Structure Determination
by X-Ray Crystallography

Structure Determination by X-Ray Crystallography

Second Edition

M. F. C. Ladd
University of Surrey
Guildford, England

and

R. A. Palmer
Birkbeck College
University of London
London, England

Plenum Press · New York and London

7308.1944

CHEMISTRY

Library of Congress Cataloging in Publication Data

Ladd, M. F. C. (Marcus Frederick Charles).
  Structure determination by X-ray crystallography.

  Includes bibliographies and index.
  1. X-ray crystallography. I. Palmer, R. A. (Rex Alfred), 1936-   . II. Title.
QD945.L32 1985   548′.83   84-24811
ISBN 0-306-41878-9

© 1985 Plenum Press, New York
A Division of Plenum Publishing Corporation
233 Spring Street, New York, N.Y. 10013

Printed in the United States of America

Foreword

X-ray crystallography provides us with the most accurate picture we can get of atomic and molecular structures in crystals. It provides a hard bedrock of structural results in chemistry and in mineralogy. In biology, where the structures are not fully crystalline, it can still provide valuable results and, indeed, the impact here has been revolutionary. It is still an immense field for young workers, and no doubt will provide yet more striking developments of a major character. It does, however, require a wide range of intellectual application, and a considerable ability in many fields.

This book will provide much help. It is a very straightforward and thorough guide to every aspect of the subject. The authors are experienced both as research workers themselves and as teachers of standing, and this is shown in their clarity of exposition. There are plenty of illustrations and worked examples to aid the student to obtain a real grasp of the subject. The practical side is encouraged by the very clarity of the theory. The examples chosen as illustrations cover the various branches of chemistry and there is useful guidance even as far as the protein field. In the later chapters dealing with the really tricky area of "getting the structure out," the treatment is well balanced, and this should help even an experienced worker to choose the most likely approach in each particular case. We seem to have passed beyond the stage at which one method is favored to the neglect of the others, and the book seems to mark a "coming of age" of X-ray crystallography.

I wish the book the great success it undoubtedly deserves.

University of Edinburgh   C. A. Beevers

Preface to the Second Edition

We have not changed the original plan and scope of this book because the first edition was, in general, well received. Additional space has been allocated to certain topics, notably experimental techniques, direct methods of phasing, and least-squares refinement. We have tried to respond to the helpful comments that we have received from reviewers and readers, and we are grateful for the interest that has been accorded to the first edition. In particular, we wish to acknowledge the kindness of Dr. C. A. Beevers for writing the Foreword to this edition.

University of Surrey   M. F. C. Ladd
Birkbeck College, London   R. A. Palmer

Preface to the First Edition

Crystallography may be described as the science of the structure of materials, using this word in its widest sense, and its ramifications are apparent over a broad front of current scientific endeavor. It is not surprising, therefore, to find that most universities offer some aspects of crystallography in their undergraduate courses in the physical sciences. It is the principal aim of this book to present an introduction to structure determination by X-ray crystallography that is appropriate mainly to both final-year undergraduate studies in crystallography, chemistry, and chemical physics, and introductory postgraduate work in this area of crystallography. We believe that the book will be of interest in other disciplines, such as physics, metallurgy, biochemistry, and geology, where crystallography has an important part to play.

In the space of one book, it is not possible either to cover all aspects of crystallography or to treat all the subject matter completely rigorously. In particular, certain mathematical results are assumed in order that their applications may be discussed. At the end of each chapter, a short bibliography is given, which may be used to extend the scope of the treatment given here. In addition, reference is made in the text to specific sources of information.

We have chosen not to discuss experimental methods extensively, as we consider that this aspect of crystallography is best learned through practical experience, but an attempt has been made to simulate the interpretive side of experimental crystallography in both examples and exercises.

During the preparation of this book, we have tried to keep in mind that students meeting crystallography for the first time are encountering a new discipline, and not merely extending a subject studied previously. In consequence, we have treated the geometry of crystals a little more fully than is usual at this level, for it is our experience that some of the difficulties which students meet in introductory crystallography lie in the unfamiliarity of its three-dimensional character.

We have limited the structure-determining techniques to the three that are used most extensively in present-day research, and we have described them in depth, particularly from a practical point of view. We hope that this treatment will indicate our belief that crystallographic methods can reasonably form part of the structural chemist's repertoire, like quantum mechanics and nmr spectroscopy.

Each chapter is provided with a set of problems, for which answers and notes are given. We recommend the reader to tackle these problems; they will provide a practical involvement which should be helpful to the understanding of the subject matter of the book. From experience in teaching this subject, the authors are aware of many of the difficulties encountered by students of crystallography, and have attempted to anticipate them in both these problems and the text. For any reader who has access to crystallographic computing facilities, the authors can supply copies of the data used to solve the structures described in Chapters 6 and 8. Certain problems have been marked with an asterisk. They are a little more difficult than the others and may be omitted at a first reading.

The Hermann–Mauguin system of symmetry notation is used in crystallography, but, unfortunately, this notation is not common to other disciplines. Consequently, we have written the Schoenflies symbols for point groups on some of the figures that depict point-group and molecular symmetry in three dimensions, in addition to the Hermann–Mauguin symbols. The Schoenflies notation is described in Appendix 3. General symbols and constants are listed in the Notation section.

We wish to acknowledge our colleague, Dr. P. F. Lindley, of Birkbeck College, London, who undertook a careful and critical reading of the manuscript and made many valuable suggestions. We acknowledge an unknown number of past students who have worked through many of the problems given in this book, to our advantage and, we hope, also to theirs. We are grateful to the various copyright holders for permission to reproduce those figures that carry appropriate acknowledgments. Finally, we thank the Plenum Publishing Company for both their interest in this book and their ready cooperation in bringing it to completion.

University of Surrey   M. F. C. Ladd
Birkbeck College, London   R. A. Palmer

Contents

8/17

Chapter 3

Preliminary Examination of Crystals by Optical and X-Ray Methods 105

Chapter 4

Intensity of Scattering of X-Rays by Crystals 153

$8/$

Chapter 5

Methods in X-Ray Structure Analysis. I 195

Chapter 6

Methods in X-Ray Structure Analysis. II   213

Chapter 7

Direct Methods and Refinement 295

Chapter 8

8/31

Notation

These notes provide a key to the main symbols and constants used throughout the book. Inevitably, some symbols have more than one use. This feature arises partly from general usage in crystallography, and partly from a desire to preserve a mnemonic character in the notation wherever possible. It is our belief that, in context, no confusion will arise. Where several symbols are closely linked, they are listed together under the first member of the set.

$A'(hkl), B'(hkl)$.	Components of the structure factor, measured along the real and imaginary axes, respectively, in the complex plane (Argand diagram)
$A(hkl), B(hkl)$.	Components of the geometric structure factor, measured along the real and imaginary axes, respectively, in the complex plane
A	A-face-centered unit cell; absorption correction factor
Å	Angstrom unit; $1 \text{ Å} = 10^{-8} \text{ cm} = 10^{-10} \text{ m}$
a, b, c	Unit-cell edges parallel to the x, y, and z axes, respectively, of a crystal; intercepts made by the parametral plane on the x, y, and z axes, respectively; glide planes with translational components of $a/2$, $b/2$, and $c/2$, respectively
a, b, c	Unit-cell edge vectors parallel to the x, y, and z axes, respectively
a^*, b^*, c^* . . .	Edges in the reciprocal unit cell associated with the x^*, y^*, and z^* axes, respectively
a*, b*, c* . . .	Reciprocal unit-cell vectors associated with the x^*, y^*, and z^* axes, respectively
B	B-face-centered unit cell; overall isotropic temperature factor
B_j	Isotropic temperature factor for the jth atom
C	C-face-centered unit cell
$\not{C}$	"Not constrained by symmetry to equal"

c	Velocity of light (2.9979×10^{-8} m s^{-1}); as a subscript: calculated, as in $\lvert F_c \rvert$
D_m	Experimentally measured crystal density
D_c	Calculated crystal density
d	Interplanar spacing
$d(hkl)$	Interplanar spacing of the (hkl) family of planes
d^*	Distance in reciprocal space
$d^*(hkl)$ . . .	Distance from the origin to the hklth reciprocal lattice point
$E, E(hkl)$ . . .	Normalized structure factor (centrosymmetric crystals)
$\mathscr{E}(hkl)$	Total energy of the hklth diffracted beam from one unit cell
e	Electron charge (1.6021×10^{-19}C); exponential factor
$\mathbf{F}(hkl)$	Structure factor for the hklth spectrum referred to one unit cell
$\mathbf{F}^*(hkl)$ . . .	Conjugate vector of $\mathbf{F}(hkl)$.
$\lvert F \rvert$	Modulus, or amplitude, of any vector $\mathbf{F}$
f	Atomic scattering factor
$f_{j,\theta}, f_j$	Atomic scattering factor for the jth atom
g	Glide line in two-dimensional space groups
g_j	Atomic scattering factor for the jth atom, in a crystal, corrected for thermal vibrations
H	Hexagonal (triply primitive) unit cell
$(hkl), (hkil)$ . .	Miller, Miller–Bravais indices associated with the x, y, and z axes or the x, y, u, and z axes, respectively; any single index containing two digits has a comma placed *after* such an index
$\{hkl\}$	Form of (hkl) planes
hkl	Reciprocal lattice point corresponding to the (hkl) family of planes
$\mathbf{h}$	Vector with components h, k, l in reciprocal space.
h	Planck's constant (6.6256×10^{-34} J s)
I	Body-centered unit cell; intensity of reflection
$I(hkl)$	Intensity of reflection from the (hkl) planes referred to one unit cell
$\mathscr{I}$	Imaginary axis in the complex plane
i	$\sqrt{-1}$; an operator that rotates a vector in the complex plane through 90° in a right-handed (anticlockwise) sense

$J(hkl)$ . . . . . .   Integrated reflection

K . . . . . .   Reciprocal lattice constant; scale factor for $|F_o(hkl)|$ data

L . . . . . .   Lorentz correction factor

M_r . . . . . .   Relative molecular weight (mass)

m . . . . . .   Mirror plane

N . . . . . .   Number of atoms per unit cell

n . . . . . .   Glide plane, with translational component of $(a+b)/2$, $(b+c)/2$, or $(c+a)/2$

n_1, n_2, n_3 . . .   Principal refractive indices in a biaxial crystal

o . . . . . .   subscript: observed, as in $|F_o(hkl)|$

P . . . . . .   Probability; Patterson function

$P(u, v, w)$ . . .   Patterson function at the fractional coordinates u, v, w in the unit cell

p . . . . . .   Polarization correction factor

R . . . . . .   Rhombohedral unit cell; rotation axis (of degree R); reliability factor

$\bar{R}$ . . . . . .   Inversion axis

$\mathscr{R}$ . . . . . .   Real axis in the complex plane

RU . . . . .   Reciprocal lattice unit

$s, s(hkl), s(\mathbf{h})$.   Sign of a centric reflection

$T_{j,\theta}$ . . . . .   Thermal vibration parameter for the jth atom

$[UVW]$ . . .   Zone or direction symbol

$\langle UVW \rangle$ . . .   Form of zone axes or directions

u . . . . . .   Atomic mass unit $(1.66057 \times 10^{-27}\,\text{kg})$

(u, v, w) . . .   Components of a vector in Patterson space

$\overline{U^2}$ . . . . .   Mean square amplitude of vibration

V_c . . . . .   Volume of a unit cell

w . . . . . .   Weight factor

$x, y, z;$ Spatial coordinates, in absolute measure, of a point,

   x, y, u, z . .   parallel to the x, y, (u), and z axes, respectively

x, y, z . . . .   Spatial fractional coordinates in a unit cell

x_j, y_j, z_j . . .   Spatial fractional coordinates of the jth atom in a unit cell

$[x, \beta, \gamma]$ . . .   Line parallel to the x axis and intersecting the y and z axes at β and γ, respectively

(x, y, γ) . . .   Plane normal to the z axis and intersecting it at γ

$\pm\{x, y, z; \ldots\}$. $x, y, z; \bar{z}, \bar{y}, \bar{z}; \ldots$

Z . . . . . .   Number of formula-entities of weight M_r per unit cell

Z_j . . . . . .   Atomic number of the jth atom in a unit cell

α, β, γ . . . .   Angles between the pairs of unit-cell edges bc, ca, and ab, respectively

$\alpha^*, \beta^*, \gamma^*$ . . .   Angles between the pairs of reciprocal unit-cell edges b^*c^*, c^*a^*, and a^*b^*, respectively

δ . . . . . .   Path difference

$\varepsilon, \varepsilon(hkl)$ . . .   Statistical weight of a reflection

ε, ω . . . .   Principal refractive indices for a uniaxial crystal

θ . . . . . .   Bragg angle

λ . . . . . .   Wavelength

μ . . . . . .   Linear absorption coefficient

ν . . . . . .   Frequency

ν_n . . . . . .   Spacing between the zeroth- and nth-layer lines

ρ . . . . . .   Radius of stereographic projection

$\rho(x, y, z)$ . . .   Electron density at the point x, y, z

Φ . . . . . .   Interfacial (internormal) angle

$\phi(hkl), \phi(h), \varphi$.   Phase angle associated with a structure factor

χ, ψ, ω . . . . $(\cos \chi, \cos \psi, \cos \omega)$ direction cosines of a line with respect to the x, y, and z axes

ω . . . . . .   Angular frequency

Ω . . . . . .   Azimuthal angle in experimental methods

Structure Determination by X-Ray Crystallography

Second Edition

Crystal Geometry. I

1.1   Introduction

Crystallography grew up as a branch of mineralogy, and involved mainly the recognition, description, and classification of naturally occurring crystal species. X-ray crystallography is a relatively new discipline, dating from the discovery in 1912 of the diffraction of X-rays by crystals. This year marked the beginning of the experimental determination of crystal structures. Figure 1.1 illustrates the structure of sodium chloride, which was among the first crystals to be studied by the new X-ray techniques.

Nowadays, we can probe the internal structure of crystals by X-ray methods and determine with certainty the actual atomic arrangement in space. Figure 1.2 shows a three-dimensional contour map of the electron density in euphenyl iodoacetate, $C_{32}H_{53}IO_2$. The contour lines join points of equal electron density in the structure; hydrogen atoms are not revealed in this map because of their relatively small scattering power for X-rays. If we

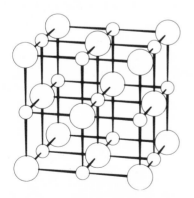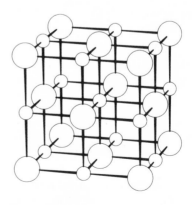

FIGURE 1.1. Stereoview of a unit cell of the crystal structure of sodium chloride: $\bigcirc = Na^+$, $\bigcirc = Cl^-$.

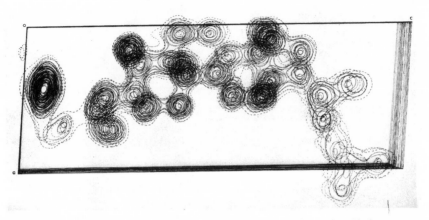

FIGURE 1.2. Three-dimensional electron density contour map for euphenyl iodoacetate, as seen along the *b* direction of the unit cell.

assume that the centers of atoms are located at the maxima in the electron density map, we can deduce the molecular model in Figure 1.3; the conventional chemical structural formula is shown for comparison. The iodine atom is represented by the large number of contours in the elongated peak at the extreme left of the figure. The carbon and oxygen atoms are depicted by approximately equal numbers of contours, except for the atoms in the side chain, shown on the extreme right of the figure. Thermal vibrations of the atoms are most severe in this portion of the molecule, and they have the effect of smearing out the electron density, so that its gradient, represented by the contour intervals, is less steep than in other parts of the molecule.

Molecules of much greater complexity than this example are now being investigated; the structures of proteins, enzymes, and nucleic acids—the "elements" of life itself—are being revealed by powerful X-ray diffraction techniques.

1.2 The Crystalline State

A crystalline substance may be defined as a homogeneous solid having an ordered internal atomic arrangement and a definite, though not necessarily stoichiometric, overall chemical composition. In addition to the more

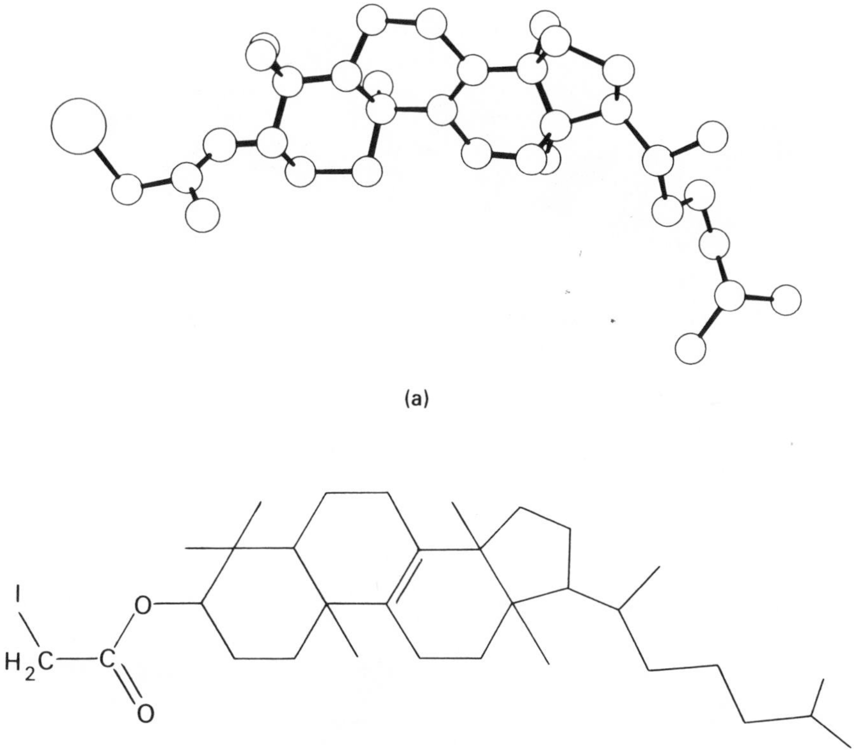

(a)

FIGURE 1.3. Euphenyl iodoacetate: (a) molecular model, excluding hydrogen atoms; (b) chemical structural formula.

obvious manifestations of crystalline material, other substances, such as cellophane sheet and fibrous asbestos, which reveal different degrees of long-range order (extending over many atomic dimensions), may be described as crystalline.

Fragments of glass and of quartz look similar to each other to the unaided eye, yet quartz is crystalline and glass is noncrystalline, or amorphous. Glass may be regarded as a supercooled liquid, with an atomic arrangement that displays only very short-range order (extending over few atomic dimensions). Figure 1.4 illustrates the structures of quartz and silica glass; both contain the same atomic group, the tetrahedral SiO_4 structural unit.

(a)

(b)

FIGURE 1.4. SiO$_4$ structural unit (the darker spheres represent Si): (a) α-quartz; (b) silica glass. [Crown copyright. Reproduced from *NPL Mathematics Report Ma62* by R. J. Bell and P. Dean, with the permission of the Director, National Physical Laboratory, Teddington, Middlesex, England.]

A crystal may be defined as a substance that is crystalline in three dimensions and is bounded by plane faces. The word crystal is derived from the Greek κρνσταλλoσ, meaning *ice*, used to describe quartz, which once was thought to be water permanently congealed by intense cold. We have made the useful distinction that crystalline substances exhibit long-range order in three dimensions or less, whereas crystals have three-dimensional regularity and plane bounding faces.

1.2.1 Reference Axes

The description of crystals and their external features presents situations which are associated with coordinate geometry. In two dimensions, a straight line AB may be referred to rectangular axes (Figure 1.5) and described by the equation

$$y = mx + b \qquad (1.1)$$

where m $(= \tan \Phi)$ is the slope of the line and b is the intercept made by AB on the y axis. Any point $P(x, y)$ on the line satisfies (1.1). If the line had been referred to oblique axes (Figure 1.6), its equation would have been

$$y = Mx + b \qquad (1.2)$$

where b has the same value as before, and M is given by

$$M = (\tan \Phi \sin \gamma - \cos \gamma) \qquad (1.3)$$

Evidently, oblique axes are less convenient in this case.

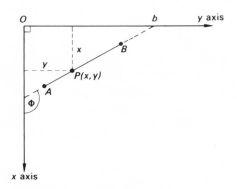

FIGURE 1.5. Line AB referred to rectangular axes.

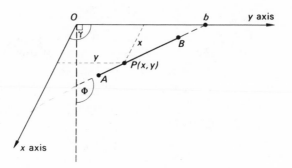

FIGURE 1.6. Line AB referred to oblique axes.

We may describe the line in another way. Let AB intersect the x axis at a and the y axis at b (Figure 1.7) and have slope m. At $x = a$, we have $y = 0$, and, using (1.1),

$$ma + b = 0 \qquad\qquad (1.4)$$

whence

$$y = -(bx/a) + b \qquad\qquad (1.5)$$

or

$$(x/a) + (y/b) = 1 \qquad\qquad (1.6)$$

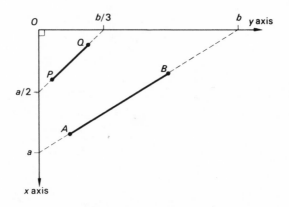

FIGURE 1.7. Lines AB and PQ referred to rectangular axes.

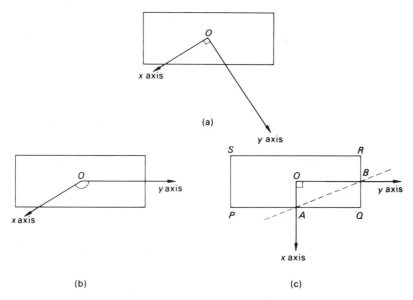

FIGURE 1.8. Rectangle referred to rectangular and oblique axes.

Equation (1.6) is the intercept form of the equation of the straight line AB. This line will be used as a reference, or parametral, line. Consider next any other line, such as PQ; let its intercepts on the x and y axes be, for example, $a/2$ and $b/3$, respectively. The line may be identified by two numbers h and k defined such that h is the ratio of the intercepts made on the x axis by the parametral line and the line PQ, and k is the corresponding ratio for the y axis. Thus

$$h = a/(a/2) = 2 \qquad (1.7)$$

$$k = b/(b/3) = 3 \qquad (1.8)$$

PQ is described as the line (23)—two-three. It follows that AB is (11). Although the values of a and b are not specified, once the parametral line is chosen, any other line can be defined uniquely by its indices h and k.

In the analysis of a plane figure, reference axes are again useful. In Figure 1.8, common sense (and convention, as we shall see) dictates the choice (c) for reference axes x and y; these lines are parallel to the perimeter lines, which are important features of the rectangle. If AB is (11), then PQ, QR, RS, and SP are (10), (01), ($\bar{1}$0),* and (0$\bar{1}$), respectively. A zero

---

* Read as "bar-one zero," or "one-bar zero" in the USA.

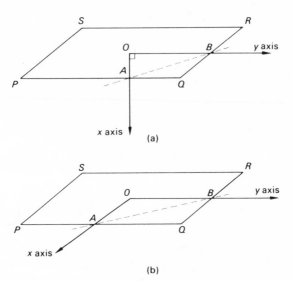

FIGURE 1.9. Parallelogram: (a) referred to rectangular
axes; (b) referred to oblique axes.

value for h or k indicates parallelism of the line with the corresponding axis
(its intercept is at infinity); a negative value for h or k, indicated by a bar over
the symbol, implies an intercept on the negative side of the reference axis.

This simple description of the lines in *PQRS* is not obtained with either
the orientation (a) or the oblique axes (b) in Figure 1.8. In considering a
parallelogram, however, oblique axes are the more convenient for our
purposes. It is left as an exercise for the reader to show that, if AB in Figure
1.9 is (11), then PQ, QR, RS, and SP are again (10), (01), ($\bar{1}$0), and (0$\bar{1}$),
respectively, provided that the reference axes are chosen parallel to the sides
of the figure.

Crystallographic Axes

Three reference axes are needed for crystal description (Figure 1.10).
An extension of the above arguments leads to the adoption of x, y, and z
axes parallel to important directions in the crystal. We shall see later that
these directions (crystal edges or possible crystal edges) are related closely to
the symmetry of the crystal; in some cases, a choice of oblique axes then will
arise naturally.

It is usual to work with right-handed axes. In Figure 1.11, $+y$ and $+z$
are in the plane of the paper, as shown, and $+x$ is directed forward; the

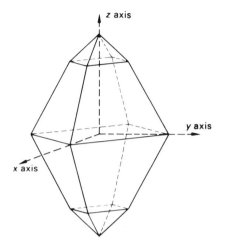

FIGURE 1.10. Idealized tetragonal crystal with the crystallographic axes drawn in.

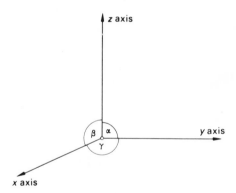

FIGURE 1.11. Right-handed crystallographic axes and interaxial angles.

succession $x \rightarrow y \rightarrow z$ simulates a right-handed screw motion. Notice the selection of the interaxial angles α, β, and γ, and the mnemonic connection between their positions and the directions of the x, y, and z axes.

1.2.2 Equation of a Plane

In Figure 1.12, the plane ABC intercepts the x, y, and z axes at A, B, and C, respectively. ON is the perpendicular from the origin O to the

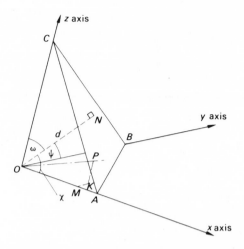

FIGURE 1.12. Plane *ABC* in three-
dimensional space.

plane; it has the length d, and its direction cosines are $\cos \chi$, $\cos \psi$, and $\cos \omega$
with respect to *OA*, *OB*, and *OC*, respectively. *OA*, *OB*, and *OC* have
the lengths a, b, and c, respectively, and P is any point X, Y, Z in the plane
ABC. Let *PK* be parallel to *OC* and meet the plane *AOB* at K, and let *KM*
be parallel to *OB* and meet *OA* at M. Then the lengths of *OM*, *MK*, and *KP*
are X, Y, and Z, respectively. Since *ON* is the projection of *OP* onto *ON*, it
is equal to the sum of the projections *OM*, *MK*, and *KP* all onto *ON*. Hence,

$$d = X \cos \chi + Y \cos \psi + Z \cos \omega \qquad (1.9)$$

In $\triangle OAN$, $d = OA \cos \chi = a \cos \chi$. Similarly, $d = b \cos \psi = c \cos \omega$, and,
hence,

$$(X/a)+(Y/b)+(Z/c) = 1 \qquad (1.10)$$

Equation (1.10) is the intercept form of the equation of the plane *ABC*, and
may be compared with (1.6).

1.2.3   Miller Indices

The faces of a crystal are planes in three dimensions. Once the crystallographic axes are chosen, a parametral plane may be defined and any other
plane described in terms of three numbers h, k, and l. It is an experimental
fact that, in crystals, if the parametral plane is designated by integral values

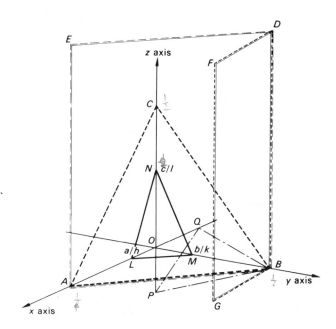

FIGURE 1.13. Miller indices of planes: $OA = a$, $OB = b$, $OC = c$.

of h, k, and l, usually (111), then the indices of all other crystal faces are integers. This notation for describing the faces of a crystal was introduced first by Miller in 1839, and h, k, and l are called Miller indices.

In Figure 1.13, let the parametral plane (111) be ABC, making intercepts a, b, and c on the crystallographic axes. Another plane, LMN (the first of its type from the origin), makes intercepts a/h, b/k, and c/l along the x, y, and z axes, respectively. Its Miller indices are expressed by the ratios of the intercepts of the parametral plane to those of the plane LMN. If in the figure, $a/h = a/4$, $b/k = b/3$, and $c/l = c/2$, then LMN is (432). If fractions occur in h, k, or l, they are cleared by multiplication throughout by the lowest common denominator. Conditions of parallelism to axes and intercepts on the negative sides of the axes lead respectively to zero or negative values for h, k, and l. Thus, $ABDE$ is (110), $BDFG$ is (010), and PBQ is ($\bar{2}1\bar{3}$). It may be noted that it has not been necessary to assign numerical values to either a, b, and c or α, β, and γ in order to describe the crystal faces. In the next chapter we shall identify a, b, and c with the edges of the crystal unit cell in a lattice, but this relationship is not needed at present.

The preferred choice of the parametral plane leads to small numerical values for the Miller indices of crystal faces. Rarely are h, k, and l greater than 4. If LMN had been chosen as (111), then ABC would have been (346). Summarizing, we may say that the plane (hkl) makes intercepts a/h, b/k, and c/l along the crystallographic x, y, and z axes, respectively, where a, and c are the corresponding intercepts made by the parametral plane.

From (1.10), the intercept equation of the plane (hkl) may be written as

$$(hx/a) + (ky/b) + (lz/c) = 1 \qquad (1.11)$$

The equation of the parallel plane passing through the origin is

$$(hx/a) + (ky/b) + (lz/c) = 0 \qquad (1.12)$$

since it must satisfy the condition $x = y = z = 0$.

Miller–Bravais Indices

In crystals that exhibit sixfold symmetry (see Table 1.3), four axes of reference are used. The axes are designated x, y, u, and z; the x, y, and u axes lie in one plane, at $120°$ to one another, and the z axis is perpendicular to the xyu plane (Figure 1.14). As a consequence, planes in these crystals are described by four numbers, the Miller–Bravais indices h, k, i, and l. The

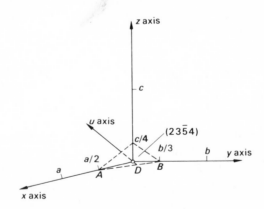

FIGURE 1.14. Miller–Bravais indices $(hkil)$. The crystallographic axes are labeled x, y, u, z, and the plane $(23\bar{5}4)$ is shown; the parametral plane is $(11\bar{2}1)$.

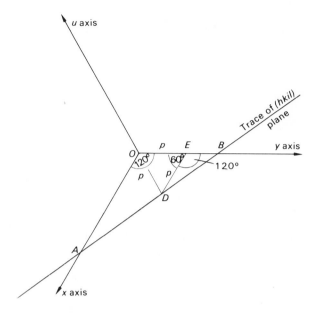

FIGURE 1.15. Equivalence of i and $-(h+k)$.

index i is not independent of h and k: thus, if the plane ABC in Figure 1.14 intercepts the x and y axes at $a/2$ and $b/3$, for example, then the u axis is intercepted at $-u/5$. If the z axis is intercepted at $c/4$, the plane is designated $(23\bar{5}4)$. In general, $i = -(h+k)$, and the parametral plane is $(11\bar{2}1)$.

We can show that $i = -(h+k)$ with reference to Figure 1.15. From the definition of Miller indices,

$$OA = a/h \qquad (1.13)$$

$$OB = b/k \qquad (1.14)$$

Let the intercept of $(hkil)$ on the u axis be p. Draw DE parallel to AO. Since OD bisects $\widehat{AOB}$,

$$\widehat{AOD} = 60° \qquad (1.15)$$

Hence, $\triangle ODE$ is equilateral, and

$$OD = DE = OE = p \qquad (1.16)$$

Triangles EBD and OBA are similar. Hence,

$$\frac{EB}{ED} = \frac{OB}{OA} = \frac{b/k}{a/h} \tag{1.17}$$

But

$$EB = b/k - p \tag{1.18}$$

Hence,

$$\frac{b/k - p}{p} = \frac{b/k}{a/h} \tag{1.19}$$

or

$$p = \frac{ab}{ak + bh} \tag{1.20}$$

Since $a = b = u$ in crystals with sixfold symmetry (hexagonal), and writing p as $-u/i$, we have

$$i = -(h + k) \tag{1.21}$$

An alternative, geometric, approach to this result consists in drawing the traces of any family of ($hkil$) planes from the origin to $+u$, whence it will be clear that $i = -(h + k)$. This construction may be appreciated fully after a study of Chapter 2.

1.2.4 Axial Ratios

If both sides of (1.12) are multiplied by b, we obtain

$$\frac{hx}{a/b} + ky + \frac{lz}{c/b} = 0 \tag{1.22}$$

The quantities a/b and c/b are termed axial ratios; they can be deduced from the crystal morphology.

1.2.5 Zones

Most well-formed crystals have their faces arranged in groups of two or more with respect to certain directions in the crystal. In other words, crystals exhibit symmetry; this feature is an external manifestation of the ordered arrangement of atoms in the crystal. Figure 1.16 illustrates zircon, $ZrSiO_4$, an example of a highly symmetric crystal. It is evident that several faces have a given direction in common. Such faces are said to lie in a zone, and the common direction is called the zone axis. Two faces, $(h_1k_1l_1)$ and $(h_2k_2l_2)$, define a zone. The zone axis is the line of intersection of the two planes, and is given by the solution of the equations

$$(h_1x/a)+(k_1y/b)+(l_1z/c)=0 \qquad (1.23)$$

and

$$(h_2x/a)+(k_2y/b)+(l_2z/c)=0 \qquad (1.24)$$

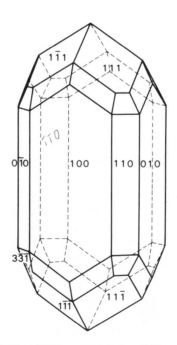

FIGURE 1.16. A highly symmetric crystal (zircon, $ZrSiO_4$).

that is, by the line

$$\frac{x}{a(k_1 l_2 - k_2 l_1)} = \frac{y}{b(l_1 h_2 - l_2 h_1)} = \frac{z}{c(h_1 k_2 - h_2 k_1)} \qquad (1.25)$$

which passes through the origin. It may be written as

$$x/(aU) = y/(bV) = z/(cW) \qquad (1.26)$$

where $[UVW]$ is called the zone symbol.

   If any other face (hkl) lies in the same zone as $(h_1 k_1 l_1)$ and $(h_2 k_2 l_2)$, then it may be shown, from (1.12) and (1.26), that

$$hU + kV + lW = 0 \qquad (1.27)$$

which is a symbolic expression of the Weiss zone law.

   In the zircon crystal, the vertical (prism) faces lie in one zone. If the prism faces are indexed in the usual manner (see Figure 1.16), then, from (1.25) and (1.26), the zone symbol is [001]. From (1.27), we see that $(\bar{1}\bar{1}0)$ is a face in the same zone, but (111) is not. In the manipulation of (1.25) and (1.26), it may be noted that the zone axis is described by $[UVW]$, the simplest symbol; the directions that may be described as $[nU, nV, nW]$ $(n = 0, \pm1, \pm2, \ldots)$ are coincident with $[UVW]$.

Interfacial Angles

   The law of constant interfacial angles states that in all crystals of the same substance, angles between corresponding faces have a constant value. Interfacial angles are measured by a goniometer, the simplest form of which is the contact goniometer (Figure 1.17). In using this instrument, large crystals are needed, a condition not easily obtainable in practice.

   An improvement in technique was brought about by the reflecting goniometer. The principle of this instrument is shown in Figure 1.18, and forms the basis of modern optical goniometers. A crystal is arranged to rotate about a zone axis O, which is set perpendicular to a plane containing the incident and crystal-reflected light beams. Parallel light reflected from the face AB is received by a telescope. If the crystal is rotated in a clockwise direction, a reflection from the face BC is received next when the crystal has been turned through the angle Φ and the interfacial angle is $180 - \Phi$ degrees.

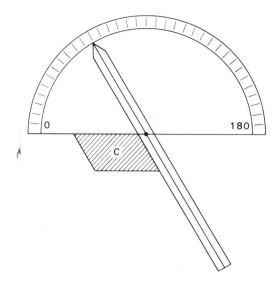

FIGURE 1.17. Contact goniometer with a crystal (C) in the measuring position.

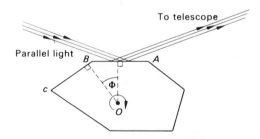

FIGURE 1.18. Principle of the reflecting goniometer.

Accurate goniometry brought a quantitative significance to observable angular relationships in crystals.

1.3 Stereographic Projection

The general study of the external features of crystals is called crystal morphology. The analytical description of planes and zones given above is inadequate for a simultaneous appreciation of the many faces exhibited by a

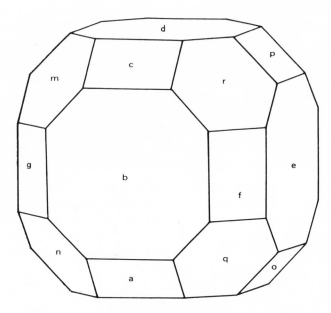

FIGURE 1.19. Cubic crystal showing three forms of planes: cube—*b*, *e*, *d*, and parallel faces; octahedron—*r*, *m*, *n*, *q*, and parallel faces; rhombic dodecahedron—*f*, *g*, *p*, *o*, *c*, *a*, and parallel faces.

crystal. It is necessary to be able to represent a crystal by means of a two-dimensional drawing, while preserving certain essential properties. For a study of crystal morphology, the interfacial angles, which are a fundamental feature of crystals, must be maintained in plane projection, and the stereographic projection is useful for this purpose. Furthermore, with imperfectly formed crystals, the true symmetry may not be apparent by inspection. In favorable cases, the symmetry may be completely revealed by a stereographic projection of the crystal. We shall develop this projection with reference to the crystal shown in Figure 1.19.

This crystal belongs to the cubic system (page 34): The crystallographic reference axes *x*, *y*, and *z* are mutually perpendicular, and the parametral plane (111) makes equal intercepts ($a = b = c$) on these axes. The crystal shows three forms of planes. In crystallography, a form of planes, represented by $\{hkl\}$, refers to the set of planes that are equivalent under the point-group symmetry of the crystal (see page 25*ff*). The crystal under discussion shows the cube form $\{100\}$—six faces (100), ($\bar{1}$00), (010), (0$\bar{1}$0), (001), and (00$\bar{1}$); the octahedron $\{111\}$—eight faces; and the rhombic dodecahedron {110}—12 faces. Each face on the crystal drawing has a related parallel face on the actual crystal, for example, b(shown) and b'. The reader may care to list the sets of planes in the latter two cubic forms; the answer will evolve from the discussion of the stereographic projection of the crystal.

From a point within the crystal, lines are drawn normal to the faces of the crystal. A sphere of arbitrary radius is described about the crystal, its center being the point of intersection of the normals which are then produced to cut the surface of the sphere (Figure 1.20).

In Figure 1.21, the plane of projection is $ABCD$, and it intersects the sphere in the primitive circle. The portion of the plane of projection enclosed by the primitive circle is the primitive plane, or primitive. The point of intersection of each normal with the upper hemisphere is joined to the lowest point P on the sphere. The intersection of each such line with the primitive is the stereographic projection, or pole, of the corresponding face on the crystal, and is indicated by a dot on the stereographic projection, or stereogram.

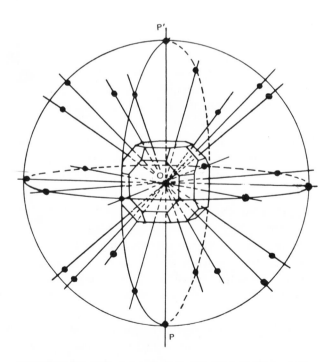

FIGURE 1.20. Spherical projection of the crystal in Figure 1.19.

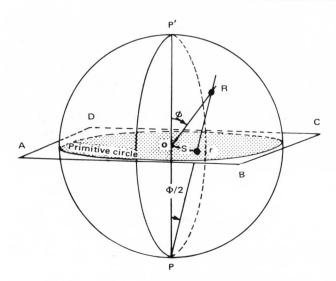

FIGURE 1.21. Development of the stereographic projection
(stereogram) from the spherical projection.

If the crystal is oriented such that the normal to face d (and d') coincides with PP' in the sphere, then the normals to the zone $e, f, b, \ldots, g'$ lie in the plane of projection and intersect the sphere on the primitive circle. In order to avoid increasing the size of the stereogram unduly, the intersections of the face normals with the lower hemisphere are joined to the uppermost point P' on the sphere and their poles are indicated on the stereogram by an open circle.

The completed stereogram is illustrated by Figure 1.22. The poles now should be compared with their corresponding faces on the crystal drawing. A fundamental property of the stereogram is that all circles drawn on the sphere project as circles. Thus, the curve G_1G_1' is an arc of a circle; specifically, it is the projection of a great circle that is inclined to the plane of projection. A great circle is the trace, on the sphere, of a plane that passes through the center of the sphere; it may be likened to a meridian on the globe of the world. Limiting cases of inclined great circles are the primitive circle, which lies in the plane of projection, and straight lines, such as G_2G_2', which are projections of great circles lying normal to the plane of projection. All poles on one great circle represent faces lying in one and the same zone.

Circles formed on the surface of the sphere by planes that do not pass through the center of the sphere are called small circles; they may be likened to parallels of latitude on the globe.

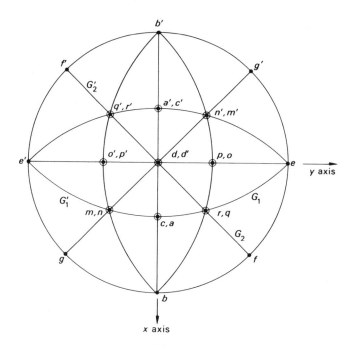

FIGURE 1.22. Stereogram of the crystal in Figure 1.19. The zone circle (great circle) $G_1 G_1'$, symbol [101], passes through e, q, a, n, e', q', a', n'; the zone circle $G_2 G_2'$, symbol [1$\bar{1}$0], passes through f, r, d, q', f', r', d', q.

In order to construct Figure 1.22, the following practical principles must be followed. The interfacial angles are measured in zones. If an optical goniometer is used, the angle Φ (see Figure 1.18) is plotted directly on the stereogram. Although Φ is the angle between the normals to planes, it is often called the interfacial angle in this context. Next, the crystal orientation with respect to the sphere is chosen: for example, let zone $b, f, e, \ldots$ be on the primitive circle, and zone b, c, $d, \ldots$ run from bottom to top of the projection. Since $\widehat{bf}$, the angle between faces b and f, is 45°, zone $f, r, d, \ldots$ can be located on the stereogram.

The distance S of the pole r from the center of the stereogram is given by

$$S = \rho \tan(\Phi/2) \qquad (1.28)$$

where ρ is the radius of the stereogram and Φ is the interfacial angle $\widehat{dr}$. This equation may be examined by means of Figure 1.23.

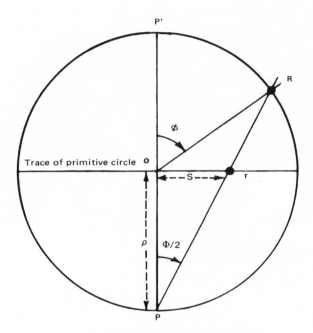

FIGURE 1.23. Evaluation of the stereographic distance S
for the pole r.

A simple graphical method, employing a Wulff net (Figure 1.24), is often sufficiently accurate to locate poles on a stereogram. On this net, the curves running from top to bottom are projected great circles; curves running from left to right are projected small circles. In use, the center of the net is pivoted at the center of the stereogram, the interfacial angle measured along the appropriate great circle, and the pole plotted. The pole of face r lies at the intersection of two zone circles, $e, r, c, \ldots$ and $f, r, d, \ldots$; if $\widehat{dc}$ and $\widehat{bf}$, for example, are known, r can be located. Interfacial angles may be measured on a stereogram by aligning a Wulff net in the manner described, rotating it until the poles in question lie on the same great circle, and then reading directly the angle.

The completed stereogram (Figure 1.22) may now be indexed. The parametral plane is chosen as face r (the parametral plane must intersect all three crystallographic axes), and the remaining faces are then allocated h, k, and l values (Figure 1.25). It is not necessary to write the indices for both poles at the same point on the stereogram. If the dot is hkl, then we know

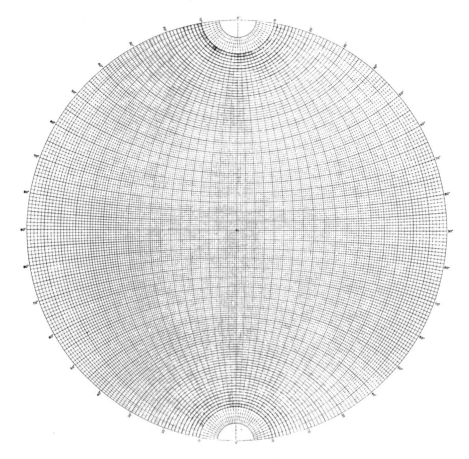

FIGURE 1.24. Wulff net.

that the open circle is, in general, $hk\bar{l}$. Figure 1.26 shows the crystal of Figure 1.19 again, but with the Miller indices inserted for direct comparison with its stereogram.

We shall not be concerned here with any further development of the stereogram.* The angular truth of the stereographic projection makes it very suitable for representing not only interfacial angles, but also symmetry directions, point groups, and bond directions in molecules and ions.

* For a fuller discussion, see Bibliography.

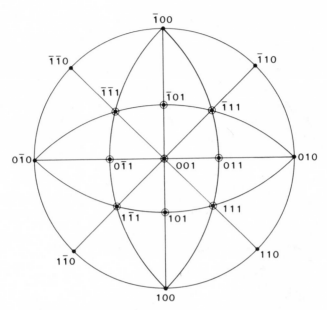

FIGURE 1.25. Stereogram in Figure 1.22 indexed, taking *r* as 111. The zone containing (100) and (111) is [0$\bar{1}$1], and that containing (010) and (001) is [100]; the face *p* common to these two zones is (011)—see (1.25) to (1.27).

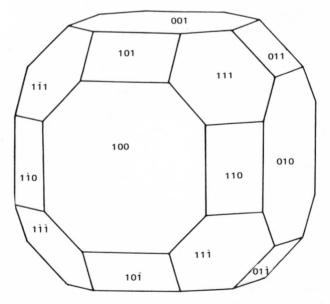

FIGURE 1.26. Crystal of Figure 1.19 with Miller indices inserted.

1.4 External Symmetry of Crystals

Often, the faces on a crystal are arranged in groups of two or more in a similar orientation with respect to some line or plane in the crystal. In other words, the crystal exhibits symmetry. The crystal drawing of zircon in Figure 1.16 shows several sets of symmetrically arranged faces.

Symmetry is a property by means of which an object is brought into self-coincidence by a certain operation. For our purposes, the operation will be considered to take place in space and to represent an action about a symmetry element. A symmetry element may be thought of as a geometric entity which generates symmetry operations, that is, a point, line, or plane with respect to which symmetry operations may be performed. The object under examination may contain more than one symmetry element. A collection of interacting symmetry elements is called a point group, which may be defined as a set of symmetry operations, all of which leave at least one point unmoved. This point is the origin, and all symmetry elements in a point group pass through this point. We shall introduce the ideas of symmetry elements first in two dimensions. Self-coincidence will be judged by appearance, by performing or imagining the symmetry operation, or by measuring interfacial angles.

It should be noted that the symmetry of a crystal may be different with respect to different physical properties, such as optical refraction, magnetism, or photoelasticity. We are concerned with the symmetry of crystals as revealed by optical or X-ray goniometry. It may be argued that, because of imperfections in real crystals, self-coincidence can be obtained *only* by a 360° rotation about any line in the crystal, the equivalent of doing nothing, and that when we speak of self-coincidence, we mean an apparent self-coincidence judged by the measuring property. Rather than enter the hypothetical realm of conceptual, geometrically perfect crystals, we shall note that for most practical purposes, the effects of crystal imperfections on symmetry observations are of a very small order, and we shall employ the term self-coincidence with this understanding.

1.4.1 Two-Dimensional Point Groups

If we examine various two-dimensional objects (Figure 1.27), we can discover two types of symmetry element that can bring an object into self-coincidence: Parts (a)–(e) of Figure 1.27 depict rotational symmetry, whereas (f) shows reflection symmetry.

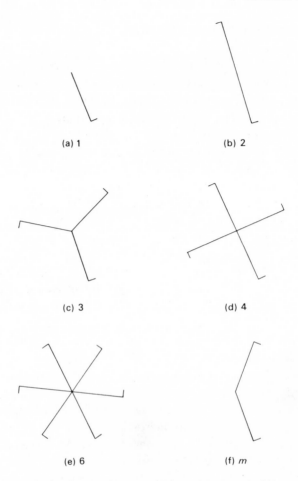

FIGURE 1.27. Two-dimensional objects and their point groups. The pattern is generated from the asymmetric unit (a) by operating on it according to the symmetry elements of the point group.

Rotation Symmetry

An object possesses rotational symmetry of degree R (or R-fold symmetry) if it is brought into self-coincidence for each rotation of $(360/R)$ degrees about the symmetry point, or rotation point. Figures 1.27a–e illustrate the rotational symmetry elements R of 1, 2, 3, 4, and 6, respectively. The onefold element is the identity element, and is trivial; every object has onefold symmetry.

Reflection Symmetry

An object possesses reflection symmetry, symbol m, in two dimensions if it is brought into self-coincidence by reflection across a line. Reflection (mirror) symmetry is a nonperformable operation: we can construct an object with the symmetry of Figure 1.27d and rotate it into self-coincidence. On the other hand, we cannot physically reflect Figure 1.27f across a symmetry line bisecting the figure. We can imagine, however, that this line divides the figure into its asymmetric unit, $\diagdown$, and the mirror image or enantiomorph of this part, $\diagup\diagdown$, which situation is characteristic of reflection symmetry.

Each of the objects in Figure 1.27 has a symmetry pattern which can be described by a two-dimensional point group, and it is convenient to illustrate these point groups by stereograms. Figure 1.28 shows stereograms for the

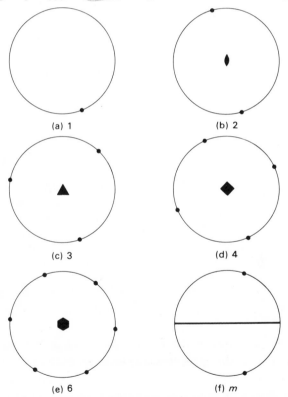

FIGURE 1.28. Stereograms of the point groups of the objects in Figure 1.27; the conventional graphic symbols for R and m are shown.

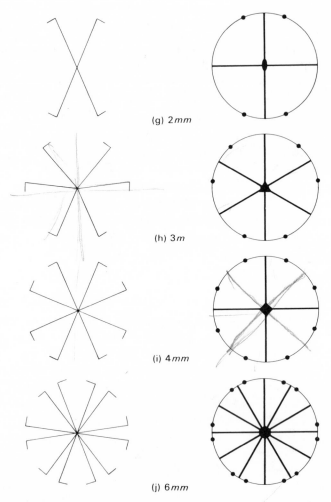

FIGURE 1.29. Other two-dimensional objects with their
stereograms and point groups.

two-dimensional point groups 1, 2, 3, 4, 6, and m. It should be noted that in using stereogram-like drawings to illustrate two-dimensional symmetry, the representative points (poles) are placed on the perimeter; such situations may represent special forms (page 37) on stereograms of three-dimensional objects.

Combinations of R and m lead to four more point groups; they are illustrated in Figure 1.29. We have deliberately omitted point groups in which $R = 5$ and $R \geq 7$; the reason for this choice will be discussed later.

TABLE 1.1 Two-Dimensional Point Groups

System	Point groups	Symbol meaning, appropriate to position occupied		
		First position	Second position	Third position
Oblique	1, 2	Rotation about a point	—	—
Rectangular	1m[a]	Rotation about a point	$m \perp x$	—
	2mm		$m \perp x$	$m \perp y$
Square	4	Rotation about a point	—	—
	4mm		$m \perp x, y$	m at 45° to x, y
Hexagonal	3		—	—
	3m	Rotation about a point	$m \perp x, y, u$	—
	6		—	—
	6mm		$m \perp x, y, u$	m at 30° to x, y, u

[a] This point group is usually called m, but the full symbol is given here in order to clarify the location of the positions.

It is convenient to allocate the ten two-dimensional point groups to two-dimensional systems, and to choose reference axes in close relation to the directions of the symmetry elements. Table 1.1 lists these systems, together with the meanings of the three positions in the point-group symbols. It should be noted that combinations of m with R ($R \geqslant 2$) introduce additional reflection lines of a different crystallographic form. In the case of $3m$ however, these additional m lines are coincident with the first set; the symbol $3mm$ is not used.

It is important to *remember* the relative orientations of the symmetry elements in the point groups, and the variations in the meanings of the positions in the different systems. In the two-dimensional hexagonal system, three axes are chosen in the plane; this selection corresponds with the use of Miller–Bravais indices in three dimensions (page 12).

1.4.2 Three-Dimensional Point Groups

The symmetry elements encountered in three dimensions are rotation axes (R), inversion axes ($\bar{R}$), and a reflection (mirror) plane (m). A center of symmetry can be invoked also, although neither this symmetry element nor the m plane is independent of $\bar{R}$.

The operations of rotation and reflection are similar to those in two dimensions, except that the geometric extensions of the operations are now increased to rotation about a line and reflection across a plane.

Inversion Axes

An object is said to possess an inversion axis $\bar{R}$ (read as bar-R), if it is brought into self-coincidence by the combination of a rotation of $(360/R)$ degrees and inversion through the origin. Like the mirror plane, the inversion axis depicts a nonperformable symmetry operation, and it may be represented conveniently on a stereogram. It is a little more difficult to envisage this operation than those of rotation and reflection. Figure 1.30 illustrates a crystal having a vertical $\bar{4}$ axis: the stereoscopic effect can be created by using a stereoviewer or, with practice, by the unaided eyes (see Appendix A1).

In representing three-dimensional point groups, it is helpful to indicate the third dimension on the stereogram, and, in addition, to illustrate the change-of-hand relationship that occurs with $\bar{R}$ (including m) symmetry elements. For example, referring to Figure 1.31, the element 2 lying in the plane of projection, and the element $\bar{4}$ normal to the plane of projection, when acting on a point derived from the upper hemisphere (symbol ●) both move the point into the lower hemisphere region (symbol ○). Both operations involve a reversal of the sign of the vertical coordinate, but only $\bar{4}$ involves also a change of hand, and this distinction is not clear from the conventional notation. Consequently, we shall adopt a symbolism, common to three-dimensional space groups, which will effect the necessary distinction.

A representative point in the positive hemisphere will be shown by $\bigcirc^+$, signifying, for example, the face (hkl). A change of hemisphere, to $(hk\bar{l})$, will be indicated by $\bigcirc^-$, and a change of hand by reflection or inversion by $\odot^+$ or $\odot^-$ (see Figure 1.32). This notation may appear to nullify the purpose of a stereogram. However, although the stereogram is a two-dimensional diagram, it should convey a three-dimensional impression, and the notation is used as an aid to this purpose.

Crystal Classes

There are 32 crystal symmetry classes, each characterized by a point group. They comprise the symmetry elements R and $\bar{R}$, taken either singly or in combination, with R restricted to the values 1, 2, 3, 4, and 6. A simple explanation for this restriction is that figures based only on these rotational symmetries can be stacked together to fill space completely, as Figure 1.33 shows. A further discussion of these values for R is given in Section 2.5.

We shall not be concerned here to derive the crystallographic point groups—and there are several ways in which it can be done—but to give,

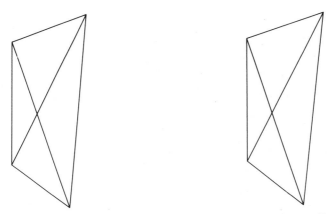

FIGURE 1.30. Stereoview of an idealized tetragonal crystal; the $\bar{4}$ axis is in the vertical direction.

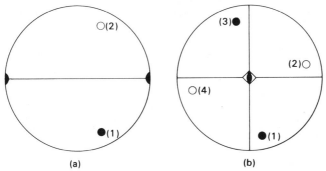

FIGURE 1.31. Stereograms of general forms: (a) point group 2 (axis horizontal and in the plane of the stereogram); (b) point group $\bar{4}$ (axis normal to the plane of the stereogram). In (a), the point ● is rotated through 180° to ○: (1) → (2). In (b), the point ● is rotated through 90° and then inverted through the origin to ○; this combined operation generates, in all, four symmetry-equivalent points: (1) → (4) → (3) → (2).

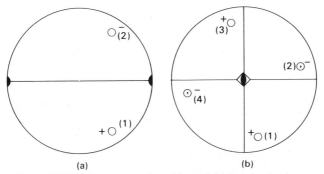

FIGURE 1.32. Stereograms from Figure 1.31 in the revised notation; the different natures of points (2) in (a) and (2) and (4) in (b), all with respect to point (1), are clear.

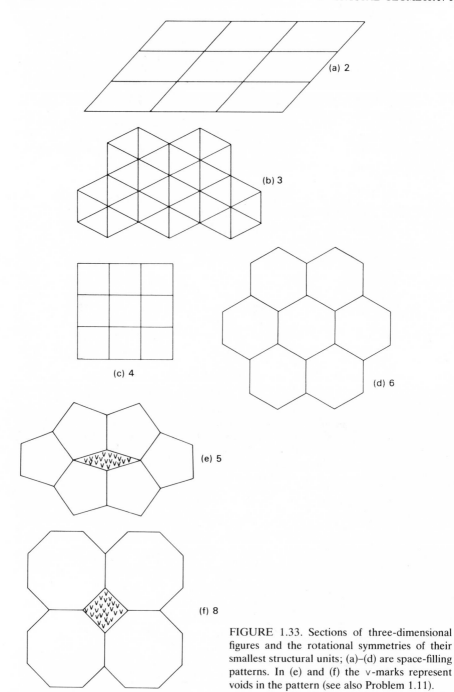

FIGURE 1.33. Sections of three-dimensional figures and the rotational symmetries of their smallest structural units; (a)–(d) are space-filling patterns. In (e) and (f) the ∨-marks represent voids in the pattern (see also Problem 1.11).

TABLE 1.2. Three-Dimensional Symmetry Symbols

Symbol	Name	Action for self-coincidence	Graphic symbol		
1	Monad	360° rotation; identity	None		
2	Diad	180° rotation	⬤ ⊥ projection	❙ ‖ projection	
3	Triad	120° rotation	▲ ⊥ or inclined to projection		
4	Tetrad	90° rotation	◆ ⊥ projection	■ ‖ projection	
6	Hexad	60° rotation	⬢ ⊥ projection		
$\bar{1}$	Inverse monad	Inversion[a]	○		
$\bar{3}$	Inverse triad	120° rotation + inversion	△ ⊥ or inclined to projection		
$\bar{4}$	Inverse tetrad	90° rotation + inversion	◈ ⊥ projection	◫ ‖ projection	
$\bar{6}$	Inverse hexad	60° rotation + inversion	⬠ ⊥ projection		
$m = \bar{2}$	Mirror plane[b]	Reflection across plane	— ⊥ projection	❨ ‖ projection	

[a] $\bar{R}$ is equivalent to R plus $\bar{1}$ only where R is an odd number: $\bar{1}$ represents the center of symmetry, but $\bar{2}, \bar{4},$ and $\bar{6}$ are not centrosymmetric point groups.

[b] The symmetry elements m and $\bar{2}$ produce an equivalent operation.

instead, a scheme which allows them to be worked through simply and adequately for present purposes.

The symbols for rotation and reflection are similar to those used in two dimensions. Certain additional symbols are required in three dimensions, and Table 1.2 lists them all.

Figure 1.34a shows a stereogram for point group m. The inverse diad is lying normal to the m plane. A consideration of the two operations in the given relative orientations shows that they produce equivalent actions. It is conventional to use the symbol m for this operation, although sometimes it is helpful to employ $\bar{2}$ instead. Potassium tetrathionate (Figure 1.34b) crystallizes in point group m.

Crystal Systems and Point-Group Scheme

The broadest classification of crystals is carried out in terms of rotation axes and inversion axes. Crystals are grouped into seven systems according

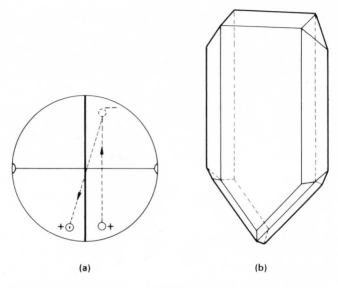

(a)   **(b)**

FIGURE 1.34. Point group m: (a) stereogram showing equivalence of m and $\bar{2}$ (the symbol () for $\bar{2}$ is not conventional); (b) crystal of potassium tetrathionate ($K_2S_4O_6$), point group m.

TABLE 1.3. The Seven Crystal Systems

System	Characteristic symmetry axes, with their orientation	Parametral plane intercepts and interaxial angles, assuming the simplest indexing of faces[a,b]
Triclinic	None	$a \mathcal{C} b \mathcal{C} c$: $\alpha \mathcal{C} \beta \mathcal{C} \gamma \mathcal{C}$ 90°, 120°
Monoclinic	One 2 or $\bar{2}$ axis along y	$a \mathcal{C} b \mathcal{C} c$; $\alpha = \gamma = 90°$; $\beta \mathcal{C}$ 90°, 120°
Orthorhombic	Three mutually perpendicular 2 or $\bar{2}$ axes[c] along x, y, and z	$a \mathcal{C} b \mathcal{C} c$; $\alpha = \beta = \gamma = 90°$
Tetragonal	One 4 or $\bar{4}$ axis along z	$a = b \mathcal{C} c$; $\alpha = \beta = \gamma = 90°$
Trigonal[d]	One 3 axis along z	$a = b \mathcal{C} c$; $\alpha = \beta = 90°$; $\gamma = 120°$
Hexagonal	One 6 or $\bar{6}$ axis along z	
Cubic	Four 3 axes inclined at 54.74° ($\cos^{-1} \frac{1}{\sqrt{3}}$) to x, y, and z	$a = b = c$; $\alpha = \beta = \gamma = 90°$

[a] We shall see in Chapter 2 that the same constraints apply to conventional unit cells in lattices.

[b] The special symbol $\mathcal{C}$ should be read as "not constrained by symmetry to equal."

[c] It must be remembered that $\bar{2}$ is equivalent to an m plane normal to the $\bar{2}$ axis.

[d] For convenience, the trigonal system is referred to hexagonal axes.

TABLE 1.4. Point-Group Scheme[a]

Type	Triclinic	Monoclinic	Trigonal	Tetragonal	Hexagonal	Cubic[b]
R	1	2	3	4	6	23
$\bar{R}$	$\bar{1}$	m	$\bar{3}$	$\bar{4}$	$\bar{6}$	$m3$
R + center		$2/m$		$4/m$	$6/m$	
		Orthorhombic				
$R2$		222	32	422	622	432
Rm		$mm2$	$3m$	$4mm$	$6mm$	$\bar{4}3m$
$\bar{R}m$			$\bar{3}m$	$\bar{4}2m$	$\bar{6}m2$	$m3m$
$R2$ + center		mmm		$\frac{4}{m}mm$	$\frac{6}{m}mm$	

$4 \perp m$

[a] The reader should consider the implication of the unfilled spaces in this table.
[b] The cubic system is characterized by its threefold axes; R refers to the element 2, 4, or $\bar{4}$ here, but 3 is always present.

to characteristic symmetry, as listed in Table 1.3. The characteristic symmetry refers to the minimum necessary for classification of a crystal in a system; a given crystal may contain more than the characteristic symmetry of its system. The conventional choice of crystallographic reference axes leads to special relationships between the intercepts of the parametral plane (111) and between the interaxial angles α, β, and γ.

A crystallographic point-group scheme is given in Table 1.4, under the seven crystal systems as headings. The main difficulty in understanding point groups lies not in knowing the action of the individual symmetry elements, but rather in appreciating both the relative orientation of the different elements in a point-group symbol and the fact that this orientation changes among the crystal systems according to the principal symmetry axis.* These orientations must be learned: They are the key to point-group and space-group studies.

Table 1.5 lists the meanings of the three positions in the three-dimensional point-group symbols. Tables 1.4 and 1.5 should be studied carefully in conjunction with Figure 1.39.

The reader should not be discouraged by the wealth of convention which surrounds this part of the subject. It arises for two main reasons. There are many different, equally correct ways of describing crystal geometry. For example, the unique axis in the monoclinic system could be

* Rotational axis of highest degree, R.

TABLE 1.5. Three-Dimensional Point Groups

System	Point groups	Symbol meaning, appropriate to position occupied		
		First position	Second position	Third position
Triclinic	$1, \bar{1}$	One symbol position only, denoting all directions in the crystal		
Monoclinic[a,b]	$2, m, 2/m$	One symbol position only: 2 or $\bar{2}$ along y		
Orthorhombic	$222, mm2,$ mmm	2 and/or $\bar{2}$ along x	2 and/or $\bar{2}$ along y	2 and/or $\bar{2}$ along z
Tetragonal	$4, \bar{4}, 4/m$ $422, 4mm,$ $\bar{4}2m, \dfrac{4}{m}mm$	4 and/or $\bar{4}$ along z	2 and/or $\bar{2}$ along x, y	2 and/or $\bar{2}$ at 45° in xy plane
Trigonal[c]	$3, \bar{3}$ $32, 3m, \bar{3}m$	3 or $\bar{3}$ along z	— 2 and/or $\bar{2}$ along x, y, u	— —
Hexagonal	$6, \bar{6}, 6/m$ $622, 6mm,$ $\bar{6}m2, \dfrac{6}{m}mm$	6 and/or $\bar{6}$ along z	— 2 and/or $\bar{2}$ along x, y, u	— 2 and/or $\bar{2}$ at 30° to x, y, u in the xyu plane
Cubic	$23, m3$	2 and/or $\bar{2}$ along x, y, z	3 or $\bar{3}$ at $54.74°$[d] to x, y, z	—
	$432, \bar{4}3m,$ $m3m$	4 and/or $\bar{4}$ along x, y, z		— 2 and/or $\bar{2}$ at 45° to x, y, z in $xy, yz,$ and zx planes

[a] In the monoclinic system, the y axis is taken as the unique 2 or $\bar{2}$ axis. Since $\bar{2} \equiv m$, then if $\bar{2}$ is along y, the m plane represented by the same position in the point-group symbol is perpendicular to y. The latter comment applies *mutatis mutandis* in other crystal systems. (It is best to specify the orientation of a plane by that of its normal.)

[b] R/m occupies a single position in a point-group symbol.

[c] For convenience, the trigonal system is referred to hexagonal axes.

[d] Actually $\cos^{-1}(1/\sqrt{3})$.

chosen along x or z instead of y, or even in none of these directions. Second, a strict system of notation is desirable for the purposes of concise, unambiguous communication of crystallographic material. With familiarity, the conventions cease to be a problem.

We shall now consider two point groups in a little more detail in order to elaborate the topics discussed so far.

Point Group mm2. We shall see that once we fix the orientations of two of the symmetry elements in this point group, the third is introduced.

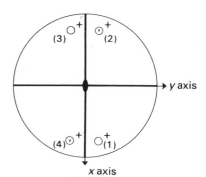

FIGURE 1.35. Stereogram of point group $mm2$.

Referring to Figure 1.35, we start with mm as shown. Point (1), in a general position, is reflected across the m plane perpendicular to the x axis (m_X) to give point (2). This point is now reflected across the second m plane to (3). Then either (3) across m_X or (1) across m_Y produces (4). It is evident now that (1) and (3), and (2) and (4), are related by the twofold rotation axis along z.

Point Group 4mm. If we start the 4 along z and m perpendicular to x, we see straightaway that another m plane (perpendicular to y) is required (Figures 1.36a and b); the fourfold axis acts on other symmetry elements in the crystal as well as faces. A general point operated on by the symmetry $4m$ produces eight points in all (Figure 1.36c). The stereogram shows that a second form of m planes, lying at 45° to the first set,* is introduced (Figure 1.36d). No further points are introduced by the second set of m planes: a fourfold rotation, (1) → (2), followed by reflection across the mirror plane normal to the x axis, (2) → (3), is equivalent to reflection of the original point across the mirror at 45° to x, (1) → (3). The reader should now look again at Table 1.5 for the relationship between the positions of the symmetry elements and the point-group symbols, particularly for the tetragonal and orthorhombic systems, from which these detailed examples have been drawn.

In this discussion, we have used a general form, which we may think of as $\{hkl\}$, to represent the point group. Each symmetry-equivalent point lies in a general position (point-group symmetry 1) on the stereogram. Certain crystal planes may coincide with symmetry planes or lie normal to symmetry axes. These planes constitute special forms, and their poles lie in special positions on the stereogram; the forms $\{110\}$ and $\{001\}$ in $4mm$ are examples of special forms. The need for the general form in a correct description of a

---

* This description is not strict; more fully, we may say that the normals to the two forms of m planes are at 45° to one another.

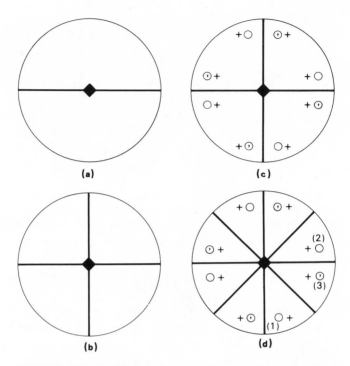

FIGURE 1.36. Intersecting symmetry elements: (a) one m plane intersecting 4 is inconsistent; (b) consistent; (c) points generated by $4m$; (d) complete stereogram, point group $4mm$.

point group is illustrated in Figure 1.37. The poles of the faces on each of the two stereograms shown are identical, although they may be derived for crystals in different classes, $4mm$ and $\bar{4}2m$ in this example. Figure 1.38 shows crystals of these two classes with the {110} form, among others, developed. In Figure 1.38b, the presence of only special forms led originally to an incorrect deduction of the point group of this crystal.

The stereograms for the 32 crystallographic point groups are shown in Figure 1.39. The conventional crystallographic axes are drawn once for each system. Two comments on the nomenclature are necessary at this stage. The symbol $^{-}\oplus^{+}$ indicates two points, O^{+} and $\odot^{-}$, related by a mirror plane in the plane of projection. In the cubic system, the four points related by a fourfold axis in the plane of the stereogram lie on a small circle (Figure 1.40). In general, two of the points are projected from the upper hemisphere and the other two points from the lower hemisphere. We can distinguish them readily by remembering that 2 is a subgroup (page 45) of both 4 and $\bar{4}$.

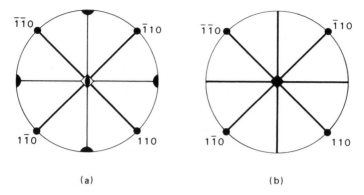

FIGURE 1.37. {110} form in tetragonal point groups: (a) point group $\bar{4}2m$; (b) point group $4mm$.

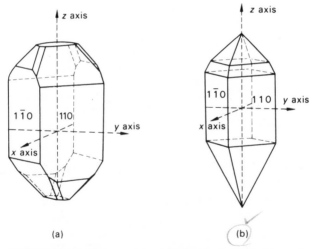

FIGURE 1.38. Tetragonal crystals showing, among others, the {110} form: (a) copper pyrites ($\bar{4}2m$); (b) iodosuccinimide, point group $4mm$; X-ray photographs revealed that the true point group is 4.

Appendix A2 describes a scheme for the study and recognition of the crystallographic point groups. Appendix A3 discusses the Schoenflies symmetry notation for point groups. Because this system is also used, we have written the Schoenflies symbols in Figure 1.39, in parentheses, after the Hermann–Mauguin symbol.

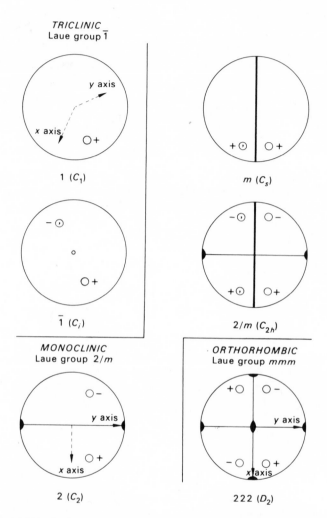

FIGURE 1.39. Stereograms showing both the symmetry elements and the general form $\{hkl\}$ in the 32 crystallographic point groups. The arrangement is by system and common Laue group. The crystallographic axes are named once for each system and the Z axis is chosen normal to the stereogram. The Schoenflies symbols are given in parentheses.

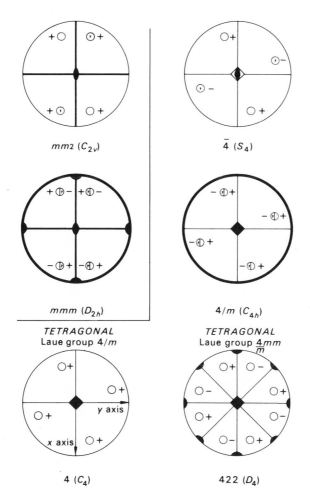

FIGURE 1.39.—*cont.*

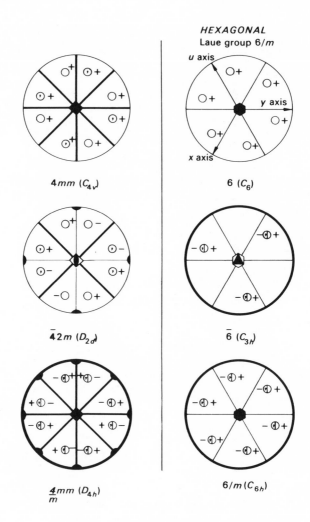

FIGURE 1.39.—cont.

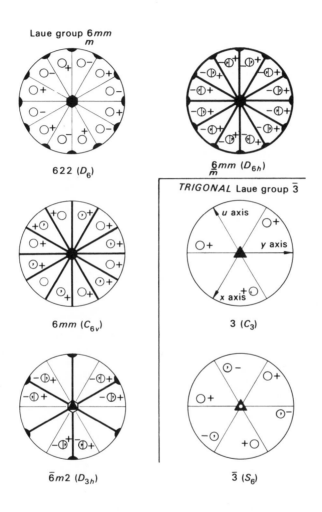

FIGURE 1.39.—*cont.*

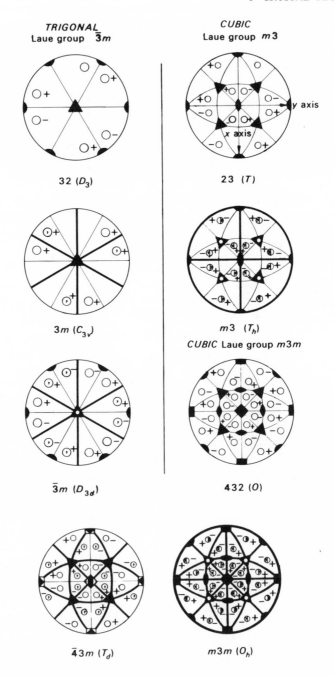

FIGURE 1.39.—cont.

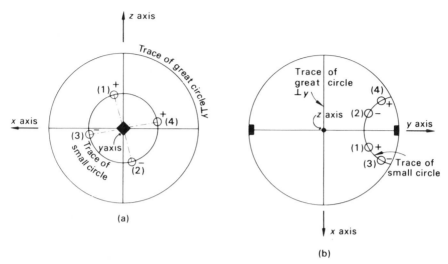

FIGURE 1.40. Stereogram nomenclature for points related by a fourfold axis (y) lying in the plane of a stereogram. The ± signs refer to the z-axis direction: (a) vertical section normal to the y axis, (b) corresponding stereogram; the pairs of points (1)–(2) and (3)–(4) are related by twofold symmetry (subgroup of 4).

Subgroups and Laue Groups

A subgroup of a given point group is a point group of lower symmetry than the given group, contained within it and capable of separate existence as a point group. For example, 32 is a subgroup of $\bar{3}m$, 622, $\bar{6}m2$, $\frac{6}{m}mm$, 432, and $m3m$, whereas $\bar{4}$ is a subgroup of $\frac{4}{m}$, $\bar{4}2m$, $\frac{4}{m}mm$, $\bar{4}3m$, and $m3m$. The subgroup principle provides a rationale for some of the graphic symbols for symmetry elements. Thus, $\bar{4}$ is shown by a square (fourfold rotation), unshaded (to distinguish it from 4), and with a twofold rotation symbol inscribed (2 is a subgroup of $\bar{4}$).

Point group $\bar{1}$ and point groups that have $\bar{1}$ as a subgroup are centrosymmetric. Since, as we shall see, X-ray diffraction effects are, in general, centrosymmetric, the symmetry pattern of X-ray diffraction spots on a flat-plate film, obtained from any crystal, can exhibit only the symmetry that would be obtained from a crystal having the corresponding centrosymmetric point group.

There are 11 such point groups; they are called Laue groups, since symmetry is often investigated by the Laue X-ray method (page 126). Neither the Laue photograph, however, nor any other X-ray photograph can show directly the presence (or absence) of a center of symmetry in a crystal. In Table 1.6, the point groups are classified according to their Laue group, and the symmetry of the Laue flat-plate film photographs is given for directions of the X-ray beam normal to the crystallographic forms listed. The Laue-projection symmetry corresponds to one of the ten two-dimensional point groups.

What is the Laue-projection symmetry on {120} of a crystal of point group 422? This question can be answered with the stereogram of the

TABLE 1.6. Laue Groups and Laue-Projection Symmetry

System	Point groups	Laue group	Laue-projection symmetry normal to the given form		
			{100}	{010}	{001}
Triclinic	$1, \bar{1}$	$\bar{1}$	1	1	1
Monoclinic	$2, m, 2/m$	$2/m$	m	2	m
Orthorhombic	$222, mm2,$ mmm	mmm	$2mm$	$2mm$	$2mm$
			{001}	{100}	{110}
Tetragonal	$4, \bar{4}, 4/m$	$4/m$	4	m	m
	$422, 4mm,$ $\bar{4}2m, \frac{4}{m}mm$	$\frac{4}{m}mm$	$4mm$	$2mm$	$2mm$
			{0001}	{10$\bar{1}$0}	{11$\bar{2}$0}
Trigonal[a]	$3, \bar{3}$	$\bar{3}$	3	1	1
	$32, 3m, \bar{3}m$	$\bar{3}m$	$3m$	m	2
Hexagonal	$6, \bar{6}, 6/m$	$6/m$	6	m	m
	$622, 6mm,$ $\bar{6}m2, \frac{6}{m}mm$	$\frac{6}{m}mm$	$6mm$	$2mm$	$2mm$
			{100}	{111}	{110}
Cubic	$23, m3$	$m3$	$2mm$	3	m
	$432, \bar{4}3m,$ $m3m$	$m3m$	$4mm$	$3m$	$2mm$

[a] Referred to hexagonal axes.

corresponding Laue group, $\frac{4}{m}$ mm. Reference to the appropriate diagram in Figure 1.39 shows that an X-ray beam traveling normal to $\{hk0\}$ $(h \neq k)$ encounters only m symmetry. The entries in Table 1.6 can be deduced in this way. The reader should refer again to Table 1.5 and compare it with Table 1.6.

Noncrystallographic Point Groups

We have seen that in crystals the elements R and $\bar{R}$ are limited to the numerical values 1, 2, 3, 4, and 6. However, there are molecules that exhibit symmetries other than those of the crystallographic point groups. Indeed, R can, in principle, take any integer value between one and infinity. The statement $R = \infty$ implies cylindrical symmetry; the molecule of carbon monoxide has an ∞-axis along the C—O bond, if we assume spherical atoms.

In biscyclopentadienyl ruthenium (Figure 1.41) a fivefold symmetry axis is present, and the point-group symbol may be written as $\overline{10}m2$, or $\frac{5}{m}$ m. The stereogram of this point group is shown in Figure 1.42; the symbol for $\frac{5}{m}$ $(\overline{10})$ is not standard.

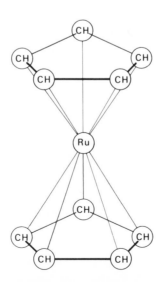

FIGURE 1.41. Biscyclopentadienyl ruthenium, $(C_5H_5)_2Ru$.

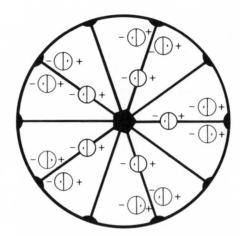

FIGURE 1.42. Stereogram of the noncrystallo-
graphic point group $\overline{1}0m2$ (D_{5h}) showing the general
form (20 poles) and a special form of 10 poles, lying
on the m planes, which can be used to represent the
CH groups in $(C_5H_5)_2Ru$.

Other examples of noncrystallographic point groups will be encountered among chemical molecules, and always a stereogram can be used to
represent the point-group symmetry. In every such example, however, the
substance will crystallize in one of the seven crystal systems and the crystals
will belong to one of the 32 crystal classes.

Bibliography

General and Historical Study of Crystallography

BRAGG, W. L., *A General Survey* (*The Cyrstalline State*, Vol. I), London,
Bell (1949).

EWALD, P. P. (Editor), *Fifty years of X-Ray Diffraction*, Utrecht, Oosthoek
(1962).

Crystal Morphology and Stereographic Projection

PHILLIPS, F. C., *An Introduction to Crystallography*, London, Longmans
(1971).

Crystal Symmetry and Point Groups

HAHN, T. (Editor), *International Tables for Crystallography*, Vol. A, Dordrecht, D. Reidel (1983).

HENRY, N. F. M., and LONSDALE, K. (Editors), *International Tables for X-Ray Crystallography*, Vol. I, Birmingham, Kynoch Press (1965).

Problems

1.1. The line AC (Figure P1.1) may be indexed as (12) with respect to the rectangular axes x and y. What are the "indices" of the same line with respect to the axes x' and y, where the angle $x'Oy = 120°$? PQ is the parametral line for both sets of axes, and $OB/OA = 2$.

1.2. Write the Miller indices for planes that make the intercepts given below:

(a) $a, -b/2, \|c$.   (b) $2a, b/3, c/2$.
(c) $\|a, \|b, -c$.   (d) $a, -b, 3c/4$.
(e) $\|a, -b/4, c/3$.   (f) $-a/4, b/2, -c/3$.

1.3. Evaluate zone symbols for the pairs of planes given below:

(a) $(123), (0\bar{1}1)$.   (b) $(20\bar{3}), (111)$.
(c) $(41\bar{5}), (1\bar{1}0)$.   (d) $(\bar{1}1\bar{2}), (001)$.

1.4. What are the Miller indices of the plane that lies in both of the zones $[123]$ and $[\bar{1}1\bar{1}]$? Why are there, apparently, two answers to this problem and to each part of **1·3**?

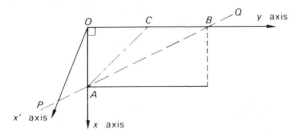

FIGURE P1.1

*1.5. Anatase, TiO_2, is tetragonal, and the following goniometric measurements have been recorded in degrees for the zones shown.

Zones $a, m, a, \ldots$		Zones $m, s, p, \ldots$		Zones $a, r, r, \ldots$	
$\widehat{am}$	45.0	$\widehat{ms}$	11.2	$\widehat{ar}$	53.9
$\widehat{aa}$	90.0	$\widehat{sp}$	10.4	$\widehat{rr}$	72.3
		$\widehat{pr}$	11.9		
Zones $a, s, s, \ldots$		$\widehat{rz}$	16.5	Zones $a, e, a, \ldots$	
		$\widehat{zz}$	79.9		
$\widehat{as}$	46.0			$\widehat{ae}$	29.4
$\widehat{ss}$	87.8	Zones $a, p, p, \ldots$		$\widehat{ee}$	121.3
		$\widehat{ap}$	48.9		
Zones $a, z, z, \ldots$		$\widehat{pp}$	82.2	Zones $m, e, e, \ldots$	
$\widehat{az}$	63.0			$\widehat{me}$	51.9
$\widehat{zz}$	54.0			$\widehat{ee}$	76.1

(a) Each result is the mean of eight measurements, except for aa where four results have been averaged. Study the morphology of the anatase crystal illustrated in Figure P1.2. Plot the goniometric measurements on a stereogram, radius 3 in. Let zones a, m, a, ... and a, e, e, ... be on the primitive circle and running from bottom to top, respectively.

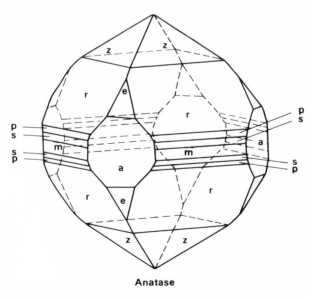

Anatase

FIGURE P1.2

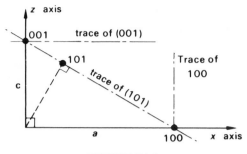

FIGURE P1.3

(b) From a consideraton of the construction in Figure P1.3, determine the axial ratio c/a.

(c) Index the poles on the stereogram.

(d) List the indices of the general forms and the special forms present on this crystal.

(e) Write the point group of anatase.

1.6. Take the cover of a matchbox (Figure P1.4a).

(a) Ignore the label, and write down its point group.

Squash it diagonally (Figure P1.4b).

(b) What is the point group now?

(c) In each case, what is the point group if the label is not ignored?

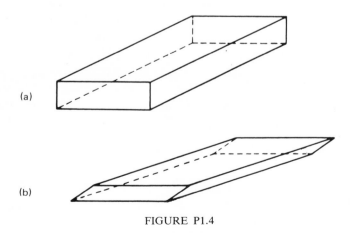

(a)

(b)

FIGURE P1.4

1.7. Draw stereograms to show the general form in each of the point groups deduced in Problems 1.6a and 1.6b. Satisfy yourself that in 1.6a three, and in 1.6b two, symmetry operations carried out in sequence produce a resultant action that is equivalent to another operation in the group.

1.8. How many planes are there in the forms $\{010\}$, $\{\bar{1}10\}$, and $\{11\bar{3}\}$ in each of the point groups $2/m$, $\bar{4}2m$, and $m3$?

1.9. What symmetry would be revealed by Laue flat-film photographs where the X-ray beam is normal to a plane in the form given in each of the examples below?

	Point group	Orientation		Point group	Orientation
(a)	$\bar{1}$	$\{100\}$	(f)	$3m$	$\{11\bar{2}0\}$
(b)	$mm2$	$\{011\}$	(g)	$\bar{6}$	$\{0001\}$
(c)	m	$\{010\}$	(h)	$\bar{6}m2$	$\{0001\}$
(d)	422	$\{120\}$	(i)	23	$\{111\}$
(e)	3	$\{10\bar{1}0\}$	(j)	432	$\{110\}$

In some examples, it may help to draw stereograms.

1.10. (a) What is the nontrivial symmetry of the figure obtained by packing a number of equivalent but irregular quadrilaterals in one plane?

(b) What is the symmetry of the Dobermann in Fig. P1.5? This example illustrates how one can study symmetry by means of everyday objects.

1.11. Name each species of molecule or ion in the 10 drawings of Figure P1.6 and write down its point-group symbol in both the Hermann–Mauguin and Schoenflies nomenclatures. Study the stereograms in Figure 1.39, and suggest the form $\{hkl\}$, the normals to faces of which may be identified with the following bond directions: (a) C—O (three bonds), (b) Ni—C (four bonds), (c) Pt—Cl (six bonds), (d) C—H (four bonds). In some examples, it may help to make ball-and-spoke models.*

* A simple apparatus for this purpose is marketed by Morris Laboratory Instruments, 480 Bath Road, Slough, Berks., England. [See also M. F. C. Ladd, Crystal Structure Models, *Education in Chemistry* **5**, 186 (1968); A.Walton (1978) *Molecular and Crystal Structure Models* (Ellis Horwood Ltd.)].

FIGURE P1.5. "Vijentor Seal of Approval at Valmara."

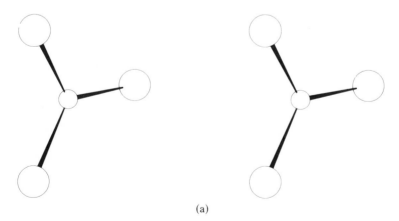

(a)

FIGURE P1.6. (a) $[CO_3]^{2-}$. Planar. (b) $[Ni(CN)_4]^{2-}$. Planar. (c) $[PtCl_6]^{2-}$. All Cl–Pt–Cl angles are 90°. (d) CH_4. All H–C–H angles are 109.47°. (e) $CHCl_3$. Pyramidal. (f) CHBrClI. (g) C_6H_6. Planar. (h) C_6H_5Cl. Planar. (i) $C_6H_4Cl_2$. Planar. (j) C_6H_4BrCl. Planar.

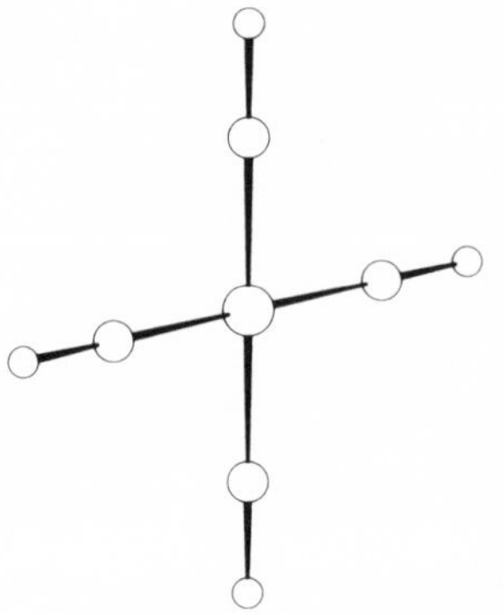

(b)

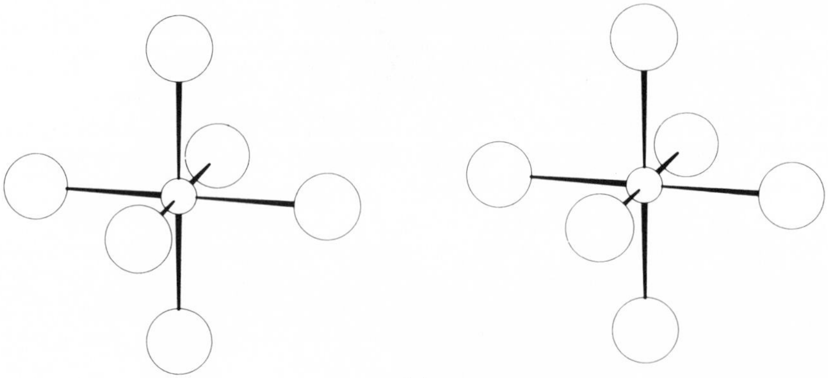

(c)

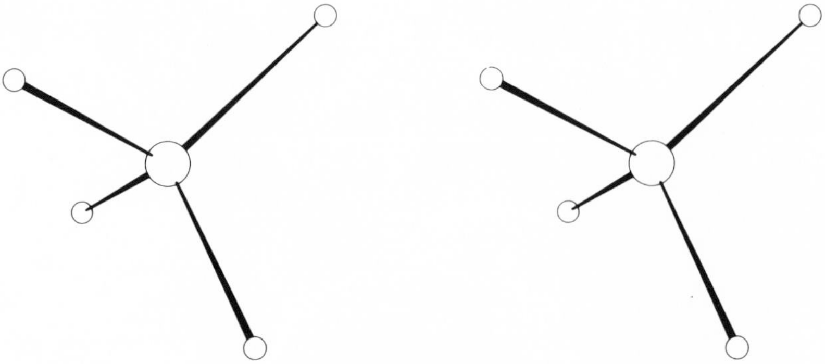

(d)

FIGURE P1.6—*cont.*

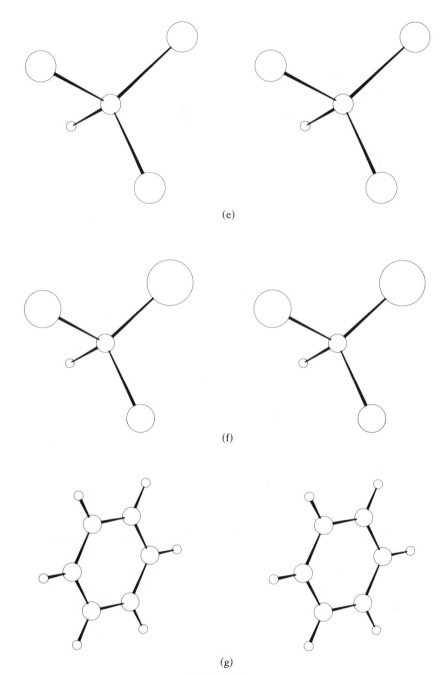

(e)

(f)

(g)

FIGURE P1.6—*cont.*

(h)

(i)

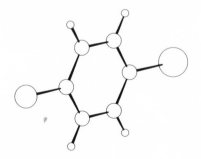

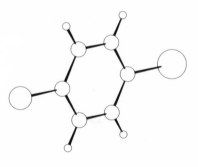

(j)

FIGURE P1.6—*cont.*

2

Crystal Geometry. II

2.1 Introduction

In this chapter, we continue our study of crystal geometry by investigating the internal arrangements of crystalline materials. Crystals are characterized by periodicities in three dimensions. An atomic grouping is repeated over and over again by a certain symmetry mechanism so as to build up a crystal, and thus we are led to a consideration of space-group symmetry. A space-group pattern in its simplest form may be considered to be derived by repeating a motif having a crystallographic point-group symmetry by the translations of a lattice, and so it is appropriate to examine lattices next.

2.2 Lattices

Every crystal has a lattice as its geometric basis. A lattice may be described as a regular, infinite arrangement of points in which every point has the same environment as any other point. This description is applicable, equally, in one-, two-, and three-dimensional space.

Lattice geometry is described in relation to three basic repeat, or translation, vectors $\mathbf{a}$, $\mathbf{b}$, and $\mathbf{c}$. Any point in the lattice may be chosen as an origin, whence a vector $\mathbf{r}$ to any other lattice point is given by

$$\mathbf{r} = U\mathbf{a} + V\mathbf{b} + W\mathbf{c} \tag{2.1}$$

where U, V, and W are positive or negative integers or zero, and represent the coordinates of the lattice point. The direction, or directed line, joining the origin to the points $U, V, W; 2U, 2V, 2W; \ldots; nU, nV, nW$ defines the row $[UVW]$. A set of such rows, or directions, related by the lattice symmetry constitutes a form of directions $\langle UVW \rangle$ (compare zone symbols, page 16).

2.2.1 Two-Dimensional Lattices

We begin our study of lattices in two dimensions rather than three. A two-dimensional lattice is called a net; it may be imagined as being formed by aligning, in a regular manner, rows of equally spaced points (Figure 2.1). The net is the array of points; the connecting lines are a convenience, drawn to aid our appreciation of the lattice geometry.

Since nets exhibit symmetry, they can be allocated to the two-dimensional systems (page 29). The most general net is shown in Figure 2.1b. A sufficient representative portion of the lattice is the unit cell, outlined by the vectors **a** and **b**; an infinite number of such unit cells stacked side by side builds up the net.

The net under consideration exhibits twofold rotational symmetry about each point; consequently, it is placed in the oblique system. The chosen unit cell is primitive (symbol p), which means that one lattice point is associated with the area of the unit cell; each point is shared equally by four adjacent unit cells. In the oblique unit cell, $a \neq b$, and $\gamma \neq 90°$ or $120°$; angles of $90°$ and $120°$ may imply symmetry higher than 2.

Consider next the stacking of unit cells in which $a \neq b$ but $\gamma = 90°$ (Figure 2.2). The symmetry at every point is $2mm$, and this net belongs to the rectangular system. The net in Figure 2.3 may be described by a unit cell in which $a' = b'$ and $\gamma' \neq 90°$ or $120°$. It may seem at first that this net is oblique, but careful inspection shows that each point has $2mm$ symmetry, and so this net, too, is allocated to the rectangular system.

In order to display this fact clearly, a centered (symbol c) unit cell is chosen, shown in Figure 2.3 by the vectors **a** and **b**. This cell has two lattice points per unit cell area. It is left as an exercise to the reader to show that a

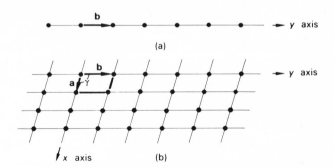

FIGURE 2.1. Formation of a net: (a) row of equally spaced
points, (b) regular stack of rows.

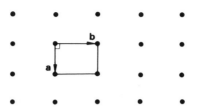

FIGURE 2.2. Rectangular net with a
p unit cell drawn in.

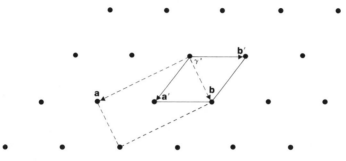

FIGURE 2.3. Rectangular net with p and c unit cells drawn in.

centered, oblique unit cell does not represent a net which is fundamentally
different from that in Figure 2.1.

2.2.2 Choice of Unit Cell

From the foregoing discussion, it will be evident that the choice of unit
cell is somewhat arbitrary. We shall follow a universal crystallographic
convention in choosing a unit cell: The unit cell is the smallest repeat unit for
which its delineating vectors are parallel to, or coincide with, important
symmetry directions in the lattice. Returning to Figure 2.3, the centered cell
is preferred because **a** and **b** coincide with the symmetry (m) lines in the net.
The primitive unit cell $(\mathbf{a}', \mathbf{b}')$ is, of course, a possible unit cell, but it does not,
in isolation, reveal the lattice symmetry clearly. The symmetry is still there;
it is invariant under choice of unit cell, as is shown by the following
equations:

$$a'^2 = a^2/4 + b^2/4 \qquad (2.2)$$

$$b'^2 = a^2/4 + b^2/4 \qquad (2.3)$$

the value of γ' depends only on the ratio a/b.

TABLE 2.1. Two-Dimensional Lattices

System	Unit cell symbol(s)	Symmetry at lattice points	Unit cell edges and angles
Oblique	p	2	$a \neq b$, $\gamma \neq 90°, 120°$
Rectangular	p, c	$2mm$	$a \neq b$, $\gamma = 90°$
Square	p	$4mm$	$a = b$, $\gamma = 90°$
Hexagonal	p	$6mm$	$a = b$, $\gamma = 120°$

Two other nets are governed by the unit cell relationships $a = b$, $\gamma = 90°$ and $a = b$, $\gamma = 120°$; their study constitutes the first problem at the end of this chapter. The five two-dimensional lattices are summarized in Table 2.1.

2.2.3 Three-Dimensional Lattices

The three-dimensional lattices, or Bravais lattices, may be imagined as being developed by regular stacking of nets. There are 14 ways in which this can be done, and the Bravais lattices are distributed, unequally, among the seven crystal systems, as shown in Figure 2.4. Each lattice is represented by a unit cell, outlined by three noncoplanar vectors* **a**, **b**, and **c**. In accordance with convention, these vectors are chosen so that they both form a parallelepipedon of smallest volume in the lattice and are parallel to, or coincide with, important symmetry directions in the lattice. In three dimensions, we encounter unit cells centered on a pair of opposite faces, body-centered, or centered on all faces. Table 2.2 lists the unit cell types and their nomenclature.

Triclinic Lattice

If oblique nets are stacked in a general, but regular, manner, a triclinic lattice is obtained (Figure 2.5). The unit cell is characterized by $\bar{1}$ symmetry at each lattice point, with the conditions $a \neq b \neq c$ and $\alpha \neq \beta \neq \gamma \neq 90°, 120°$. This unit cell is primitive (symbol P),† which means that one lattice point is associated with the unit cell volume; each point is shared equally by eight adjacent unit cells in three dimensions (see Figure 2.6). There is no symmetry direction to constrain the choice of the unit cell vectors, and a parallelepipedon of smallest volume can always be chosen conventionally. The angles α, β, and γ are selected to be oblique, as far as is practicable.

---

* The magnitudes of the edges a, b, and c and the angles α, β, and γ provide an equivalent description.

† Capital letters are used in three dimensions.

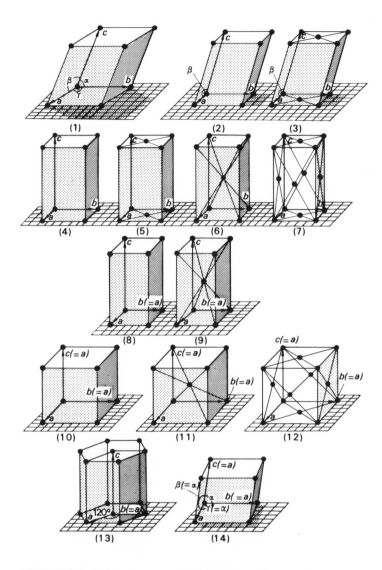

FIGURE 2.4. Unit cells of the 14 Bravais lattices; interaxial angles are
90° unless indicated otherwise by a numerical value or symbol: (1) triclinic
P, (2) monoclinic P, (3) monoclinic C, (4) orthorhombic P, (5) orthorhombic C, (6) orthorhombic I, (7) orthorhombic F, (8) tetragonal P, (9)
tetragonal I, (10) cubic P, (11) cubic I, (12) cubic F, (13) hexagonal P, (14)
trigonal R. It should be noted that (13) shows three P hexagonal unit
cells. A hexagon of lattice points (without the central point in the basal
planes shown) does not lead to a lattice. Why?

TABLE 2.2. Unit Cell Nomenclature

Centering site(s)	Symbol	Miller indices of centered faces of the unit cell	Fractional coordinates[a] of centered sites
None	P	—	—
bc faces	A	100	$0, \frac{1}{2}, \frac{1}{2}$
ca faces	B	010	$\frac{1}{2}, 0, \frac{1}{2}$
ab faces	C	001	$\frac{1}{2}, \frac{1}{2}, 0$
Body center	I	—	$\frac{1}{2}, \frac{1}{2}, \frac{1}{2}$
All faces	F	$\begin{cases} 100 \\ 010 \\ 001 \end{cases}$	$\begin{cases} 0, \frac{1}{2}, \frac{1}{2} \\ \frac{1}{2}, 0, \frac{1}{2} \\ \frac{1}{2}, \frac{1}{2}, 0 \end{cases}$

[a] A fractional coordinate x is given by x/a, where x is the coordinate in absolute measure (Å) and a is the unit-cell repeat distance in the same direction and in the same units.

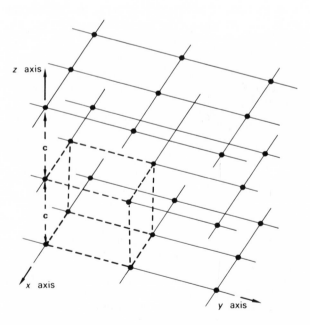

FIGURE 2.5. Oblique nets stacked to form a triclinic lattice.

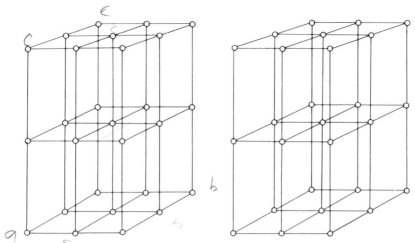

FIGURE 2.6. Stereoview showing eight adjacent P unit cells in a monoclinic lattice. The sharing of corner points can be seen readily by focusing attention on the central lattice point in the drawings.

Monoclinic Lattices

The monoclinic system is characterized by one diad (rotation or inversion), with the y axis (and b) chosen parallel to it. The conventional unit cell is specified by the conditions $a \not\subset b \not\subset c$, $\alpha = \gamma = 90°$, and $\beta \not\subset 90°$, $120°$. Figure 2.6 illustrates a steroscopic pair of drawings of a monoclinic lattice, showing eight unit cells; according to convention, the β angle is chosen to be oblique.

Reference to Figure 2.4 shows that there are two conventional monoclinic lattices, symbolized by the unit cell types P and C.

A monoclinic unit cell centered on the A faces is equivalent to that described as C; the choice of the b axis* is governed by symmetry, but a and c are interchangeable labels.

The centering of the B faces is illustrated in Figure 2.7. In this situation a new unit cell, $\mathbf{a}'$, $\mathbf{b}'$, $\mathbf{c}'$, can be defined by the following equations:

$$\mathbf{a}' = \mathbf{a} \tag{2.4}$$

$$\mathbf{b}' = \mathbf{b} \tag{2.5}$$

$$\mathbf{c}' = \mathbf{a}/2 + \mathbf{c}/2 \tag{2.6}$$

If β is not very obtuse, the equivalent transformation $\mathbf{c}' = -\mathbf{a}/2 + \mathbf{c}/2$ can ensure that β' is obtuse.

* We often speak of the b axis (to mean y axis) because our attention is usually confined to the unit cell.

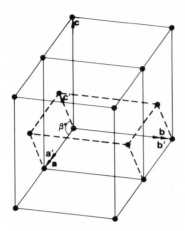

FIGURE 2.7. Monoclinic lattice
showing that $B \equiv P$.

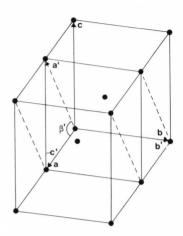

FIGURE 2.8. Monoclinic lattice
showing that $I \equiv C$.

Since c' lies in the ac plane, $\alpha' = \gamma' = 90°$, but $\beta' \not\subset 90°$ or $120°$. The new monoclinic cell is primitive; symbolically we may write $B \equiv P$. Similarly, it may be shown that $I \equiv F \equiv C$ (Figures 2.8 and 2.9).

If the C cell (Figure 2.10) is reduced to primitive, it no longer displays the characteristic monoclinic symmetry clearly (see Table 2.3), and we may conclude that there are two monoclinic lattices, described by the unit cell types P and C.

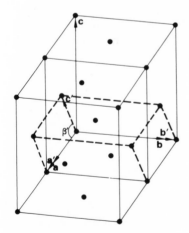

FIGURE 2.9. Monoclinic lattice
showing that $F \equiv C$.

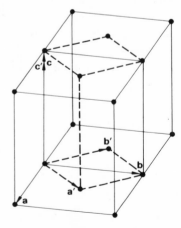

FIGURE 2.10. Monoclinic lattice
showing that $C \not\equiv P$.

It may be necessary to calculate the new dimensions of a transformed unit cell. Consider the example $B \rightarrow P$, (2.4)–(2.6). Clearly, $a' = a$ and $b' = b$. Taking the scalar product* of (2.6) with itself, we obtain

$$\mathbf{c}' \cdot \mathbf{c}' = (\mathbf{a}/2 + \mathbf{c}/2) \cdot (\mathbf{a}/2 + \mathbf{c}/2) \tag{2.7}$$

Hence,

$$c'^2 = a^2/4 + c^2/4 + ac(\cos \beta)/2 \tag{2.8}$$

The new angle β' is given by†

$$\cos \beta' = \mathbf{a}' \cdot \mathbf{c}'/a'c' \tag{2.9}$$

Using (2.6), we obtain

$$\cos \beta' = [a/2 + c(\cos \beta)/2]/c' \tag{2.10}$$

where c' is given by (2.8). These calculations can be applied to all crystal systems, giving due consideration to any nontrivial relationships between a, b, and c and between α, β, and γ.

Orthorhombic Lattices

The monoclinic system was treated in some detail. It will not be necessary to give such an extensive discussion for either this system or the remaining crystal systems. Remember always to think of the unit cell as a representative portion of its lattice and not as a finite body.

The orthorhombic system is characterized by three mutually perpendicular diads (rotation and/or inversion); the unit cell vectors are chosen to be parallel to, or coincide with, these symmetry axes. The orthorhombic unit cell is specified by the relationships $a \not\subset b \not\subset c$ and $\alpha = \beta = \gamma = 90°$. It will not be difficult for the reader to verify that the descriptions P, C, I, and F are necessary and sufficient in this system. One way in which this exercise may be carried out is as follows. After centering the P unit cell, three questions must be asked, in the following order:

1. Does the centered cell represent a true lattice?
2. If it is a lattice, is the symmetry changed?
3. If the symmetry is unchanged, does it represent a new lattice, and has the unit cell been chosen correctly?

We answered these questions implicitly in discussing the monoclinic lattices.

---

*The scalar (dot) product of two vectors $\mathbf{p}$ and $\mathbf{q}$ is denoted by $\mathbf{p} \cdot \mathbf{q}$ and is equal to $pq(\cos \widehat{pq})$, where $\widehat{pq}$ represents the angle between the positive directions of $\mathbf{p}$ and $\mathbf{q}$.
† To make β' obtuse, it may be necessary to use $-\mathbf{a}/2$ in (2.7).

It should be noted that the descriptions A, B, and C do not remain equivalent for orthorhombic space groups in the class $mm2$; it is necessary to distinguish C from A (or B). The reader may like to consider now, or later, why this distinction is necessary.

Tetragonal Lattices

The tetragonal system is characterized by one tetrad (rotation or inversion along z (and c); the unit cell conditions are $a = b \not C c$ and $\alpha = \beta = \gamma = 90°$. There are two tetragonal lattices, specified by the unit cell symbols P and I (Figure 2.4); C and F tetragonal unit cells may be transformed to P and I, respectively.

Cubic Lattices

The symmetry of the cubic system is characterized by four triad axes at angles of $\cos^{-1}(1/3)$ to one another; they are the body diagonals $\langle 111 \rangle$ of a cube. The threefold axes in this orientation introduce twofold axes along $\langle 100 \rangle$. There are three cubic Bravais lattices (Figure 2.4) with conventional unit cells P, I, and F.

Hexagonal Lattice

The basic feature of a hexagonal lattice is that it should be able to accommodate a sixfold symmetry axis. This requirement is achieved by a lattice based on a P unit cell, with $a = b \not C c$, $\alpha = \beta = 90°$, and $\gamma = 120°$, the c direction being parallel to a sixfold axis in the lattice.

Lattices in the Trigonal System

A two-dimensional unit cell in which $a = b$ and $\gamma = 120°$ is compatible with either sixfold or threefold symmetry (see Figure 2.22, plane groups $p6$ and $p3$). For this reason, the hexagonal lattice (P unit cell) is used for certain crystals which belong to the trigonal system. However, as shown in Figure 2.11, the presence of two threefold axes within a unit cell, with x, y coordinates of $\frac{2}{3}, \frac{1}{3}$ and $\frac{1}{3}, \frac{2}{3}$, respectively, and parallel to the Z axis, introduces the possibility of a crystal structure which belongs to the trigonal system but has a triply primitive hexagonal unit cell R_{hex} with centering points at $\frac{2}{3}, \frac{1}{3}, \frac{1}{3}$ and $\frac{1}{3}, \frac{2}{3}, \frac{2}{3}$ in the unit cell.

Thus for some trigonal crystals the unit cell will be P, and for others it will be R_{hex}, the latter being distinguished by systematically absent X-ray

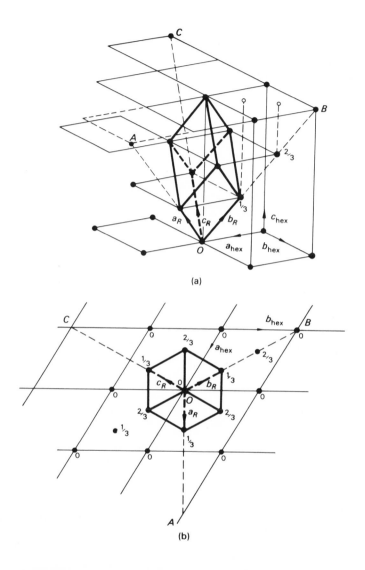

(a)

(b)

FIGURE 2.11. Trigonal lattice (the fractions refer to values of c_{hex}): (a) rhombohedral (R) unit cell (obverse setting) developed from a triply primitive hexagonal (R_{hex}) unit cell. (In the *reverse* setting, the rhombohedral lattice and unit cell are rotated about [111] 60° clockwise with respect to the R_{hex} axes.) The ratio of the volumes of any two unit cells in one and the same lattice is equal to the ratio of the numbers of lattice points in the two unit cell volumes. (b) Plan view of (a) as seen along c_{hex}.

TABLE 2.3. Fourteen Bravais Lattices

System	Unit cell(s)	Symmetry at lattice points	Axial relationships
Triclinic	P	$\bar{1}$	$a \neq b \neq c; \alpha \neq \beta \neq \gamma \neq 90°, 120°$
Monoclinic	P, C	$2/m$	$a \neq b \neq c; \alpha = \gamma = 90°; \beta \neq 90°, 120°$
Orthorhombic	P, C, I, F	mmm	$a \neq b \neq c; \alpha = \beta = \gamma = 90°$
Tetragonal	P, I	$\frac{4}{m}mm$	$a = b \neq c; \alpha = \beta = \gamma = 90°$
Cubic	P, I, F	$m3m$	$a = b = c; \alpha = \beta = \gamma = 90°$
Hexagonal	P	$\frac{6}{m}mm$	$a = b \neq c; \alpha = \beta = 90°; \gamma = 120°$
Trigonal[a]	R or P	$\bar{3}m$	$a = b = c; \alpha = \beta = \gamma \neq 90°, <120°$

[a] On hexagonal axes, column 4 is the same as for the hexagonal system, but the symmetry at each lattice points remains $\bar{3}m$.

reflections (Table 4.1). The R_{hex} cell can be transformed to a primitive rhombohedral unit cell R, with $a = b = c$ and $\alpha = \beta = \gamma \neq 90°, <120°$; the threefold axis is then along [111]. The R cell may be thought of as a cube extended or squashed along one of its threefold axes.

The lattice based on an R unit cell is the only truly exclusive trigonal lattice, the lattice based on a P unit cell being borrowed from the hexagonal system (Table 2.3).

2.3 Families of Planes and Interplanar Spacings

Figure 2.12 shows one unit cell of an orthorhombic lattice projected onto the ab plane. The trace of the (110) plane nearest the origin O is indicated by a dashed line, and the perpendicular distance of this plane from O is $d(110)$. By repeating the operation of the translation $\pm\mathbf{d}(110)$ on the plane (110), a series, or family, of parallel, equidistant planes is generated, as shown in Figure 2.13. Our discussion of the external symmetry of crystals led to a description of the external faces of crystals by Miller indices, which are by definition prime to one another. In discussing X-ray diffraction effects, however, it is necessary to consider planes for which the indices h, k, and l may contain a common factor while still making intercepts a/h, b/k, and c/l on the X, Y, and Z axes, respectively, as required by the definition of Miller indices. It follows that the plane with indices (nh, nk, nl) makes

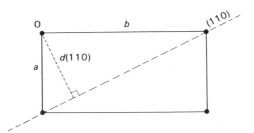

FIGURE 2.12. One unit cell in an orthorhombic
lattice as seen in projection along c.

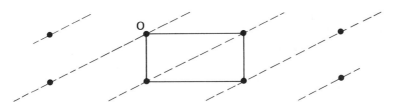

FIGURE 2.13. Family of (110) planes.

intercepts a/nh, b/nk, and c/nl along x, y, and z, respectively, and that this
plane is nearer to the origin by a factor of $1/n$ than is the plane (hkl). In other
words, $d(nh, nk, nl) = d(hkl)/n$. In general, we denote a family of planes as
(hkl) where h, k, and l may contain a common factor. For example, the (220)
family of planes is shown in Figure 2.14 with interplanar spacing $d(220) =
d(110)/2$; alternate (220) planes therefore coincide with (110) planes. It
should be noted that an external crystal face normal to $d(hh0)$ would always
be designated (110), since external observations reveal the shape but not the
size of the unit cell.

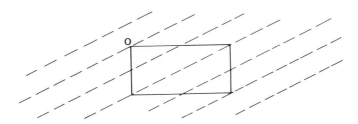

FIGURE 2.14. Family of (220) planes.

2.4   Reciprocal Lattice

We introduce the reciprocal lattice concept in this chapter; it will be needed in the study of X-ray diffraction from crystals. For each direct (Bravais) lattice, a corresponding reciprocal lattice may be postulated. It has the same symmetry as the direct lattice, and may be derived from it graphically. Let Figure 2.15a represent a monoclinic direct lattice as seen in a direction normal to the (010) plane. From the origin O of the P unit cell, lines are drawn normal to families of planes (hkl) in direct space. It may be noted in passing that the normal to a plane (hkl) does not, in general, coincide with the direction of the same indices $[hkl]$. Along each line, reciprocal lattice points hkl (no parentheses) are marked off such that the distance from the origin to the first point in any line is inversely proportional to the corresponding interplanar spacing $d(hkl)$.†

In three dimensions, we refer to $d^*(100)$, $d^*(010)$, and $d^*(001)$ as a^*, b^*, and c^*, respectively, and so define a unit cell in the reciprocal lattice. In general,

$$d^*(hkl) = K/d(hkl) \qquad (2.11)$$

where K is a constant. Hence, for the monoclinic system,

$$a^* = K/d(100) = K/(a \sin \beta) \qquad (2.12)$$

From Figure 2.15a, the scalar product $\mathbf{a} \cdot \mathbf{a}^*$ is given by

$$\mathbf{a} \cdot \mathbf{a}^* = aa^* \cos(\beta - 90°) = aK\frac{\cos(\beta - 90°)}{a \sin \beta} = K \qquad (2.13)$$

The mixed scalar products, such as $\mathbf{a} \cdot \mathbf{b}^*$ are identically zero, because the angle between a and b^* is 90° (see Figure 2.15a).

The reciprocal lattice points form a true lattice with a representative unit cell outlined by $\mathbf{a}^*$, $\mathbf{b}^*$, and $\mathbf{c}^*$, and, therefore, involving six reciprocal cell parameters in the most general case—three sides a^*, b^*, and c^*, and three angles α^*, β^*, and γ^*. The size of the reciprocal cell is governed by the choice of the constant K. In practice, K is frequently taken as the wavelength λ of the X-radiation used; reciprocal lattice units are then dimensionless.

† The correspondence between directions of reciprocal lattice vectors containing no common factor and those in a spherical projection (see Figure 1.20) should be noted.

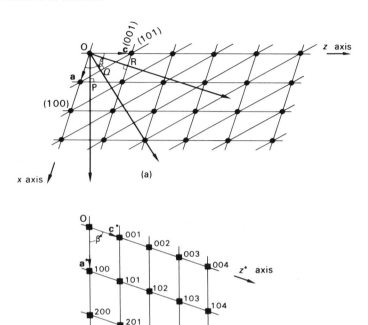

FIGURE 2.15. Direct and reciprocal lattices: (a) monoclinic P, as seen in projection along b, showing three families of planes; (b) corresponding reciprocal lattice.

A reciprocal lattice row hkl; $2h,2k,2l; \ldots$ may be considered to be derived from the families of planes (nh, nk, nl) with $n = 1, 2, \ldots$, since $d(nh, nk, nl) = d(hkl)/n$. Hence,

$$d^*(nh, nk, nl) = nd^*(hkl) \qquad (2.14)$$

where $d^*(hkl)$ is the distance of the reciprocal lattice point hkl from the origin, expressed in reciprocal lattice units (RU). The vector $\mathbf{d}^*(hkl)$ is given by

$$\mathbf{d}^*(hkl) = h\mathbf{a}^* + k\mathbf{b}^* + l\mathbf{c}^* \qquad (2.15)$$

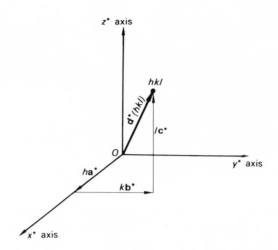

FIGURE 2.16. Vector components of $\mathbf{d}^*(hkl)$ in a reciprocal lattice. Following the appropriate vector paths parallel to the three reciprocal lattice axes x^*, y^*, and z^* gives the result, $\mathbf{d}^*(hkl) = h\mathbf{a}^* + k\mathbf{b}^* + l\mathbf{c}^*$.

This equation is illustrated in Figure 2.16. Equation (2.15) provides a straightforward method for deriving expressions for d^* and d. Thus, generally, from (2.15)

$$\mathbf{d}^*(hkl) \cdot \mathbf{d}^*(hkl) = d^{*2}(hkl) = h^2 a^{*2} + k^2 b^{*2} + l^2 c^{*2} + 2kl b^* c^* \cos \alpha^* \\ + 2lh c^* a^* \cos \beta^* + 2hk a^* b^* \cos \gamma^*$$

$$(2.16)$$

Now $d(hkl)$ may be obtained from (2.11) and (2.16). Simplifications of (2.16) arise through symmetry constraints on the unit cell vectors in different crystal systems. The reader should check the entries in Table 2.4, starting with Table 2.3 and equation (2.16).

It should be noted that the relationship between the direct and reciprocal lattices requires that $\mathbf{a}^*$ be perpendicular to both $\mathbf{b}$ and $\mathbf{c}$, and, conversely, that $\mathbf{a}$ be perpendicular to both $\mathbf{b}^*$ and $\mathbf{c}^*$, and so on for the other cell constants. Hence, in converting from direct to reciprocal space and vice versa, the crystal system is preserved. For an orthogonal lattice, the direct and reciprocal axes are coincident.

We give, without proof, the following general formulae for the triclinic system:

$$a^* = Kbc(\sin \alpha)/V_c \qquad (2.17)$$

TABLE 2.4. Expressions for $d^*(hkl)$ and $d(hkl)$ in the Seven Crystal Systems[a]

System	$d^{*2}(hkl)$	$d^2(hkl)$
Triclinic	$h^2a^{*2}+k^2b^{*2}+l^2c^{*2}+2klb^*c^*\cos\alpha^*$ $+2lhc^*a^*\cos\beta^*+2hka^*b^*\cos\gamma^*$	$K^2/d^{*2}(hkl)$
Monoclinic	$h^2a^{*2}+k^2b^{*2}+l^2c^{*2}+2hla^*c^*\cos\beta^*$	$\left\{\dfrac{1}{\sin^2\beta}\left[\dfrac{h^2}{a^2}+\dfrac{l^2}{c^2}-\dfrac{2hl\cos\beta}{ac}\right]+\dfrac{k^2}{b^2}\right\}^{-1}$
Orthorhombic	$h^2a^{*2}+k^2b^{*2}+l^2c^{*2}$	$\left\{\dfrac{h^2}{a^2}+\dfrac{k^2}{b^2}+\dfrac{l^2}{c^2}\right\}^{-1}$
Tetragonal	$(h^2+k^2)a^{*2}+l^2c^{*2}$	$\left\{\dfrac{h^2+k^2}{a^2}+\dfrac{l^2}{c^2}\right\}^{-1}$
Hexagonal and trigonal (P)	$(h^2+k^2+hk)a^{*2}+l^2c^{*2}$	$\left\{\dfrac{4(h^2+k^2+hk)}{3a^2}+\dfrac{l^2}{c^2}\right\}^{-1}$
Trigonal (R) (rhombohedral)	$[h^2+k^2+l^2+2(hk+kl+hl)(\cos\alpha^*)]a^{*2}$	$a^2(TR)^{-1}$, where $T=h^2+k^2+l^2+2(hk+kl+hl)[(\cos^2\alpha-\cos\alpha)/\sin^2\alpha]$ and $R=(\sin^2\alpha)/(1-3\cos^2\alpha+2\cos^3\alpha)$
Cubic	$(h^2+k^2+l^2)a^{*2}$	$\left\{\dfrac{h^2+k^2+l^2}{a^2}\right\}^{-1}=\dfrac{a^2}{h^2+k^2+l^2}$

[a] In the monoclinic system, $d(100)=a\sin\beta$, $d(001)=c\sin\beta$, and hence $a=K/(a^*\sin\beta^*)$ and $c=K/(c^*\sin\beta^*)$.
In the hexagonal system (and trigonal P), $a=b=K/(a^*\sin\gamma^*)=K/(a^*\sqrt3/2)$.
In general, the expressions for d^{*2} are simpler in form than the corresponding expressions for d^2.

and similarly for b^* and c^*;

$$\cos \alpha^* = (\cos \beta \cos \gamma - \cos \alpha)/(\sin \beta \sin \gamma) \qquad (2.18)$$

and similarly for $\cos \beta^*$ and $\cos \gamma^*$; V_c is the unit cell volume, given by

$$V_c = abc(1 - \cos^2\alpha - \cos^2\beta - \cos^2\gamma + 2 \cos \alpha \cos \beta \cos \gamma)^{1/2} \quad (2.19)$$

Hence, the volume of the reciprocal unit cell V^* is given by

$$V^* = K^3/V_c \qquad (2.20)$$

Simplifications of the expressions (2.17–2.19) arise for other crystal systems. A vector algebraic treatment of the reciprocal lattice is given in Appendix A8.

2.5   Rotational Symmetries of Lattices

We can now discuss analytically the permissible rotational symmetries in lattices, already stated to be of degrees 1, 2, 3, 4, and 6. In Figure 2.17, let A and B represent two adjacent lattice points, of repeat distance t, in any row. An R-fold rotation axis is imagined to act at each point and to lie normal to the plane of the diagram. An anticlockwise rotation of Φ about A maps B onto B', and a clockwise rotation of Φ about B maps A onto A'. Lines AB' and BA' are produced to meet in Q.

Since triangles ABQ and $A'B'Q$ are similar, $A'B'$ is parallel to AB. From the properties of lattices, it follows that $A'B' = Jt$, where J is an integer.

Lines $A'S$ and $B'T$ are drawn perpendicular to AB, as shown. Hence,

$$A'B' = TS = AB - (AT + BS) \qquad (2.21)$$

or

$$Jt = t - 2t \cos \Phi \qquad (2.22)$$

whence

$$\cos \Phi = (1 - J)/2 = M/2 \qquad (2.23)$$

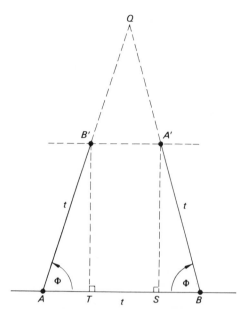

FIGURE 2.17. Rotational symmetry in crystal
lattices. Permissible values of Φ are 90°, 60°,
120°, 360°, 180° (4-fold, 6-fold, 3-fold, 1-fold,
2-fold).

where M is another integer. Since $-1 \le \cos\Phi \le 1$, and, from (2.23), the only
admissible values for M are $0, \pm1, \pm2$, these values give rise to the rotational
symmetries already discussed. This treatment gives a quantitative aspect to
the packing considerations mentioned previously (page 30).

2.6 Space Groups

In order to extend our study of crystal geometry into the realm of
atomic arrangements, we must consider now the symmetry of extended,
ideally infinite, patterns in space. We recall that a point group describes the
symmetry of a finite body, and that a lattice constitutes a mechanism for
repetition, to an infinite extent, by translations parallel to three noncoplanar
directions. We may ask, therefore, what is the result of repeating a point-
group pattern by the translations of a Bravais lattice. We shall see that it is
like an arrangement of atoms in a crystal.

A space group may be described as a set of symmetry elements, the
operation of any of which brings the infinite array of points to which they

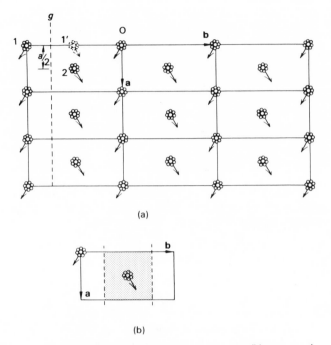

(a)

(b)

FIGURE 2.18. Wallpaper pattern: (a) extended pattern, (b) asymmetric unit.

belong into self-coincidence. We may apply space-group rules to crystals
because the dimensions of crystals used in experimental investigations are
large with respect to the repeat distances of the pattern.

A space group may be considered to be made up of two parts, a pattern
unit and a repeat mechanism. An analogy may be drawn with a wallpaper
(Victorian style), a simple example of which is shown in Figure 2.18a. We
shall analyze this pattern.

The conventional unit cell for this pattern is indicated by the vectors **a**
and **b**. If we choose a pattern unit consisting of two flowers (Figure 2.18b)
and continue it indefinitely by the repeat vectors **a** and **b**, the plane pattern
is generated. However, we have ignored the symmetry between the two
flowers in the chosen pattern unit. If one flower (1) is reflected across the
dashed line (g) to (1′) and then translated by **a**/2, it then occupies the
position of the second flower (2), or the pattern represented by Figure 2.18a
is brought into self-coincidence by the symmetry operation. This operation
takes place across a glide line, a symmetry element that occurs in some
extended two-dimensional patterns.

We say that the necessary and sufficient pattern unit is a single flower, occupying the asymmetric unit—the shaded (or unshaded) portion of Figure 2.18b. If the single flower is repeated by both the glide-line symmetry and the unit cell translations, then the extended pattern is again generated. Thus, to use our analogy, if we know the asymmetric unit of a crystal structure, which need not be the whole unit cell contents, and the space-group symbol for the crystal, we can generate the whole structure.

2.6.1 Two-Dimensional Space Groups

Our discussion leads naturally into two-dimensional space groups, or plane groups. Consider a pattern motif showing twofold symmetry (Figure 2.19a)—the point-group symbolism is continued into the realm of space groups. Next, consider a primitive oblique net (Figure 2.19b); it is of infinite extent in the plane, and the framework of lines divides the field into a number of identical primitive (p) unit cells. An origin is chosen at any lattice point.

Now, let the motif be repeated around each point in the net, and in the same relationship, with the twofold rotation points of the motif and the net in coincidence (Figure 2.19c). It will be seen that additional twofold rotation points are introduced, at the fractional coordinates (see footnote to Table 2.2) $0, \frac{1}{2}; \frac{1}{2}, 0$; and $\frac{1}{2}, \frac{1}{2}$ in each unit cell. We must always look for such "additional" symmetry elements after the point-group motif has been operated on by the unit cell repeats. This plane group is given the symbol $p2$.

In general, we shall not need to draw several unit cells; one cell will suffice provided that the pattern motif is completed around all lattice points intercepted by the given unit cell. Figure 2.20 illustrates the standard drawing of $p2$: The origin is taken on a twofold point, the x axis runs from top to bottom, and the y axis runs from left to right. Thus, the origin is considered to be in the top left-hand corner of the cell, but each corner is an equivalent position; we must remember that the drawing is a representative portion of an infinite array, whether in two or three dimensions.

The asymmetric unit, represented by $\bigcirc$, may be placed anywhere in the unit cell (for convenience, near the origin), and then repeated by the symmetry $p2$ to build up the picture, taking care to complete the arrangement around each corner. The additional twofold points can then be identified. The reader should now carry out this construction.

The list of fractional coordinates in Figure 2.20 refers to symmetry-related sites in the unit cell. The maximum number of sites generated by

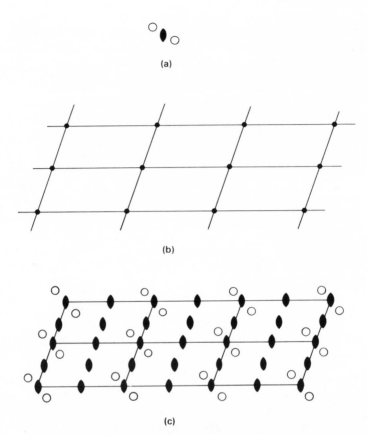

(a)

(b)

(c)

FIGURE 2.19 Plane group $p2$: (a) twofold symmetry motif, (b) oblique net with p unit cells, (c) extended pattern of plane group $p2$.

the space-group symmetry are called the general equivalent positions. In $p2$ they are given the coordinates x, y and $\bar{x}$, $\bar{y}$. We could use $1-x$, $1-y$ instead of $\bar{x}$, $\bar{y}$, but it is more usual to work with a set of coordinates near one and the same origin.

Each coordinate line in the space-group description lists, in order from left to right, the number of positions in each set, the Wyckoff* notation (for reference purposes only), the symmetry at each site in the set, and the coordinates of all sites in the set.

In a conceptual two-dimensional crystal, or projected real atomic arrangement, the asymmetric unit may contain either a single atom or a

* See Bibliography to Chapter 8.

Origin at 2

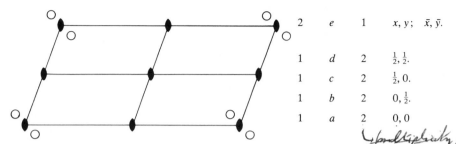

2	e	1	$x, y;$ $\bar{x}, \bar{y}.$
1	d	2	$\frac{1}{2}, \frac{1}{2}.$
1	c	2	$\frac{1}{2}, 0.$
1	b	2	$0, \frac{1}{2}.$
1	a	2	$0, 0$

FIGURE 2.20. Standard drawing and description of plane group $p2$. The lines which divide the unit cell into four quadrants are drawn for convenience only.

group of atoms. If it consists of part (half, in this plane group) of one molecule, then the whole molecule, as seen in projection at least, must contain twofold rotational symmetry, or a symmetry of which 2 is a sub-group. There are four unique twofold points in the unit cell; in the Wyckoff notation they are the sets (a), (b), (c), and (d), and they constitute the sets of special equivalent positions in this plane group. Notice that general positions have symmetry 1, whereas special positions have a higher crystallographic point-group symmetry. Where the unit cell contains fewer (an integral submultiple) of a species than the number of general equivalent positions in its space group, then it may be assumed that the species are occupying special equivalent positions and have the symmetry of the special site, at least. Exceptions to this rule arise in disordered structures, but this topic will not be discussed in this book.

We move now to the rectangular system, which includes point groups m and $2mm$, and both p and c unit cells. We shall consider first plane groups pm and cm.

The formation of these plane groups may be considered along the lines already described for $p2$, and we refer immediately to Figure 2.21. The origin is chosen on m, but its y coordinate is not defined by this symmetry element. In pm, the general equivalent positions are two in number, and there are two sets of special equivalent positions on m lines.

Plane group cm introduces several new features. The coordinate list is headed by the expression $(0, 0; \frac{1}{2}, \frac{1}{2}) +$; this means that two translations—$0, 0$ and $\frac{1}{2}, \frac{1}{2}$—are added to all the listed coordinates. Hence, the full list of general positions reads

$$x, y; \quad \bar{x}, y; \quad \tfrac{1}{2}+x, \tfrac{1}{2}+y; \quad \tfrac{1}{2}-x, \tfrac{1}{2}+y$$

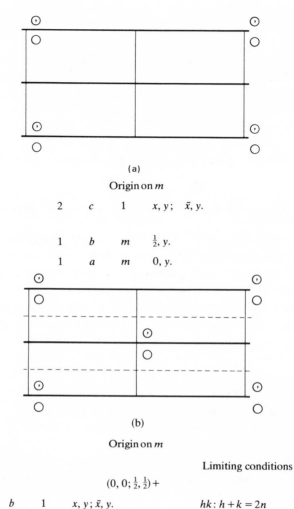

(a)

Origin on m

2	c	1	x, y; $\bar{x}, y$.
1	b	m	$\frac{1}{2}, y$.
1	a	m	$0, y$.

(b)

Origin on m

Limiting conditions

$(0, 0; \frac{1}{2}, \frac{1}{2}) +$

4	b	1	x, y; $\bar{x}, y$.	$hk: h + k = 2n$
2	a	m	$0, y$.	As above

FIGURE 2.21. Plane groups in the rectangular system: (a) *pm*, (b) *cm*.

Given x, the distance $\frac{1}{2} - x$, for example, is found by first moving $\frac{1}{2}$ along the a axis from the origin and then moving back along the same direction by the amount x.

The centering of the unit cell in conjunction with the m lines introduces the glide-line symmetry element (symbol g and graphic symbol - - -). The glide lines interleave the mirror lines, and their action is a combination of reflection and translation. The translational component is one-half of the repeat distance in the direction of the glide line. Thus, the pair of general positions x, y and $\frac{1}{2} - x$, $\frac{1}{2} + y$ are related by the g line. We shall encounter glide lines in any centered unit cell where m lines are present, and in certain other plane groups. For example, we may ask if there is any meaning to the symbol pg, a glide-symmetry motif repeated by the lattice translations. The answer is that pg is a possible plane group; in fact, it is the symmetry of the pattern in Figure 2.18.

There is only one set of special positions in cm, in contrast to two sets in pm. This situation arises because the centering condition in cm requires that both mirror lines in the unit cell be included in one and the same set. If we try to postulate two sets, by analogy with pm, we obtain

$$0, y; \quad \tfrac{1}{2}, \tfrac{1}{2} + y \tag{2.24}$$

and

$$\tfrac{1}{2}, y; \quad 0 \text{ (or 1)}, \tfrac{1}{2} + y \tag{2.25}$$

Expressions (2.24) and (2.25) differ only in the value of the variable y and therefore do not constitute two different sets of special equivalent positions.

We could refer to plane group cm by the symbol cg. If we begin with the origin on g and draw the general positions as before, we should find the glide lines interleaved with m lines. Two patterns that differ only in the choice of origin or the values attached to the coordinates of the equivalent positions do not constitute different space groups. The reader can illustrate this statement by drawing cg, and also, by drawing pg, can show that pm and pg are different. The glide line, or, indeed, any translational symmetry element is not encountered in point groups; it is a property of infinite patterns. The 17 plane groups are illustrated in Figure 2.22. The asymmetric unit is represented by a scalene triangle instead of the usual circle.

2.6.2 Limiting Conditions Governing X-Ray Reflection

Our main reason for studying space-group symmetry is that it provides information about the repeat patterns of atoms in crystal structures. X-ray diffraction spectra are characterized partly by the indices of the families of planes from which, in the Bragg treatment of diffraction (page 120), the X-rays are considered to be reflected. The pattern of indices reveals information about the space group of the crystal. Where a space group

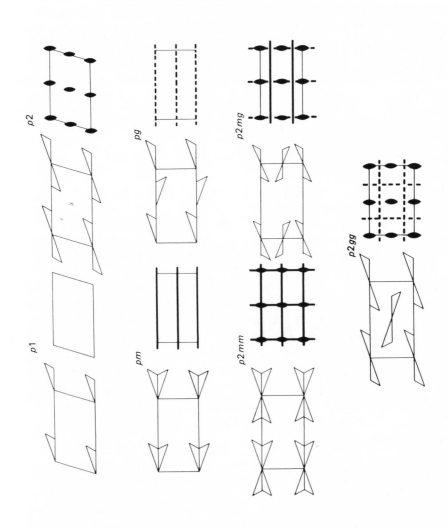

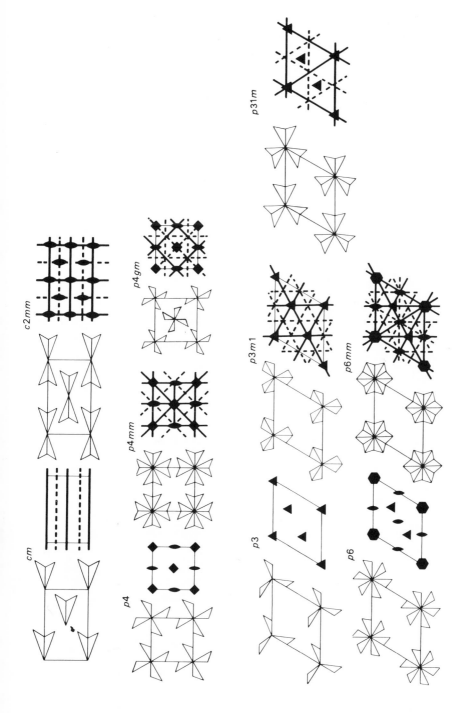

FIGURE 2.22. Unit cells of the 17 plane groups.

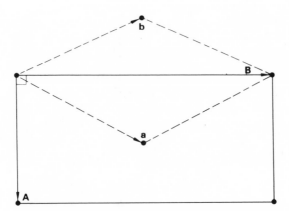

FIGURE 2.23. Centered rectangular unit cell ($\mathbf{A}$, $\mathbf{B}$)
and primitive unit cell ($\mathbf{a}$, $\mathbf{b}$) within the same lattice.

contains translational symmetry,* certain sets of reflections will be systematically absent from the experimental diffraction data record. We meet this situation for the first time in *cm* (Figure 2.21b); a reflection *hk* is absent unless the sum $h + k$ is an even number.

Figure 2.23 illustrates a rectangular lattice. Two unit cells are depicted on this lattice, a centered cell with vectors $\mathbf{A}$ and $\mathbf{B}$, and a primitive cell with vectors $\mathbf{a}$ and $\mathbf{b}$. The relationship between them is summarized by the equations

$$\mathbf{A} = \mathbf{a} - \mathbf{b} \tag{2.26}$$

$$\mathbf{B} = \mathbf{a} + \mathbf{b} \tag{2.27}$$

It is shown in Appendix A.6 that Miller indices of planes transform in the same way as unit cell vectors. Hence,

$$h' = h - k \tag{2.28}$$

$$k' = h + k \tag{2.29}$$

Adding (2.28) and (2.29), we obtain

$$h' + k' = 2h \tag{2.30}$$

which is even for all values of *h*.

Limiting conditions describe circumstances in which reflections can occur; systematic absences refer to conditions under which reflections cannot arise. Both terms are in common use, and we must distinguish between them carefully. Limiting conditions are discussed more fully in Chapter 4.

---

* With translational components of less than unity.

Point group $2mm$ belongs to the rectangular system, and, as a final example in two dimensions, we shall study plane group $p2gg$. It is often helpful to recall the parent point group of any space group. All that we need to do is to ignore the unit cell symbol, and replace any translational symmetry elements by the corresponding nontranslational symmetry elements. Thus, pg is derived from point group m, and $p2gg$ from $2mm$.

In $p2gg$, we are not at liberty to choose the actual positions of 2 and the two g lines freely. In studying point groups, we saw that the symmetry elements in a given symbol have a definite relative orientation with respect to the crystallographic axes; this is preserved in the corresponding space groups. Thus, we know that the g lines are normal to the x and y axes, and we can take an origin, initially, at their intersection (Figure 2.24a). In Figure 2.24b the general equivalent positions have been inserted; this diagram reveals the positions of the twofold points, inserted in Figure 2.24c, together with the additional g lines in the unit cell. The standard orientation of $p2gg$ places the twofold point at the origin; Figure 2.24d shows this setting and the description of this plane group. We see again that two interacting symmetry elements lead to a combined action which is equivalent to that of a third symmetry element, but their positions must be chosen correctly. This question did not arise in point groups because, by definition, all symmetry elements pass through a point—the origin.

There are two sets of special equivalent positions in $p2gg$; the pairs of twofold rotation points must be selected correctly. One way of ensuring proper selection is by inserting the coordinate values of the point-group symmetry element constituting a special position into the coordinates of the general positions. Thus, by taking $x = y = 0$ for one of the twofold points, we obtain a set of special positions with coordinates 0, 0 and $\frac{1}{2}, \frac{1}{2}$. If we had chosen 0, 0 and 0, $\frac{1}{2}$ as a set, the resulting pattern would not have conformed to $p2gg$ symmetry, but to pm, as Figure 2.25 shows. Special positions form a subset of the general positions, under the same space-group symmetry.

The general equivalent positions give rise to two limiting conditions, because the structure is "halved" with respect to b for the reflections $0k$, and with respect to a for the reflections $h0$. The special positions take both of these conditions, and the extra conditions shown because occupancy of the special positions* in this plane group gives rise to centered arrangements (Figure 2.26). After the development of the structure factor (page 166), some of the different limiting conditions will be derived analytically.

---

* The entities occupying special positions must, themselves, conform to the space group symmetry.

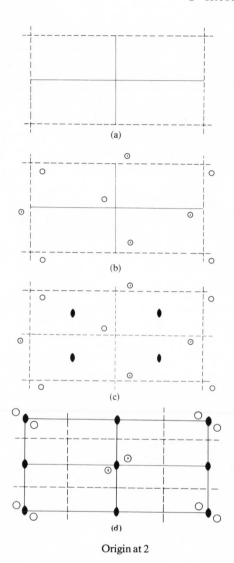

Origin at 2

4	c	1	$x, y;$ $\bar{x}, \bar{y};$ $\frac{1}{2}+x, \frac{1}{2}-y;$ $\frac{1}{2}-x, \frac{1}{2}+y.$	Limiting conditions

Limiting conditions

hk: None

$h0: h = 2n$

$0k: k = 2n$

| 2 | b | 2 | $\frac{1}{2}, 0;$ $0, \frac{1}{2}.$ | } As above + |
| 2 | a | 2 | $0, 0;$ $\frac{1}{2}, \frac{1}{2}.$ | } $hk: h + k = 2n$ |

FIGURE 2.24. Formation and description of $p2gg$.

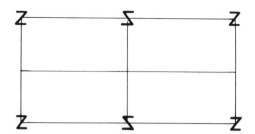

FIGURE 2.25. Occupation of the special positions 0, 0 and 0, $\frac{1}{2}$ leads to *pm* symmetry.

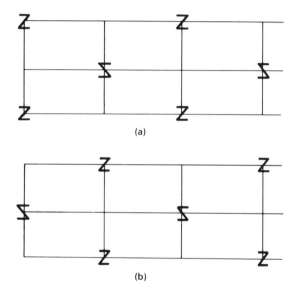

(a)

(b)

FIGURE 2.26. Special equivalent positions in *p2gg*; both sets (a) and (b) give rise to centered arrangements of the entity at the center of the Z symbol.

2.6.3 Three-Dimensional Space Groups

The principles which have emerged from the discussion on plane groups can be extended to three dimensions. Whereas the plane groups are limited to 17 in number, there are 230 space groups. We shall limit our discussion to a few space groups in the monoclinic and orthorhombic systems. We believe this will prove adequate because most of the important principles will evolve and, from a practical point of view, about 90% of crystals belong to these two systems.

Monoclinic Space Groups

In the monoclinic system, the lattices are characterized by the *P* and *C* unit cell descriptions, and the point groups are 2, *m*, and 2/*m*. We consider first space groups *P*2 and *C*2.

As with plane groups, we may begin with a motif which has twofold symmetry, but now about a line (axis) in three-dimensional space. This motif is arranged in a fixed orientation with respect to the points of a monoclinic lattice. Figure 2.27 shows a stereoscopic pair of illustrations for the unit cell of *C*2, drawn with respect to the conventional right-handed axes (page 9).

In Figure 2.28, *P*2 and *C*2 are shown in projection. The standard drawing of space-group diagrams is on the *ab* plane of the unit cell, with +*x* running from top to bottom, +*y* from left to right, both in the plane of the paper, and +*z* coming up from the paper. The positive or negative signs attached to the representative points indicate the *z* coordinates, that is, in ○⁺ and ○⁻, the signs stand for *z* and *z̄*, respectively. The relationship with the chosen stereogram nomenclature will be evident here.

In both *P*2 and *C*2, the origin is chosen on 2, and is, thus, defined with respect to the *x* and *z* axes, but not with respect to *y*. The graphic symbol for a diad axis in the plane of the diagram is →.

In space group *P*2, the general and special equivalent positions may be derived quite readily. The special sets (b) and (d) should be noted carefully; they are sometimes forgotten by the beginner because symmetry elements

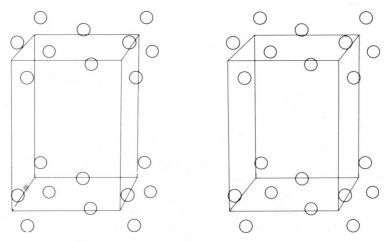

FIGURE 2.27. Stereoscopic pair of illustrations of the environs of one unit cell of space group *C*2; general equivalent positions are shown.

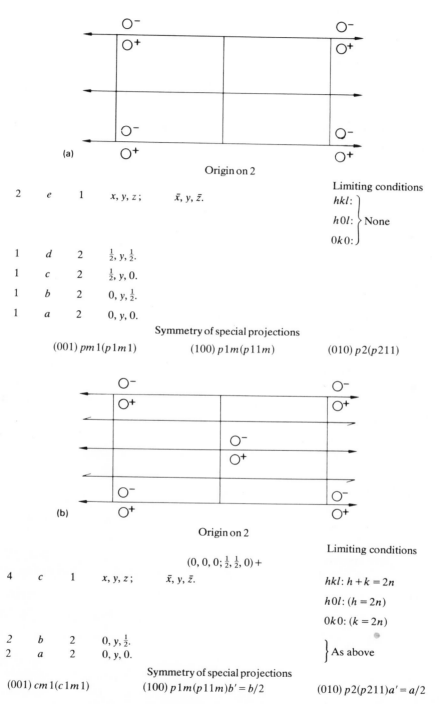

(a)

Origin on 2

2   e   1 x, y, z; $\bar{x}, y, \bar{z}$.

Limiting conditions

hkl:
h0l: } None
0k0:

1   d   2 $\frac{1}{2}, y, \frac{1}{2}$.
1   c   2 $\frac{1}{2}, y, 0$.
1   b   2 $0, y, \frac{1}{2}$.
1   a   2 $0, y, 0$.

Symmetry of special projections

(001) $pm1(p1m1)$ (100) $p1m(p11m)$ (010) $p2(p211)$

(b)

Origin on 2

$(0, 0, 0; \frac{1}{2}, \frac{1}{2}, 0)+$

4   c   1 x, y, z; $\bar{x}, y, \bar{z}$.

Limiting conditions

hkl: $h+k=2n$
h0l: $(h=2n)$
0k0: $(k=2n)$

2   b   2 $0, y, \frac{1}{2}$.
2   a   2 $0, y, 0$.

} As above

Symmetry of special projections

(001) $cm1(c1m1)$ (100) $p1m(p11m)b'=b/2$ (010) $p2(p211)a'=a/2$

FIGURE 2.28. Monoclinic space groups: (a) $P2$, (b) $C2$.

distant $c/2$ from those drawn in the ab plane are not indicated on the conventional diagrams. The diad along y and at $x = 0$, $z = \frac{1}{2}$, for example, relates x, y, z to a point at $1 - x, 1 - y, 1 - z$; its presence, and that of the diad at $x = z = \frac{1}{2}$, may be illustrated by drawing the space group in projection on the ac plane of the unit cell. The reader should make this drawing and compare it with Figure 2.28a.

It is often useful to consider a structure in projection onto one of the principal planes (100), (010), or (001). The symmetry of a projected space group corresponds with a plane group, and the symmetries of the principal projections are included with the space-group description (Figure 2.28). The full plane-group symbols, given in parentheses, indicate the orientations of the symmetry elements. In $C2$, certain projections produce more than one repeat in certain directions; the projected cell dimensions, represented by a', b', and c', may then be halved with respect to their original values.

The projection of $C2$ onto (100) is shown by Figure 2.29 in three stages, starting from the y and z coordinates of the set of general equivalent positions. The question is sometimes asked, how do two points of the same hand, such as x, y, z and $\bar{x}, y, \bar{z}$ in $C2$, become of opposite hand, such as y, z and $y, \bar{z}$ in $p1m$, after projection? The difficulty may be associated with the use of the highly symmetric circle as a representative point. It is suggested that the reader make a drawing of $C2$, and of the stages of the projection on to (100), using either $\mathbf{9}$ instead of $\bigcirc^{+}$, and $\mathbf{9}$ instead of $\bigcirc^{-}$, or the scalene triangle shown in Figure 2.22.

Space group $C2$ may be obtained by adding the translation $\frac{1}{2}, \frac{1}{2}, 0$, that associated with a C cell (Table 2.2), to the equivalent positions of $P2$. This operation is equivalent to repeating the original twofold motif at the lattice points of the C monoclinic unit cell. This simple relationship between P and C cells is indicated by the heading $(0, 0, 0; \frac{1}{2}, \frac{1}{2}, 0)+$ of the coordinate list in $C2$; it may be compared with that for cm (Figure 2.21b).

There are four sets of special positions in $P2$, but only two sets in $C2$; the reason for this has been discussed in relation to plane groups pm and cm (page 79).

2.6.4   Screw Axes

Screw axes are symmetry elements that relate points in an infinite, three-dimensional, regular array: thus, they are not a feature of point groups. A screw axis may be thought of as a combination of rotation and

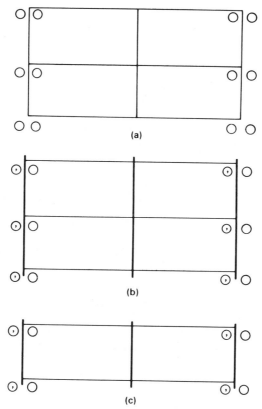

FIGURE 2.29. Projection of $C2$ onto (100): (a) y, z positions from $C2$ (Z axis left to right), (b) two-dimensional symmetry elements added, (c) one unit cell—$p1m$ ($p11m$), $b' = b/2$, $c' = c$. (Plane groups $p11m$ and $p1m1$ are equivalent because they correspond only to an interchange of the X and Y axes; 1 is a trivial symmetry element.)

translation: an infinitely long spiral staircase would give an indication of the nature of the symmetry operation.

Imagine that the bottom step is rotated by 60° and then translated upward by one sixth of the repeat distance between steps in similar orientations, so that it takes the place of the second step, which itself moves upward in a similar manner. Clearly, if this procedure were repeated six times, the bottom step would reach the position and orientation of the sixth step up shown in Figure 2.30; we symbolize this screw axis as 6_1. Infinite length is,

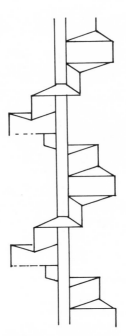

FIGURE 2.30. Spiral
staircase: illustration of 6_1
screw axis symmetry.

strictly, a requirement because as the bottom step is rotated and translated
upward, so another step, below the figure, comes up into its position in
order that self-coincidence be obtained. A similar argument must apply at
the top of the diagram. The spiral staircases of the Monument in London
and of the Statue of Liberty in New York seem to be of infinite length,
and might be considered as macroscopic examples of screw axes. Examine
them carefully on your next visit and determine their symmetry nature.

The centering of the unit cell in $C2$ introduces screw axes which
interleave the diads. A screw axis may be designated R_p ($p < R$), and the
operation consists of an R-fold rotation plus a translation parallel to the
screw axis of p/R times the repeat in that direction. Thus, in $C2$, the screw
axis is of the type 2_1 and has a translational component of $\frac{1}{2}$ parallel to b.
The general equivalent positions x, y, z and $\frac{1}{2} - x, \frac{1}{2} + y, \bar{z}$ are related by a
2_1 axis along $[\frac{1}{4}, y, 0]$.* Screw axes are present in the positions shown by
their graphic symbol $\rightarrow$.

* We use this nomenclature to describe, in this example, the line parallel to the Y axis through
  $x = \frac{1}{4}, z = 0$.

Limiting Conditions in $C2$

The limiting conditions for this space group are given in Figure 2.28b. Two of them are placed in parentheses; this notation is used to indicate that they are dependent upon a more general condition. Thus, since the hkl reflections are limited by the condition $h + k = 2n$ (even), because the cell is C-centered, it follows that $h0l$ are limited by $h = 2n$ (0 is an even number). There are several other nonindependent conditions which could have been listed. For example, $0kl$: $k = 2n$ and $h00$: $h = 2n$. However, in the monoclinic system, in addition to the hkl reflections, we are concerned particularly only with $h0l$ and $0k0$, because the symmetry plane is parallel to (010) and the symmetry axis is parallel to [010]. This feature is discussed more fully in Chapter 4.

Space Group $P2_1$

Space groups $C2$ and $C2_1$ are equivalent (compare cm and cg). On the other hand, $P2$ contains no translational symmetry, so $P2_1$ is a new space group (Figure 2.31). There are no special positions in $P2_1$. Special positions

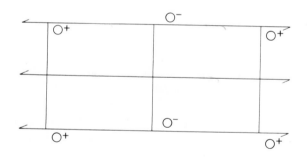

Origin on 2_1

					Limiting conditions
2	a	1	x, y, z;	$\bar{x}, \frac{1}{2}+y, \bar{z}.$	hkl: None
					$h0l$: None
					$0k0$: $k = 2n$

Symmetry of special projections

(001) $pg1(p1g1)$ (100) $p1g(p11g)$ (010) $p2(p211)$

FIGURE 2.31. Space group $P2_1$.

TABLE 2.5. Glide-Plane Notation

Symbol	Orientation	Graphic symbol		Translational component
		‖ to projection	⊥ to projection	
a	(010) or (001)	- - - - -	⌐	$a/2$
b	(100) or (001)	- - - - -	←	$b/2$
c	(100) or (010)	· · · · · · · · ·	None	$c/2$
n	(100)			$(b+c)/2$
	(010)	—·—·—·—	⟋	$(a+c)/2$
	(001)			$(a+b)/2$
d	(100)	·—·←·—·—·—·	p ↘	$(b\pm c)/4$
	(010)	—·—·—·→·—·	q ↙	$(a\pm c)/4$
	(001)			$(a\pm b)/4$

Notes:
1. In rhombohedral, tetragonal, and cubic space groups, c, n, and d glide planes have additional translations.
2. The symbols p and q are fractions differing by $c/4$.

cannot exist on a single translational symmetry element, since it would mean that the entity placed on such an element consisted of an infinite repeating pattern.

2.6.5 Glide Planes

Consider again Figure 2.18, but let each dashed line be the trace of a glide plane normal to b. In two dimensions, the direction of translation, after the reflection part of the operation, is unequivocal. In the three-dimensional case, however, there are four possibilities, although each of them will not necessarily give rise to a different space group.

In the case of a glide plane normal to b, the direction of translation could be along a (amount $a/2$), along c (amount $c/2$), along a diagonal direction n (amount $(a+c)/2$), or along a diagonal direction d (amount $(a\pm c/4$—as in the diamond allotrope of carbon). The d glide plane is not often encountered in practice, and will not be discussed in detail here.

The translations form mnemonics for the glide plane names. Thus, in Figure 2.18, in three dimensions, an a glide plane is shown. The symbol n may refer to more than one orientation (Table 2.5). However, the space-group symbol, derived from the appropriate point-group symbol in Table 1.5, provides the necessary information. Thus, if the n glide plane is normal to b, the translation component of the n glide symmetry operation must be $(a + c)/2$. It is imperative to understand fully the Hermann–Mauguin point-group notation; that for space groups follows in a logical manner.

It should be noted that the translational components for screw axes and for glide planes are integer fractions of the repeat distances. A 2_1 axis parallel to b has a component of translation of $b/2$. A 2_2, or in general an R_R, axis is trivial: it is a combination of 2 (or R) and the corresponding unit-cell translation, in this case b. The term *translation* is preferably used in the context of screw axis or glide plane, and the term *repetition* is used for the unit-cell (Bravais) translation.

If a space group is formed from the combination of a point group with m planes and a lattice of centered unit cells, glide planes are introduced into the space group. They are the three-dimensional analog of glide lines. The glide-plane operation consists of reflection across the plane plus a translation parallel to the plane. The direction of translation is indicated by the glide-plane symbol (Table 2.5). Other types of glide planes exist, particularly in the higher symmetry systems.*

As an example of a space group with a glide plane, we shall study $P2_1/c$, a space group encountered frequently in practice. This space group is derived from point group $2/m$, and must, therefore, be centrosymmetric. However, the center of symmetry does not lie at the intersection of 2_1 and c. It is convenient to take the origin on a center of symmetry† in centrosymmetric space groups, and, in this example, we must determine the correct positions of the symmetry elements in the unit cell. We shall approach the solution of this problem in two ways, the first of which is similar to our treatment of plane group $p2gg$.

Since the screw axis must intersect the glide plane, the point of intersection will be taken as an origin and the space group drawn (Figure 2.32). We see now that the centers of symmetry lie at points such as $0, \frac{1}{4}, \frac{1}{4}$. This point may be taken as a new origin, and the space group redrawn (Figure 2.33); a fraction ($\frac{1}{4}$, for example) placed next to a symmetry element indicates the position of that symmetry element with respect to the ab plane.

It is desirable, however, to be able to draw the standard space-group illustration at the outset. From a choice of origin, and using the full meaning of the space-group symbol, we can obtain the positions of the symmetry elements by means of a simple scheme.

---

*The d (diamond) glide plane, with translations of, for example, $(b \pm c)/4$ is also known but not often encountered in practice.

†Sometimes the origin will have a point symmetry higher than $\bar{1}$, for example, $2/m$ or mmm, but $\bar{1}$ is a subgroup of such point symmetries.

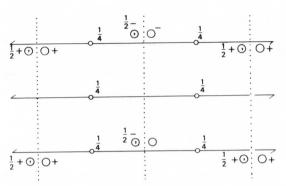

FIGURE 2.32. Space group $P2_1/c$ with the origin at an
intersection of 2_1 and c.

Let the symmetry elements be placed as follows:

$\bar{1}$ at 0, 0, 0 (choice of origin)
2_1 parallel to $[p, y, r]$ (parallel to the y axis)
c parallel to (x, q, z) (normal to the y axis)

It is important to note that we have employed only the standard choice of
origin and the information contained in the space-group symbol. Next, we
carry out the symmetry operations:

$$(1)\quad x, y, z \xrightarrow{\quad 2_1 \quad} 2p-x, \tfrac{1}{2}+y, 2r-z \quad (2)$$

$$2p-x, 2q-\tfrac{1}{2}-y, -\tfrac{1}{2}+2r-z \quad (3)$$

$$\xrightarrow{\bar{1}} -x, -y, -z \quad (4)$$

The symbol $-c$ is used to indicate that the c-glide translation of $\tfrac{1}{2}$ is
subtracted, which is crystallographically equivalent to being added.*

We now use the fact that the combined effect of two operations is
equivalent to a third operation, starting from the original point (1). Symboli-
cally, $c.2_1 \equiv \bar{1}$, or 2_1 followed by c is equivalent to $\bar{1}$. Thus, points (3) and (4)
are one and the same, whence, by comparing coordinates, $p = 0$ and
$q = r = \tfrac{1}{4}$. Comparison with Figure 2.33 shows that these conditions lead to
the desired positions of the three symmetry elements in $P2_1/c$.

* ± 1 may be added to any coordinate to give a crystallographically equivalent situation.

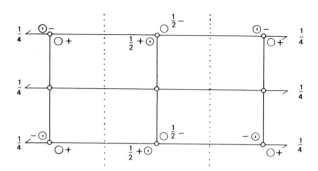

FIGURE 2.33. Space group $P2_1/c$ with the origin on $\bar{1}$ (standard setting).

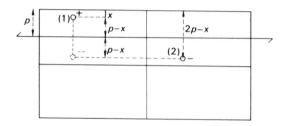

FIGURE 2.34. Operation about a 2_1 axis along the line $[p, y, 0]$: $x \to 2p - x$. A similar construction may be used for the y coordinate in a c-glide operation, for example.

The change in the x coordinate in the operation $(1) \to (2)$ is illustrated in Figure 2.34; the argument can be applied to any similar situation in monoclinic and orthorhombic space groups, and we can always consider one coordinate at a time. The completion of the details of this space group forms the basis of a problem at the end of this chapter.

We shall not discuss centered monoclinic space groups, but they do not present difficulty once the primitive space groups have been mastered. Figure 2.35 shows a stereoscopic pair of illustrations of the unit cell of diiodo-(N,N,N',N'-tetramethylethylenediamine) zinc(II), $I_2[(CH_3)_2NCH_2CH_2N(CH_3)_2]Zn$, which crystallizes in space group $C2/c$ with four molecules in the unit cell; the zinc atoms lie on twofold axes.*

Appendix A1 contains stereoscopic illustrations of the thirteen monoclinic space groups.

---

* S. Htoon and M. F. C. Ladd, *Journal of Crystal and Molecular Structure* **4**, 357 (1974).

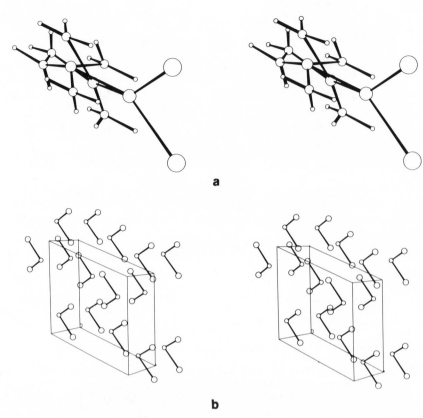

FIGURE 2.35. Stereoviews of the structure of diiodo-(N,N,N',N'-tetramethyl-ethylenediamine) zinc(II): (a) Structural formula; the circles, in decreasing order of size, represent I, Zn, N, C, and H. (b) Unit cell; for clarity, only the I and Zn atoms are shown.

Orthorhombic Space Groups

We shall consider two orthorhombic space groups, $P2_12_12_1$ and *Pnma*. The first is illustrated in Figure 2.36; it should be noted that the three mutually perpendicular 2_1 axes do *not* intersect one another in this space group. Although $P2_12_12_1$ is a noncentrosymmetric space group, the three principal projections are centrosymmetric; each has the two-dimensional space group $p2gg$.

Change of Origin. Considering the projection of $P2_12_12_1$ onto (001), we obtain from the general equivalent positions the two-dimensional set

$$x, y; \quad \tfrac{1}{2}-x, \bar{y}; \quad \tfrac{1}{2}+x, \tfrac{1}{2}-y; \quad \bar{x}, \tfrac{1}{2}+y$$

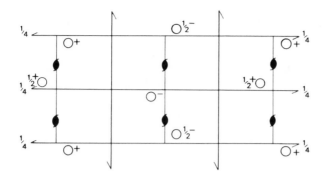

Origin halfway between three pairs of nonintersecting screw axes

Limiting conditions

4   a 1 x, y, z; $\frac{1}{2}-x, \bar{y}, \frac{1}{2}+z$; $\frac{1}{2}+x, \frac{1}{2}-y, \bar{z}$; $\bar{x}, \frac{1}{2}+y, \frac{1}{2}-z$.

hkl: ⎫
$0kl$: ⎪
        ⎬ None
$h0l$: ⎪
$hk0$: ⎭

$h00$: $h = 2n$
$0k0$: $k = 2n$
$00l$: $l = 2n$

Symmetry of special projections

(001) $p2gg$ (100) $p2gg$ (010) $p2gg$

FIGURE 2.36. Space group $P2_12_12_1$: in space-group diagrams, ⬬ represents a 2_1 axis normal to the plane of projection.

It is convenient to change the origin to a twofold rotation point, say at $\frac{1}{4}$, 0. To carry out this transformation, the coordinates of the new origin are subtracted from the original coordinates:

$$x - \tfrac{1}{4}, y; \quad \tfrac{1}{4} - x, \bar{y}; \quad \tfrac{1}{4} + x, \tfrac{1}{2} - y; \quad -x - \tfrac{1}{4}, \tfrac{1}{2} + y$$

Next, new variables x_0 and y_0 are chosen such that, for example, $x_0 = x - \frac{1}{4}$ and $y_0 = y$. Then, by substituting, we obtain

$$x_0, y_0; \quad \bar{x}_0, \bar{y}_0; \quad \tfrac{1}{2} + x_0, \tfrac{1}{2} - y_0; \quad \tfrac{1}{2} - x_0, \tfrac{1}{2} + y_0$$

If the subscript is dropped, these coordinates are exactly those given already for $p2gg$ (Figure 2.24d). This type of change of origin is useful when studying projections. In this example, the reverse transformation takes us back to the standard setting in $P2_12_12_1$.

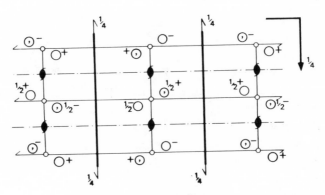

Origin at $\bar{1}$

8 d 1 x,y,z; $\frac{1}{2}+x,\frac{1}{2}-y,\frac{1}{2}-z$; $\bar{x},\frac{1}{2}+y,\bar{z}$; $\frac{1}{2}-x,\bar{y},\frac{1}{2}+z$;   Limiting conditions

        $\bar{x},\bar{y},\bar{z}$; $\frac{1}{2}-x,\frac{1}{2}+y,\frac{1}{2}+z$; $x,\frac{1}{2}-y,z$; $\frac{1}{2}+x,y,\frac{1}{2}-z$. hkl: None

                                                     $0kl$: $k+l=2n$

                                                     $h0l$: None

                                                     $hk0$: $h=2n$

                                                     $h00$: $(h=2n)$

                                                     $0k0$: $(k=2n)$

                                                     $00l$: $(l=2n)$

4 c m $x,\frac{1}{4},z$; $\bar{x},\frac{3}{4},\bar{z}$; $\frac{1}{2}-x,\frac{3}{4},\frac{1}{2}+z$; $\frac{1}{2}+x,\frac{1}{4},\frac{1}{2}-z$.   As above

4 b $\bar{1}$ $0,0,\frac{1}{2}$; $0,\frac{1}{2},\frac{1}{2}$; $\frac{1}{2},0,0$; $\frac{1}{2},\frac{1}{2},0$.   As above +

4 a $\bar{1}$ $0,0,0$; $0,\frac{1}{2},0$; $\frac{1}{2},0,\frac{1}{2}$; $\frac{1}{2},\frac{1}{2},\frac{1}{2}$. hkl: $h+l=2n$; $k=2n$

Symmetry of special projections

(001) $p2gm$ (100) $c2mm$ (010) $p2gg$

FIGURE 2.37. Space group $Pnma$.

Space group $Pnma$ is shown with the origin on $\bar{1}$ (Figure 2.37). The symbol tells us that the unit cell is primitive, with an n-glide plane normal to the x axis (see Table 2.5), an m plane normal to y, and an a-glide plane normal to z. Although this space group is derived from point group mmm, we cannot assume that the three planes in $Pnma$ intersect in a center of symmetry. We are, therefore, faced with a problem similar to that discussed with $P2_1/c$. The solution of this problem depends upon the fact that $m.m.m \equiv \bar{1}$, and is illustrated fully in Problem 12 at the end of this chapter.[*]

* See also Appendix A7.

The coordinates of the general and the special equivalent positions can be derived easily from the diagram. The translational symmetry elements n and a give rise to the limiting conditions shown. Nonindependent conditions are shown in parentheses; in the orthorhombic system, all of the classes of reflection listed should be considered, as will be discussed in Chapter 4.

It is useful to remember that among the triclinic, monoclinic, and orthorhombic space groups, at least, pairs of coordinates which have one *sign* change of x, y, or z indicate a symmetry plane normal to the axis of the coordinate with the changed sign. If two sign changes exist, a symmetry axis lies parallel to the axis of the coordinate that has *not* changed sign. Three sign changes indicate a center of symmetry. In these three systems, where any coordinate, say x, is related by symmetry to another at $t - x$, the symmetry element intersects the x axis at $t/2$.

2.6.6 Analysis of the Space-Group Symbol

In this section we consider the general interrelationship between space-group symbols and point-group symbols. On encountering a space-group symbol, the first problem is to determine the parent point group. This process has been discussed (page 85); here are a few more examples. It is not necessary to have explored all space groups in order to carry out this exercise:

$$P2_1/c \to (2_1/c) \to (2/c) \to 2/m$$
$$Ibca \to mmm$$
$$P4_12_12 \to 422$$
$$F\bar{4}3c \to \bar{4}3m$$

Next we must identify a crystal system for each point group:

$$2/m \to \text{monoclinic}$$
$$mmm \to \text{orthorhombic}$$
$$422 \to \text{tetragonal}$$
$$\bar{4}3m \to \text{cubic}$$

Now, from Table 1.5, we can associate certain crystallographic directions with each symmetry element in the space group symbol:

$P2_1/c$: Primitive, monoclinic unit cell; c-glide plane $\perp b$; 2_1 axis $\| b$; centrosymmetric.

$Ibca$: Body-centered, orthorhombic unit cell; b-glide plane $\perp a$; c-glide plane $\perp b$; a-glide plane $\perp c$; centrosymmetric.

$P4_12_12$: Primitive, tetragonal unit cell; 4_1 axis $\| c$; 2_1 axes $\| a$ and b; twofold axes at 45° to a and b; noncentrosymmetric.

$F\bar{4}3c$: Face-centered, cubic unit cell; $\bar{4}$ axes $\| a, b$, and c; threefold axes $\| \langle 111 \rangle$; c-glide planes $\perp \langle 110 \rangle$; noncentrosymmetric.

It should be noted carefully that the unique symmetry elements (where there are more than two present) given in a space-group symbol may not intersect, and the origin must always be selected with care. Appropriate procedures for the monoclinic and orthorhombic systems have been discussed; in working with higher symmetry space groups, similar rules can be evaluated.

Because of the similarities between space groups and their parent point groups, a reflection symmetry, for example, in the same orientation with respect to the crystallographic axes always produces the same changes in the *signs* of the coordinates. Thus, the m plane perpendicular to Z in point group mmm changes x, y, z to $x, y, \bar{z}$. The a-glide plane in $Pnma$ changes x, y, z to $\frac{1}{2}+x, y, \frac{1}{2}-z$; the translational components of $\frac{1}{2}$ are a feature of this space group, but the signs of x, y, and z are still $+, +$, and $-$ after the operation (see also Appendix A7).

Bibliography

Lattices and Space Groups

HAHN, T. (Editor), *International Tables for Crystallography*, Vol. A, Dordrecht, D. Reidel (1983).

HENRY, N. F. M., and LONSDALE, K. (Editors), *International Tables for X-Ray Crystallography*, Vol. I, Birmingham, Kynoch Press (1965).

Problems

2.1. Two nets are described by the unit cells (i) $a = b$, $\gamma = 90°$ and (ii) $a = b$, $\gamma = 120°$. In each case (a) what is the symmetry at each net point, (b) to which two-dimensional system does the net belong, and (c) what are the results of centering the unit cell?

2.2. A monoclinic F unit cell has the dimensions $a = 6.000$, $b = 7.000$, $c = 8.000$ Å and $\beta = 110.0°$. Show that an equivalent monoclinic C unit cell, with an *obtuse* β angle, can represent the same lattice, and calculate its dimensions. What is the ratio of the volume of the C cell to that of the F cell?

2.3. Carry out the following exercises with drawings of a tetragonal P unit cell.

(a) Center the B faces. Comment on the result.
(b) Center the A and B faces. Comment on the result.
(c) Center all faces. What conclusions can you draw now?

2.4. Calculate the length of $[31\bar{2}]$ (see page 55) for both unit cells in Problem 2.2.

2.5. The relationships $a \mathcal{C} b \mathcal{C} c$, $\alpha \mathcal{C} \beta \mathcal{C} 90°$, $120°$, and $\gamma = 90°$ may be said to define a diclinic system. Is this an eighth system? Give reasons for your answer.

2.6. (a) Draw a diagram to show the symmetry elements and general equivalent positions in $c2mm$ (origin on $2mm$). Write the coordinates and point symmetry of the general and special positions, in their correct sets, and give the conditions limiting X-ray reflection in this plane group. (b) Draw a diagram of the symmetry elements in plane group $p2mg$ (origin on 2); take care not to put the twofold point at the intersection of m and g (why?). On the diagram, insert each of the motifs P, V, and Z in turn, using the *minimum* number of motifs consistent with the space-group symmetry.

2.7. (a) Continue the study of space group $P2_1/c$ (page 95). Write the coordinates of the general and special positions, in their correct sets. Give the limiting conditions for all sets of positions, and write the plane-group symbols for the three principal projections. Draw a diagram of the space group as seen along the b axis. (b) Biphenyl, ◯–◯ , crystallizes in space group $P2_1/c$, with two molecules per unit cell. What can be deduced about both the positions of the molecules in the unit cell and the molecular conformation? (The planarity of each benzene ring in the molecule may be assumed.)

2.8. Write the coordinates of the vectors between all pairs of general equivalent positions in $P2_1/c$ with respect to the origin, and note that they are of two types. Remember that $-\frac{1}{2}$ and $+\frac{1}{2}$ in a coordinate are crystallographically equivalent, because we can always add or subtract 1 from a fractional coordinate without altering its crystallographic implication.

2.9. The orientation of the symmetry elements in the orthorhombic space group *Pban* may be written as follows:

$\bar{1}$ at 0, 0, 0 (choice of origin)
b-glide $\parallel (p, y, z)$ ⎫
a-glide $\parallel (x, q, z)$ ⎬ (from the space-group symbol)
n-glide $\parallel (x, y, r)$ ⎭

Determine p, q, and r from the following scheme, using the fact that $n.a.b \equiv \bar{1}$:

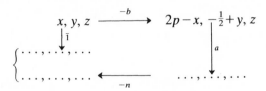

*2.10. Construct a space-group diagram for *Pbam*, with the origin at the intersection of the three symmetry planes. List the coordinates of both the general equivalent positions and the centers of symmetry. Derive the standard coordinates for the general positions by transforming the origin to a center of symmetry.

2.11. Show that space groups *Pa*, *Pc*, and *Pn* represent the same pattern, but that *Ca* is different from *Cc* (*Cn*). What is the more usual symbol for space group *Ca*? What would be the space group for *Cc* after an interchange of the *x* and *z* axes?

2.12. For each of the space groups $P2/c$, $Pca2_1$, $Cmcm$, $P\bar{4}2_1c$, $P6_322$, and $Pa3$:

(a) Write down the parent point group and crystal system.
(b) List the full meaning conveyed by the symbol.
(c) State the independent conditions limiting X-ray reflection.

2.13. Consider Figure 2.25. What would be the result of constructing this diagram with Z alone, and not using its mirror image?

3

Preliminary Examination of Crystals by Optical and X-Ray Methods

3.1 Introduction

In this chapter we shall discuss the interaction between crystals and two different electromagnetic radiations, light and X-rays. Light, with its longer wavelength (5000–6000 Å), can reveal only limited information about crystal structures, whereas X-rays with wavelengths of less than about 2 Å can be used to determine the relative positions of atoms in crystals. A preliminary examination of a crystal aims to determine its space group and unit-cell dimensions, and may be carried out by a combination of optical and X-ray techniques. The optical methods described here are simple, but, nevertheless, often very effective; they should be regarded as a desirable prerequisite to an X-ray structure determination.

3.2 Polarized Light

An ordinary light source emits wave trains, or pulses of light, vibrating in all directions perpendicular to the direction of propagation (Figure 3.1); the light is said to be unpolarized. The vibrations of interest to us are those of the electric vector associated with the waves. Any one of these random vibrations can be resolved into two mutually perpendicular components, and the resultant vibration may, therefore, be considered as the sum of all components in these two perpendicular directions. In order to study the optical properties of crystals, we need to restrict the resultant vibration of the light source to one direction only by eliminating the component at right angles to it.

105

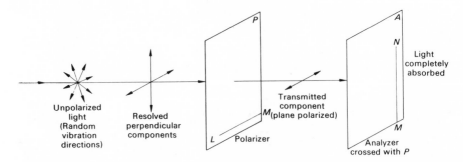

FIGURE 3.1. Production of plane-polarized light by passing unpolarized light through a sheet of Polaroid film (the polarizer, P). A second sheet of Polaroid (the analyzer, A), rotated through 90° with respect to P, completely absorbs all light transmitted by P. The lines LM and MN were parallel on the sheet from which P and A were cut.

Let us consider that a polarizer (P), consisting of a sheet of Polaroid, transmits light vibrating in the horizontal direction LM and absorbs all components vibrating in the direction perpendicular to LM. Thus, light passing through the polarizer vibrates in one plane only, and is said to be plane polarized. The plane contains the vibration direction, which is perpendicular to the direction of propagation, and the direction of propagation itself. A second Polaroid, the analyzer (A), is placed after the polarizer and rotated so that its vibration transmission direction (MN) is at 90° to that of the polarizer. It receives no component parallel to its transmission direction and, therefore, absorbs all the light transmitted by the polarizer. The two Polaroids are then said to be crossed. This effect may be demonstrated by cutting a Polaroid sheet marked with a straight line LMN into two sections, P and A (Figure 3.1). When superimposed, the two halves will not transmit light if the reference lines LM and MN are exactly perpendicular. In intermediate positions, the intensity of light transmitted varies from a maximum, where they are parallel, to zero (crossed). The production and use of plane-polarized light by this method is used in the polarizing microscope.

3.3 Optical Classification of Crystals

Crystals may be grouped, optically, under two main headings, isotropic crystals and anisotropic (birefringent) crystals. All crystals belonging to the cubic system are optically isotropic; the refractive index of a cubic crystal is

TABLE 3.1. Crystal Directions Readily Derivable from an Optical Study

Optical classification	Crystal system	Information relating to crystal axes likely to be revealed
Isotropic	Cubic	Axes may be assigned from the crystal morphology
Anisotropic, uniaxial	Tetragonal	Direction of z axis
	Hexagonal	Direction of z axis
	Trigonal[a]	Direction of z axis
Anisotropic, biaxial	Orthorhombic	Direction of at least the x, y, or z axis, possibly all three
	Monoclinic	Direction parallel to the y axis
	Triclinic	No special relationship between the crystal axes and vibration directions

[a] Referred to hexagonal axes.

independent of direction, and its optical characteristics are similar to those of glass. Noncubic crystals exhibit a dependence on direction in their interaction with light.

Anisotropic crystals are divided into two groups, uniaxial crystals, which have one optically isotropic section and include the tetragonal, hexagonal, and trigonal crystal systems, and biaxial crystals, which have two optically isotropic sections and belong to the orthorhombic, monoclinic, and triclinic crystal systems.

A preliminary optical examination of a crystal will usually show whether it is isotropic, uniaxial, or biaxial. Distinction between the three biaxial crystal systems is often possible in practice and, depending on how well the crystals are developed, a similar differentiation may also be effected for the uniaxial crystals. Even if an unambiguous determination of the crystal system is not forthcoming, the examination should, at least, enable the principal symmetry directions to be identified; Table 3.1 summarizes this information.

3.3.1 Uniaxial Crystals

As an example of the use of the polarizing microscope, we shall consider a tetragonal crystal, such as potassium dihydrogen phosphate, lying on a

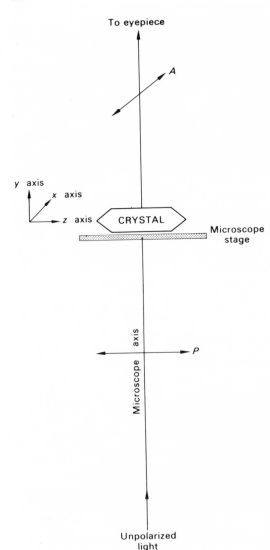

FIGURE 3.2. Schematic experimental arrangement for examining extinction directions. A tetragonal crystal is shown on the microscope stage, and the incident light is perpendicular to the z axis of the crystal.

microscope slide with its *y* axis parallel to the axis of the optical path through a microscope (Figure 3.2). The microscope is fitted with a polarizer (*P*), and an analyzer (*A*) which is crossed with respect to *P* and may be removed from the optical path. The crystal can be rotated on the microscope stage between *P* and *A*. With the Polaroids crossed and no crystal in between, the field of view is uniformly dark. However, with the crystal interposed, this situation will not necessarily be obtained.

The tetragonal crystal is lying with (010) on the microscope slide; both the x and z axes are, therefore, perpendicular to the microscope axis. In general, some of the light passing through the crystal will be transmitted by the analyzer, even though P and A are crossed. The intensity of the transmitted light varies as the crystal is rotated on the microscope stage between the polarizer and the analyzer. During a complete revolution of the stage, the intensity of transmitted light passes through four maxima and four minima. At the minimum positions, the crystal is usually only just visible. These positions are called extinction positions, and they occur at exactly 90° intervals of rotation. Maximum intensity is observed with the crystal at 45° to these directions.

These changes would be observed if the crystal itself were replaced by a sheet of Polaroid. Extinction would occur when the vibrations of the "crystal Polaroid" were perpendicular to those of P or A. A simple explanation of these effects is that the crystal behaves as a polarizer. Incident plane-polarized light from P is resolved by the crystal into two perpendicular components (Figure 3.3). In our tetragonal crystal, the vibration directions

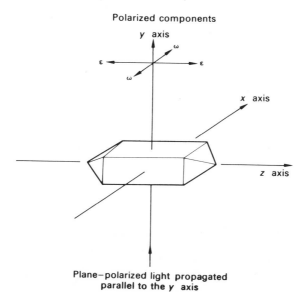

FIGURE 3.3. Resolution of incident light into components vibrating parallel to the x and z axes by a tetragonal crystal lying with its y axis parallel to the incident beam. ω and ε are the refractive indices for light vibrating, respectively, perpendicular and parallel to z.

associated with this polarizing effect are parallel to its x and z axes. Rotating the crystal on the microscope stage will, therefore, produce extinction whenever x and z are parallel to the vibration directions of P and A. The x and z axes of a tetragonal crystal correspond to its extinction directions. It should be noted that the x and y directions are equivalent under the fourfold symmetry of the crystal.

3.3.2 Birefringence

The vibration components produced by the crystal are associated with different refractive indices. With reference to Figure 3.3, a tetragonal crystal with light vibrating parallel to the fourfold symmetry axis (z) has a refractive index ε, whereas light vibrating perpendicular to z has a different refractive index, ω; the crystal is said to be birefringent, or optically anisotropic.

Figure 3.4 represents plane-polarized light incident in a general direction with respect to the crystallographic axes. It is resolved into two components, one with an associated refractive index ω and the other with an associated refractive index ε', both vibrating perpendicular to each other and to the direction of incidence. In general, the value of ε' lies between those of ω and ε. Two special cases arise: one, already discussed, where the incident light is perpendicular to z, for which $\varepsilon' = \varepsilon$; the second arises where the incident light is parallel to z, for which $\varepsilon' = \omega$. It follows that where the direction of incidence is parallel to the z axis, the refractive index is always ω for any vibration direction in the xy plane. Plane-polarized incident light parallel to the z axis will pass through the crystal unmodified. In this particular direction, the crystal is optically isotropic, and if rotated on the microscope stage between crossed Polaroids, it remains an extinction. The z direction of a uniaxial crystal is called the optic axis, and there is only one such direction in the crystal.

Identification of the z Axis of a Uniaxial Crystal

A polarizing microscope is usually fitted with eyepiece cross-wires arranged parallel and perpendicular to the vibration directions of the polarizer, and therefore we can relate the crystal vibration directions to its morphology. There are two important optical orientations for a tetragonal crystal, namely with the z axis either perpendicular or parallel to the axis of the microscope. These orientations are, in fact, important for all uniaxial crystals, and will be described in more detail.

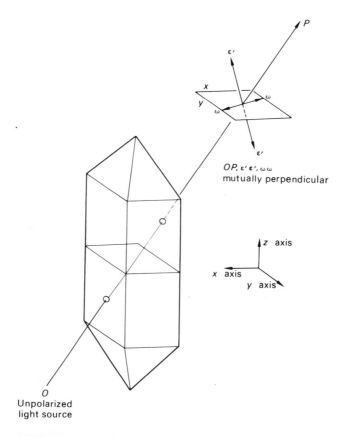

FIGURE 3.4. Uniaxial crystal showing a light ray *OP* resolved into two components. One component, with refractive index ω, vibrates in the *xy* plane, the other, with refractive index ε', vibrates parallel to both ω and the ray direction.

z Axis Perpendicular to the Microscope Axis. In this position, a birefringent orientation is always presented to the incident light beam (Figure 3.5). Extinction will occur whenever the z axis is parallel to the cross-wires, no matter how the crystal is rotated, or flipped over, *while keeping z parallel to the microscope slide.* The success of this operation depends to a large extent on having a crystal with well-developed ($hk0$) faces. The term straight extinction is used to indicate that the field of view is dark when a crystal edge is aligned with a cross-wire. A face of a uniaxial crystal for which one edge is parallel to z, an ($hk0$) face, or to its trace on a crystal face, for example, an ($h0l$) face, will show straight extinction.

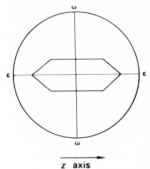

z axis

FIGURE 3.5. Extinction position for a tetragonal crystal lying with its Z axis parallel to the microscope slide. Any $[UV0]$ direction may be parallel to the microscope axis; extinction will always be straight with respect to the Z axis or its trace.

z Axis Parallel to the Microscope Axis. The crystal now presents an isotropic section to the incident light beam, and will remain extinguished for all rotations of the crystal, *while keeping z along the microscope axis.* A reasonably thin section of the crystal is required in order to observe this effect. Because of the needle-shaped habit (external development) of the crystal (KH_2PO_4), it would be necessary to cut the crystal carefully so as to obtain the desired specimen.

The section of a uniaxial crystal normal to the *z* axis, if well developed, may provide a clue to the crystal system. Tetragonal crystals often have edges at 90° to one another, whereas hexagonal and trigonal crystals often exhibit edges at 60° or 120° to one another. These angles are external manifestations of the internal symmetry; idealized uniaxial crystal sections are shown in Figure 3.6.

3.3.3 Biaxial Crystals

Biaxial crystals have two optic axes and, correspondingly, two isotropic directions. The reason for this effect lies in the low symmetry associated with

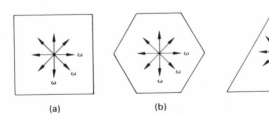

(a)   (b)   (c)

FIGURE 3.6. Idealized uniaxial crystals as seen along the *z* axis: (a) tetragonal, (b) hexagonal, (c) trigonal. The refractive index for light vibrating perpendicular to the *z* axis is always given the symbol ω, and the crystal appears isotropic in this orientation.

the orthorhombic, monoclinic, and triclinic systems, which, in turn, results in less symmetric optical characteristics. Biaxial crystals have three principal refractive indices, n_1, n_2, and n_3 ($n_1 < n_2 < n_3$), associated with light vibrating parallel to three mutually perpendicular directions in the crystal. The optic axes that derive from this property are not directly related to the crystallographic axes. We shall not concern ourselves here with a detailed treatment of the optical properties of biaxial crystals, but will concentrate on relating the vibration, or extinction, directions to the crystal symmetry.

Orthorhombic Crystals

In the orthorhombic system, the vibration directions associated with n_1, n_2, and n_3 are parallel to the crystallographic axes, but any combination of x, y, and z with n_1, n_2, and n_3 may occur. Consequently, recognition of the extinction directions facilitates identification of the directions of the crystallographic axes. For a crystal with x, y, or z perpendicular to the microscope axis, the extinction directions will be parallel (or perpendicular) to the axis in question, as shown in Figure 3.7. If the crystal is a well-developed orthorhombic prism, the three crystallographic axes may be identified by this optical method. A common alternative habit of orthorhombic crystals has one axis, x, for example, as a needle axis and the $\{0\bar{1}1\}$ form prominent. The appearance of such a crystal viewed along x is illustrated in Figure 3.8, and is an example of a symmetric extinction.

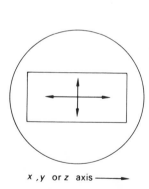

x, y or z axis ⟶

FIGURE 3.7. Extinction directions in an orthorhombic crystal viewed along the x, y, or z axis.

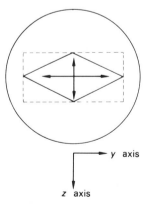

y axis

z axis

FIGURE 3.8. Extinction directions as seen along the x axis of an orthorhombic crystal with $\{011\}$ development—an example of symmetric extinction.

Monoclinic Crystals

The lower symmetry of monoclinic crystals results in a corresponding modification of the optical properties in this system. The symmetry axis *y* is chosen, conventionally, to be parallel to one of the vibration directions; *x* and *z* are related arbitrarily to the other two vibration directions. Hence, two directions are of importance in monoclinic crystals, namely, perpendicular to and parallel to the *y* axis.

When viewed between crossed Polaroids, a monoclinic crystal lying with its *y* axis perpendicular to the microscope axis will always show straight extinction, with the cross-wires parallel (and perpendicular) to *y*. Often, the *y* axis is a well-developed needle axis (Figure 3.9); rotation of the crystal about this axis while keeping it perpendicular to the microscope axis will not cause any change in the extinction positions.

If, on the other hand, a monoclinic crystal is arranged so that *y* is parallel to the microscope axis, the (010) plane will lie on the microscope slide. Extinction in this position will, in general, be oblique, as shown in Figure 3.10, thus giving further evidence for the position of the *y*-axis direction. The appearance of extinction in a monoclinic crystal in this orientation may be somewhat similar to that of an orthorhombic crystal showing prominent {011} development (compare Figures 3.8 and 3.10), and confusion may sometimes occur in practice.

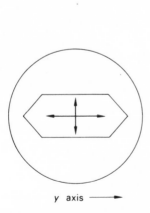

y axis ⟶

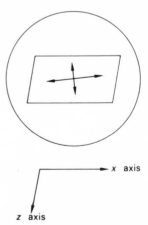

➤ x axis

z axis

FIGURE 3.9. Extinction directions in a monoclinic crystal viewed perpendicular to the *y* axis—an example of straight extinction.

FIGURE 3.10. Extinction directions in a monoclinic crystal viewed along the *y* axis—an example of oblique extinction.

Triclinic Crystals

The mutually perpendicular vibration directions associated with n_1, n_2, and n_3 are arbitrarily related to the crystallographic axes, which are selected initially from morphological and X-ray studies.

Reference again to Table 3.1 should now enable the reader to consolidate the ideas presented in the discussion of extinction directions in the seven crystal systems. Although it gives only limited information* on the optical properties of crystals, a practical study of a crystal along these lines can often provide useful information about both its system and its axial directions.

3.3.4 Interference Figures

The effects which we have discussed so far may be observed when the crystal specimen is illuminated by a more or less parallel beam of plane-polarized light. There is another technique worthy of mention, in which the crystal is examined in a convergent beam of polarized light, which produces characteristic interference figures for uniaxial and biaxial crystals. This examination may be effected, at high magnification and between crossed Polaroids, either by removing the microscope eyepiece or by inserting a Bertrand lens† into the microscope system, below the eyepiece. Figure 3.11a shows an idealized interference figure from a section of a uniaxial crystal cut perpendicular to the optic axis, while Figures 3.11a and 3.11b are interference figures for a biaxial crystal section cut perpendicular to a bisector of the two optic axes.

If optical figures of good quality can be obtained, the distinction between uniaxial and biaxial specimens may be achieved with one orientation of the crystal. It may be confirmed by rotation of the crystal specimen about the microscope axis, which causes the dark brushes or isogyres in the biaxial figure to break up, as in Figure 3.11c, while those for the uniaxial interference figure remain intact.

3.4 Direction of Scattering of X-Rays by Crystals

The first experiments involving the scattering, or diffraction, of X-rays by crystals were initiated by von Laue in 1912. It is well known that similar

---

* For a fuller discussion, see Bibliography.
† A Bertrand lens is a normal accessory with a good polarizing microscope.

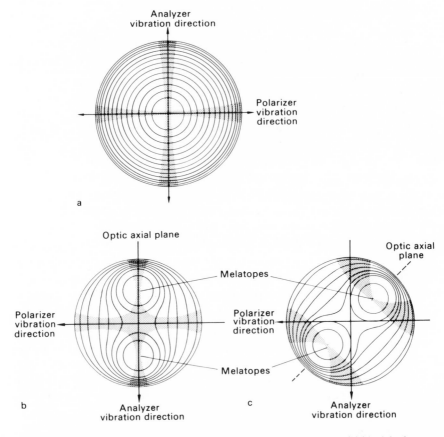

FIGURE 3.11. Interference figures: (a) uniaxial, along the optic axis; (b) biaxial, along a bisector of the optic axes and with the Polaroids crossed; (c) as in (b), but with the polarizer rotated by 45° (position of maximum transmitted intensity). [Reproduced from *An Introduction to Crystal Optics* by P. Gay, with the permission of Longmans Group Ltd., London.]

effects with visible light can be achieved with a ruled grating, provided that the rulings are spaced at about the same order of magnitude as the wavelength of the light. The analogy in a crystallographic experiment is that the wavelength of the X-rays must be of the same order of magnitude as the distance between the scattering units in the crystal. These scattering units are the electron clouds associated with the atoms in the structure, and their regularity is provided by the crystal lattice translations.

The wavelength range of X-rays used in crystallography is between about 0.7 and 2.0 Å. Crystals are, therefore, ideal materials for studying diffraction effects with an X-ray source, and, conversely, X-rays provide a

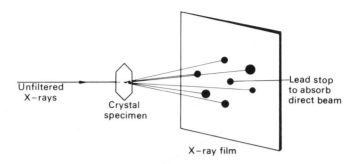

FIGURE 3.12. Experimental arrangement for taking a Laue photograph on a flat-plat film.

powerful method for investigating crystal structure. The generation and
properties of X-rays are discussed in Appendix A4.

Figure 3.12 shows, schematically, the experimental arrangement
required to produce a Laue X-ray photograph. The photograph is obtained
by irradiating a stationary single crystal with white X-radiation, which is
composed of a continuous range of wavelengths. Figure 3.13 is a Laue
photograph of Al_2O_3; it shows the symmetry of the point group $3m$. In
common with other X-ray photographs, the diagram shows two important

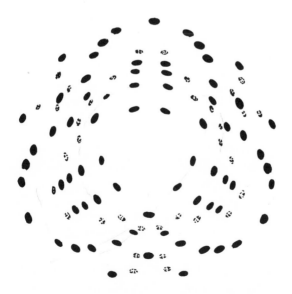

FIGURE 3.13. Sketch of a Laue photograph of α-Al_2O_3.

features: The spots occur in definite positions, which are determined by the wavelength of the X-rays and the size and orientation of the unit cell, and the intensity, or degree of blackening, varies from one spot to another. This second feature arises from both the geometry of the experiment and, most particularly, from the crystal structure itself. Thus, an X-ray photograph contains information about several aspects of the internal structure of a crystal, and is, therefore, a fingerprint of the particular specimen. X-ray structure analysis uses this information to deduce the positions of the atoms in the crystal.

This chapter is concerned with the interpretation of the positions of the spots on an X-ray photograph, that is, with the direction of scattering.

3.4.1 Laue Equations for X-Ray Scattering

Consider a row of scattering centers of regular spacing b (Figure 3.14). X-rays are incident at an angle ϕ_2 and are scattered at an angle ψ_2. The path difference between rays scattered by neighboring centers is given by

$$\delta_2 = AQ - BP \tag{3.1}$$

or

$$\delta_2 = b(\cos \psi_2 - \cos \phi_2) \tag{3.2}$$

For reinforcement of the scattered rays, the path difference must be an integral number of wavelengths, and we may write

$$b(\cos \psi_2 - \cos \phi_2) = k\lambda \tag{3.3}$$

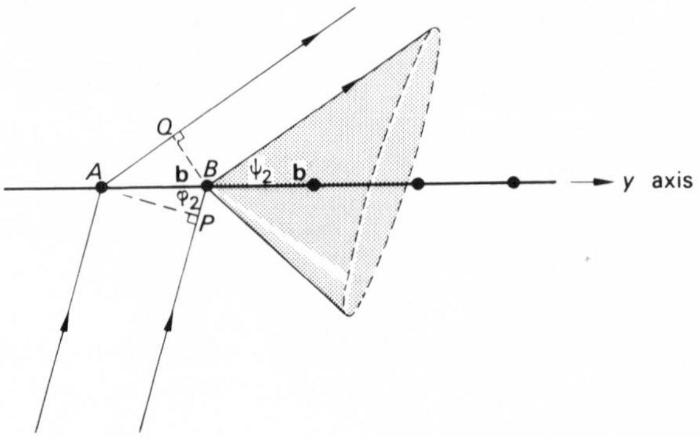

FIGURE 3.14. Diffraction from a row of scattering centers.

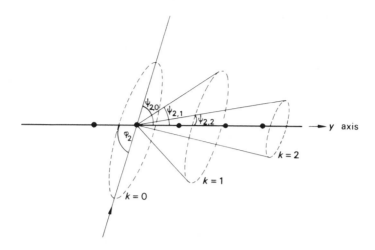

FIGURE 3.15. Several orders of diffraction from a row of scattering
centers.

This equation is satisfied by the generators of a cone which is coaxial with the
line of scattering centers and has a semi-vertical angle of ψ_2. For a series of
values of ϕ_2, there will be a number of such cones, each corresponding to an
order of diffraction k and a semi-vertical angle $\psi_{2,k}$ (Figure 3.15).

This discussion is readily extended to a net of scattering centers (Figure
3.16). For the rows parallel to the X axis, we can write, by analogy with (3.3),

$$a(\cos \psi_1 - \cos \phi_1) = h\lambda \qquad (3.4)$$

When both (3.3) and (3.4) are satisfied simultaneously, as they are along the
lines of intersection (BR and BS) of the two cones, the entire net scatters in
phase, producing hk spectra. For the particular case that BR and BS
coincide, the diffracted beam lies in the plane of the two-dimensional array
of scattering centers.

Generalizing to three dimensions, we may write down the three Laue
equations:

$$a(\cos \psi_1 - \cos \phi_1) = h\lambda$$
$$b(\cos \psi_2 - \cos \phi_2) = k\lambda \qquad (3.5)$$
$$c(\cos \psi_3 - \cos \phi_3) = l\lambda$$

Any of the three possible pairs of equations corresponds to scattering from
the corresponding net. For the particular case that the three cones intersect

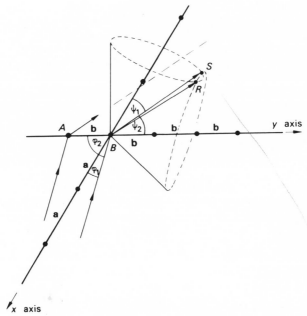

FIGURE 3.16. Diffraction from a net of scattering centers;
for clarity only one row parallel to each axis has been drawn.

in a line, the entire three-dimensional array scatters in phase, producing the
hklth spectrum.

3.4.2 Bragg's Treatment of X-Ray Diffraction

The interaction of X-rays with a crystal is a complex process, often
described as a diffraction phenomenon, although it is, strictly speaking, a
combined scattering and interference effect. The Bragg treatment of X-ray
diffraction, although an oversimplification of the complete process, gives a
clear and accurate picture of the directional features of a diffraction pattern,
and provides a valuable means for interpreting the positions of the spots on
an X-ray photograph.

Any atom in a crystal structure is repeated by the symmetry operations
of the space group. The simplest type of structure would consist of a single
atom located at the lattice points associated with a primitive unit cell. Figure
3.17a is a representation of one unit cell of such a structure, based on an
orthorhombic lattice. Figure 3.17b shows the c-axis projection of the same
structure, but rotated so as to make the traces of (110) horizontal. Any other
family of planes would be just as suitable in this discussion.

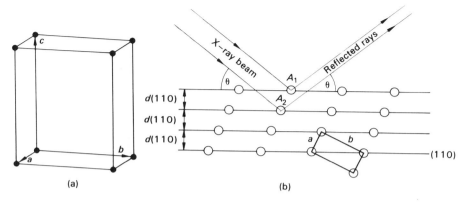

FIGURE 3.17. (a) One *P* unit cell of an orthorhombic structure; (b) X-ray beam "reflected" by the (110) planes of the orthorhombic structure; two typical rays are shown.

Bragg considered that the crystal planes behaved as though they partially reflected the X-rays, like sheets of atomic mirrors. This approach was not *ad hoc*. The early experiments with X-ray diffraction showed that if a crystal was turned from one diffracting position to another through an angle α, then the diffracted ray was rotated through an angle of 2α. The $\alpha,2\alpha$ relationship is reminiscent of the reflection of visible light from a plane mirror. The analogy breaks down with X-rays because the Bragg equation (3.14) has to be satisfied, but the reflection treatment is useful, and we shall speak of Bragg reflection, or just reflection, of X-rays by crystals. In this description, the angles of incidence and reflection (θ) are equal and are coplanar with the normal to the reflecting planes, but the angle of incidence used here is the complement of that employed in geometrical optics.

The part of the X-ray beam that is not reflected at a given level in the crystal passes on to be subjected to a similar process at the next level deeper into the crystal. In Figure 3.17b, Bragg reflection of two parallel rays is illustrated, and a special relationship between the interplanar spacing $d(hkl)$, the X-ray wavelength λ, and the Bragg angle θ exists when all planes in the (hkl) family cooperate in the scattering process.

A more complete picture is given in Figure 3.18, which demonstrates also that all rays reflected from a given level remain in phase after reflection, since no path difference is introduced. The paths *AB* and *CD* are of equal length. However, two rays reflected from neighboring planes are, in general, out of phase because they travel different path lengths. The detailed geometry of this process is shown in Figure 3.19, where the typical path

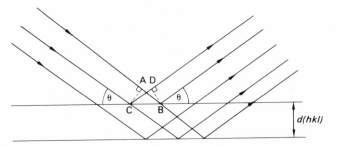

FIGURE 3.18. More complete picture of the Bragg reflection process. Two rays reflected from the same plane do not suffer any relative phase change or path difference $(AB = CD)$.

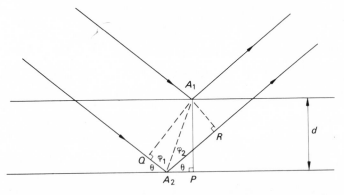

FIGURE 3.19. Detailed geometry of X-ray reflection. The path difference between the two typical rays reflected from successive planes is $(QA_2 + A_2R)$, and is equal to $2d \sin \theta$.

difference δ between two rays is given by

$$\delta = QA_2 + A_2R = A_1A_2 \cos \phi_1 + A_1A_2 \cos \phi_2 \qquad (3.6)$$

or

$$\delta = A_1A_2(\cos \phi_1 + \cos \phi_2) = 2A_1A_2 \cos[(\phi_1 - \phi_2)/2] \cos[(\phi_1 + \phi_2)/2] \qquad (3.7)$$

The three components of (3.7) may be expressed as follows

$$A_1A_2 = d/\sin(\theta + \phi_2) \qquad (3.8)$$

$$\phi_1 + \phi_2 = 180° - 2\theta \qquad (3.9)$$

whence

$$\cos[(\phi_1+\phi_2)/2]=\cos(90°-\theta)=\sin\theta \qquad (3.10)$$

$$\phi_1-\phi_2=180°-2(\theta+\phi_2) \qquad (3.11)$$

whence

$$\cos[(\phi_1-\phi_2)/2]=\cos[90°-(\theta+\phi_2)]=\sin(\theta+\phi_2) \qquad (3.12)$$

Combining terms to give δ, we obtain

$$\delta=2d\sin\theta \qquad (3.13)$$

Since δ is independent of ϕ_1 and ϕ_2, this equation applies to all rays in the bundle reflected from two adjacent planes. By the usual rules applied to the combination of waves of the same wavelength, the rays reflected by these two planes will interfere with one another, the interference being at least partially destructive unless the path difference δ is equal to an integral number of wavelengths. Thus,

$$2d\sin\theta=n\lambda \qquad (3.14)$$

which is the Bragg equation, sometimes called Bragg's law. Reflection will be obtained when this equation is satisfied, which may be achieved in practice by varying one of the four quantities θ, d, λ, or n.

In (3.14), n is the order of the Bragg reflection. We can write this expression in another form if we recall from page 69 that

$$d(hkl)/n=d(nh,nk,nl) \qquad (3.15)$$

$d(nh, hk, nl)$ is usually replaced by $d(hkl)$, with h, k, and l taking values with or without common factors. Hence, n is included in the crystallographic definition of d, and the Bragg equation is now written as

$$2d(hkl)\sin\theta(hkl)=\lambda \qquad (3.16)$$

which means that we consider each Bragg reflection from a crystal as a first-order reflection from the family (hkl), which is specified uniquely by its general Miller indices. To quantify this point further, the following example refers to reflections from planes parallel to (100) in a cube of unit cell side 12 Å.

Original Bragg formulation			Current usage	
Reflection	Order	d, Å	Reflection	d, Å
100	1	12	100	12
	2	6	200	6
	3	4	300	4
	4	3	400	3

3.4.3 Equivalence of Laue and Bragg Treatments of X-Ray Diffraction

The treatments exemplified by (3.5) and (3.16) may be shown to be equivalent. Consider any three-dimensional array of scattering centers (Figure 3.20). From (3.3), we may write

$$p(\cos \psi - \cos \phi) = n\lambda \qquad (3.17)$$

where n is an integer. Expanding (3.17), we obtain

$$-2p \sin[(\psi + \phi)/2] \sin[(\psi - \phi)/2] = n\lambda \qquad (3.18)$$

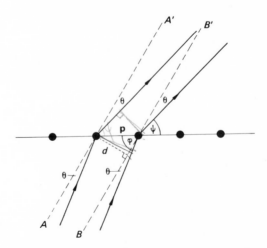

FIGURE 3.20. Equivalence of the Laue and Bragg equations.

Now AA' and BB' must be the traces of part of the family of planes that make equal angles θ with the incident and diffracted beams. From the diagram,

$$\phi - \theta = \psi + \theta \qquad (3.19)$$

or

$$\theta = (\phi - \psi)/2 \qquad (3.20)$$

Furthermore,

$$d = p \, \sin(\phi - \theta) \qquad (3.21)$$

which, from (3.20), becomes

$$d = p \, \sin[(\psi + \phi)/2] \qquad (3.22)$$

Using (3.18), (3.20), and (3.22), we obtain

$$2d \, \sin \theta = n\lambda \qquad (3.23)$$

and n can be incorporated into d in the manner described above. Hence, the two approaches to X-ray scattering by crystals are equivalent. We shall find both of them useful in our subsequent discussions. The diffraction equations used so far in this chapter are considered in terms of a vector algebraic treatment in Appendix A8.

3.5 X-Ray Techniques

The X-ray photographs in common use can be divided into two classes, single-crystal and powder photographs. If a detailed structure analysis is to be carried out, it is desirable to have well-formed single crystals of the given compound available, and the diffraction data are collected by one of the appropriate photographic methods, or with a single-crystal diffractometer (see Appendix A5). If, on the other hand, it is required only to characterize a particular substance from its X-ray pattern, then it may be possible to effect identification from powder photographs, taken with a small amount of finely powdered material. Powder photographs are of minimal value in crystal structure analysis, and are not discussed in this book.*

---

* See Bibliography.

We now continue the preliminary examination of a single crystal by X-ray methods. X-ray photographs can be used to provide the information necessary to confirm the crystal system, to measure the unit-cell dimensions, to determine the number of chemical entities in the unit cell, and to establish, at least partially, the space group.

3.5.1 Laue Method

The three variables in the Bragg equation (3.16) provide a basis for the interpretation of X-ray crystallographic experiments. In the Laue method (Figure 3.12), the Bragg equation is satisfied by effectively varying λ, using a beam of continuous (white) radiation. Since the crystal is stationary with respect to the X-ray beam, it acts as a sort of filter, selecting the correct wavelengths for each reflection according to (3.16).

The spots on a Laue photograph lie on ellipses, all of which have one end of their major axis at the center of the photographic film (Figure 3.13). All spots on one ellipse arise through reflections from planes that lie in one and the same zone. In Figure 3.21, a zone axis for a given Bragg angle θ is represented by ZZ'. A reflected ray is labeled R, and we can simulate the effect of the zone by imagining the crystal to be rotated about ZZ', taking the reflected beam with it. The rays, such as R, generate a cone, coaxial with ZZ' and with a semivertical angle θ. The lower limit, in the diagram, of R is the direction (xy) of the X-ray beam, and the general intersection of a circle with a plane is an ellipse. Hence, we can understand the general appearance of the Laue photograph. On each ellipse, discrete spots appear instead of continuous bands because only those orientations parallel to zone axes, such as ZZ', that actually exist for crystal planes can give rise to X-ray reflections.

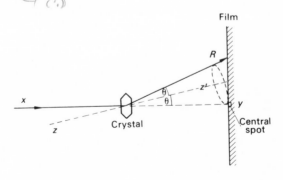

FIGURE 3.21. Basic geometry of the Laue method.

Symmetry in Laue Photographs. One of the most useful features of Laue photographs is the symmetry observable in them. The crystal orientation with respect to the X-ray beam is selected by the experimenter from morphological and optical considerations. This orientation, together with the crystal point group, controls the symmetry on the Laue photograph.

In practice, a complication arises by the introduction of a center of symmetry into the diffraction pattern, in normal circumstances, whether or not the crystal is centrosymmetric. This situation is embodied in Friedel's law, the theoretical grounds for which are discussed in Chapter 4. As a result of this law, the diffraction pattern may not reveal the true point-group symmetry of a crystal. Table 1.6 shows the classification of the 32 crystallographic point groups according to Laue, or diffraction, symmetry.

It cannot be over-emphasized that the Laue group assigned to a crystal describes the symmetry of the *complete* X-ray diffraction pattern from that crystal. No single X-ray photograph can exhibit the complete diffraction symmetry, only that of a selected portion which is a projection, along the direction of the X-ray beam, of the symmetry information that would be encountered in that direction in a crystal having the Laue group of the given crystal.

It follows that in the triclinic system, no symmetry higher than 1 is ever observable in a Laue photograph. In other crystal systems, the Laue-projection symmetry depends on the orientation of the crystal with respect to the X-ray beam. Rotation axes of any order reveal their true symmetry when the X-ray beam is parallel to the symmetry axis. Even-order rotation axes, 2, 4, or 6, give rise to mirror diffraction symmetry in the plane normal to the rotation axis when the X-ray beam is normal to that axis. A mirror plane itself shows *m* symmetry parallel to the mirror plane when the X-ray beam is contained by the plane. Various combinations of these effects may be observable, depending upon the Laue group in question.

The supplementary nature of the X-ray results to those obtained in the optical examination should be evident now. Uniaxial crystals can be allocated to their correct systems by a Laue photograph taken with the X-ray beam along the *z* axis. Figure 3.13 is an example of such a photograph.* Distinction between the monoclinic and orthorhombic systems, which is not always possible in an optical examination, is fairly straightforward with Laue photographs, as Table 1.6 shows. Cubic crystals can exhibit a variety of

---

* Laue projection symmetry 3*m*.

symmetries, but with the X-ray beam along $\langle 100 \rangle$, differentiation between Laue groups $m3$ and $m3m$ is obvious.

It should be noted that, in practice, the symmetry pattern on a Laue photograph is very sensitive to precise orientation of the crystal.* Slight deviation from the ideal position will result in a distortion of the relative positions and intensities of the spots on the photographs.

3.5.2 Oscillation Method

The oscillation method is a somewhat more sophisticated technique for recording the X-ray diffraction patterns from single crystals. Reflections are produced, in accordance with the Bragg equation, by varying the angle θ for a given wavelength λ. The variation of θ is brought about by oscillating or rotating the crystal about a crystallographic axis, and λ is "fixed" by the use of an appropriate filter (see Appendix A4) placed in the path of the incident X-ray beam.

The basic arrangement used in the oscillation method is illustrated schematically in Figure 3.22. X-ray reflections produced by the moving crystal are recorded on a cylindrical film coaxial with the axis of oscillation of the crystal. The general appearance of an oscillation photograph is illustrated in Figure 3.23. For the moment, we shall concentrate on the periodicity of the crystal parallel to the oscillation axis. Thinking of the crystal as a row of scattering centers, and following the development of (3.1) to (3.3), we see that, for normal incidence ($\phi = 90°$), the diffracted beams will lie on cones that are coaxial with the oscillation axis and intersect the film in circular traces. When the film is flattened out for inspection, the spots are found to lie on parallel straight lines.

Axial Spacings from Oscillation Photographs. The equatorial layer line, or zero-layer line, E passes through the origin O where the direct X-ray beam intersects the film (Figures 3.22 and 3.23). The layer-line spacings, measured with respect to the zero layer, are denoted by $\nu(\pm n)$, where $\nu(+n) = \nu(-n)$, and are related to the repeat distance in the crystal parallel to the oscillation axis, as is shown in the following treatment.

Let the oscillation axis be a, and let R be the radius of the film, measured in the same units as ν. Consider the crystal to be acting as a one-dimensional diffraction grating with respect to the direction of the a axis, and giving rise to spectra of order $\pm 1, \pm 2, \ldots, \pm n$; $\psi(n)$ is the scattering angle for the nth-order maximum, measured with respect to the

---

* See Bibliography.

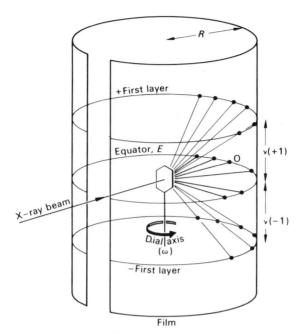

FIGURE 3.22. Basic geometry of the oscillation method, showing how diffraction spots are recorded on a cylindrical film placed around the crystal.

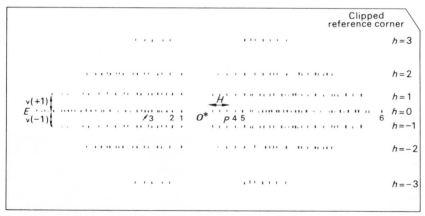

FIGURE 3.23. Sketch of a 15° oscillation photograph of an orthorhombic crystal mounted on the *a* axis ($a = 6.167$ Å); the camera radius *R* is 30.0 mm and λ (Cu $K\alpha$) = 1.542 Å. The film is flattened out and the right-hand corner, looking toward the X-ray source, is clipped in order to provide a reference mark. *P* represents any equatorial reflection at a distance *OP* (= *H* mm) from the center *O*. Reflections numbered 1–6 on the zero-level are indexed by the method given on page 136. Linear scale of the diagram is (1/1.78).

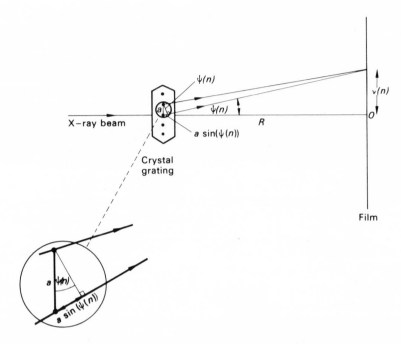

FIGURE 3.24. Diffraction grating analogy explaining the layer-line spacings on oscillation photographs. Monochromatic X-rays are incident normal to a (oscillation axis) in the crystal. The size of any spot at height such as $\nu(n)$ depends upon the experimental conditions.

direct beam. Normal beam diffraction uses the geometry of Figure 3.24. The path difference for rays scattered at an angle $\psi(n)$ by successive elements of the grating is $a \sin \psi(n)$, which, for maximum intensity, is equal to $n\lambda$. Hence, for layer n,

$$a \sin \psi(n) = n\lambda \tag{3.24}$$

where $\psi(n)$ is measured experimentally as

$$\tan \psi(n) = \nu(n)/R \tag{3.25}$$

Hence

$$a = \frac{n\lambda}{\sin\{\tan^{-1}[\nu(n)/R]\}} \tag{3.26}$$

For a known wavelength, this equation provides a convenient and reasonably accurate method for determining unit-cell spacings. In practice, we

TABLE 3.2. Symmetry Indications from Oscillation Photographs

Feature of photograph	Interpretation(s)
Horizontal m line	Horizontal m plane in the corresponding Laue group
Vertical m line[a]	m plane in Laue group of crystal, parallel to the plane defined by the oscillation axis and the beam
Twofold symmetry about the center of the photograph[a]	Twofold axis in the Laue group of the crystal, and parallel to the X-ray beam
Approximate R-fold symmetry around the central portion of the photograph[a]	R-fold axis in Laue group of crystal, and parallel to the X-ray beam

[a] Symmetric oscillation photographs.

measure the double spacing between the $\pm n$ th orders so as to enhance the precision of the result.

Symmetry in Oscillation Photographs. Oscillation photographs have several useful symmetry properties. A horizontal mirror line along the equator E of a general oscillation photograph indicates a mirror plane perpendicular to the oscillation axis in the corresponding Laue group of the crystal (Table 3.2).

Further observations on the symmetry of the Laue group can be made by arranging for a particular crystal symmetry direction to be parallel to the X-ray beam at the midpoint of the oscillation range (symmetric oscillation photograph). The situations that can arise are also summarized in Table 3.2. Note that the highest symmetry observable by the oscillation method is $2mm$, obtained from a symmetric oscillation photograph of an orthorhombic crystal mounted on a, b, or c and with one of these axes parallel to the X-ray beam at the center of the oscillation range. However, if an R-fold rotation axis $(R > 2)$ is parallel to the beam at the midpoint of the oscillation, then the central portion of the photograph will reveal an approximate R-fold symmetry pattern, particularly where the reciprocal unit cell is small. The true symmetry will not appear exactly, as it is degraded by the symmetry of the oscillation movement (mmm); the exact symmetries are subgroups of the plane point group $2mm$.

Detection of threefold, fourfold, or sixfold rotational symmetry parallel to the oscillation axis may be effected by taking a series of photographs, the first of which is taken with the crystal oscillating about an arbitrary setting (ω_1) of the dial axis (Figure 3.22). The identical appearance of succeeding photographs with the dial axis set at $\omega_1 + 60°$, $\omega_1 + 90°$, or $\omega_1 + 120°$ indicates sixfold, fourfold, or threefold (and sixfold) symmetry, respectively. A twofold axis cannot be detected by this method, because of Friedel's law.

Indexing the Zero Level of a Crystal with an Orthogonal Lattice. We discuss next the relatively small portion of the X-ray diffraction pattern produced in an oscillation photograph. It is of great importance in structure analysis to assign the correct indices *hkl* to each observed reflection. This process is known as indexing, and is reasonably straightforward. In this discussion, we shall consider the indexing of an orthogonal reciprocal lattice, using, as an example, an orthorhombic crystal mounted with *a* as the oscillation axis. It may be noted in passing that monoclinic *b*-axis and hexagonal and trigonal *c*-axis photographs can be indexed in a similar manner. Two prerequisites to indexing are the reciprocal unit-cell dimensions and the orientation of the reciprocal lattice axes perpendicular to the oscillation axis (b^* and c^* in the example) with respect to the incident X-ray beam. The reciprocal unit-cell dimensions may be derived from the corresponding direct space values through (2.17) and (2.18).

3.5.3 Ewald's Construction

The geometric interpretation of X-ray diffraction photographs is greatly facilitated by means of a device due to Ewald, and known as the Ewald sphere, or sphere of reflection. The sphere is centered on the crystal (C) and drawn with a radius of one reciprocal space unit (RU) on the X-ray beam (AQ) as diameter (Figure 3.25). The Bragg construction for reflection is superimposed, and a reflected beam *hkl* cuts the sphere in *P*. The points *A*, *P*, and *Q* lie on a circular section of the sphere which passes through the center *C*.

From the construction

$$AQ = 2 \qquad \text{(by construction)} \qquad (3.27)$$

$$\widehat{APQ} = 90° \qquad \text{(angle in a semicircle)} \qquad (3.28)$$

Hence

$$QP = AQ \sin \theta (hkl) = 2 \sin \theta (hkl) \qquad (3.29)$$

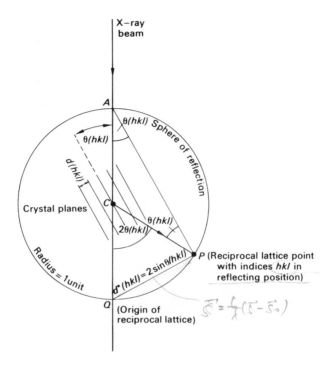

FIGURE 3.25. Ewald construction illustrating how an X-ray
reflection may be considered to arise when a reciprocal lattice
point P passes through the sphere of reflection. AP is parallel to
the (hkl) planes, and the reciprocal lattice vector QP forms a
right angle at P. (*Note*: If the rotation is about a, normal to
the circular section shown, this circle becomes the zero-layer
circle and reflected beams such as CP will all be denoted $0kl$.)

From Bragg's equation (3.16),

$$2 \sin \theta(hkl) = \lambda/d(hkl) \tag{3.30}$$

and from the definition of the reciprocal lattice (page 70), we identify the
point P with the reciprocal lattice point hkl; hence

$$QP = d^*(hkl) \tag{3.31}$$

with $K = \lambda$ [equation (2.11)], and

$$d^*(hkl) = 2 \sin \theta(hkl) \tag{3.32}$$

We now have a mechanism for predicting the occurrence of X-ray reflections and their directions in terms of the sphere of reflection and the reciprocal lattice. The origin of the reciprocal lattice is taken at Q, and, although the crystal is at C, it may be helpful to imagine a conceptual crystal at Q identical to the real crystal and moving about a parallel oscillation axis in a synchronous manner.

The condition that the crystal is in the correct orientation for a Bragg reflection hkl to take place is that the corresponding reciprocal lattice point P is on the sphere of reflection. As the crystal oscillates, an X-ray reflection flashes out each time a reciprocal lattice point cuts the sphere of reflection, and the direction of reflection is given by CP.

Ewald's construction provides an elegant illustration of the formation of layer lines on an oscillation photograph. Figure 3.26 shows an Ewald sphere and portions of several layers of an orthogonal reciprocal lattice. As the sphere and the X-ray beam oscillate about the x axis, reciprocal lattice points cut the sphere of reflection in circles because the axis of the cylindrical film is arranged to be parallel to the oscillation axis x.

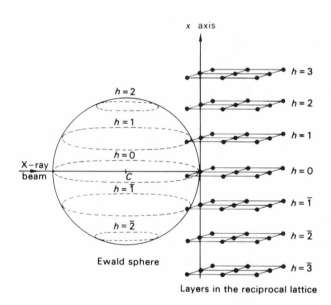

FIGURE 3.26. Formation of layer lines in terms of the Ewald construction. Layers with $|h| \geq 3$ lie outside the range of recording in this illustration. Note that except for $h = 0$, the circles labeled $\pm h$ ($h = 1, 2, 3, \ldots$) cannot be identified with the circular section of Figure 3.25.

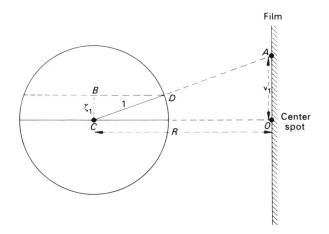

FIGURE 3.27 Relationship between reciprocal lattice spacing
and the corresponding layer-line spacing on a film.

The relationship between the reciprocal lattice spacing of the layers and
the corresponding repeat distance a in direct space is shown in Figure 3.27.
From the similar triangles AOC and BCD,

$$AO/OC = BC/BD \tag{3.33}$$

or

$$\nu_1/R = \zeta_1/(1-\zeta_1^2)^{1/2} \tag{3.34}$$

where ν_1 is the distance between the zero layer and the first layer line and R
is the radius of the film. If the lattice is orthogonal in the aspect illustrated
(x^* coinciding with x), then from (3.30) and (3.32), since ζ_1 is equivalent
to $d^*(100)$,

$$a = \lambda/\zeta_1 \tag{3.35}$$

Although (3.35) holds generally, if the lattice is not orthogonal, $\zeta_1 \neq d^*(100)$
and the appropriate expressions are a little more complicated (see page 73),
requiring a knowledge also of the interaxial angles.

Indexing Procedure for an a-Axis Oscillation Photograph. On the
zero level of an a-axis oscillation photograph, reflections are of the type $0kl$.

The relevant portion of the reciprocal lattice is the y^*z^* net which, for an orthorhombic crystal, is determined by $b^* (=\lambda/b)$, $c^* (=\lambda/c)$, and $\alpha^* (=90°)$. Without going into further detail, we note that the simplest method of determining b^* and c^* would be from the values of b and c, through (3.26) in the appropriate forms.

A drawing of the reciprocal net is prepared carefully, using a convenient scale, for example, 1 RU = 50 mm. The values of $d^*(0kl)$ are obtained from measurements of $H(0kl)$ on the zero-layer line (Figure 3.23), noting also whether the spot lies to the left or the right of the center O. Since the angular deviation of the X-ray beam is 2θ (Figure 3.25),

$$2\theta = H/R \qquad\qquad (3.36)$$

in radian measure. Using degree measure,

$$d^*(0kl) = 2 \sin \theta(hkl) = 2 \sin[180H(0kl)/2\pi R] \qquad (3.37)$$

H and R are, conveniently, measured in millimeters.

Worked Example of Indexing. The a-axis oscillation photograph of an orthorhombic crystal (Figure 3.23) was taken with $+b^*$ pointing toward the X-ray source at the start of a 15° anticlockwise oscillation. A sample of reflections recorded on the zero level of this photograph at H mm from the center is listed in Table 3.3. For the X-ray wavelength used, $b^* = 0.114$ and $c^* = 0.113$ RU. A y^*z^* reciprocal net was constructed and used to index these reflections.

The scheme for carrying out this indexing is illustrated in Figure 3.28, with the help of Table 3.4. Instead of thinking in terms of the crystal oscillating (first anticlockwise), we imagine that the sphere of reflection oscillates (first clockwise) about an axis through Q, normal to the X-ray beam, from $QA (0°)$ to $QA'(15°)$, taking the incident X-ray beam with it. The reciprocal lattice points that would intersect the sphere of reflection are

TABLE 3.3. Measurements on the Zero-Layer a-Axis Photograph

Reflection number	H, mm LHS of center	Reflection number	H, mm RHS of center
1	11.2	4	14.1
2	15.7	5	17.4
3	25.3	6	83.3

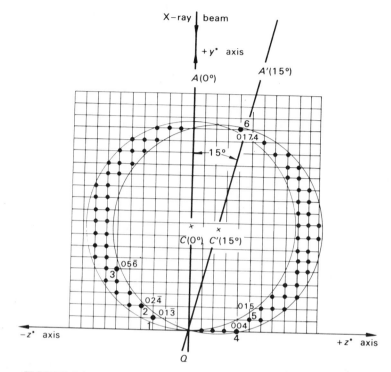

FIGURE 3.28. Indexing the 15° oscillation photograph in Figure 3.23: orthorhombic crystal, a parallel to the rotation axis, $-y^*$ along the direction of the incident X-ray beam at the start of the oscillation, scale 1 RU ≈ 27.7 mm. Possible reflections are shown as points within the lunes. Reflections 1–6 indexed in Table 3.4 are indicated.

those within the lunes swept out during this motion. Points lying within these lunes are shown in Figure 3.28, and they correspond to possible $0kl$ reflections for this oscillation movement. By measuring the d^* values from Q to reciprocal lattice points within the lunes on the left- or right-hand side,

TABLE 3.4. Indexed Reflections for the $0kl$ Layer Line

Reflection number	d^*	hkl	Reflection number	d^*	hkl
1	0.371	$01\bar{3}$	4	0.464	004
2	0.516	$02\bar{4}$	5	0.573	015
3	0.818	$05\bar{6}$	6	1.967	$017,4$

as appropriate, the required indices may be determined for the reflections that *do* occur for this crystal.

The construction of the lunes in the correct orientation greatly reduces the number of reciprocal lattice points to be considered. The indexed points are shown in both Table 3.4 and Figure 3.28. Use the diagram to determine which other reflections might have been recorded on this photograph. It should be clear that no reflection for which d^* is greater than 2 can be observed; this number is the radius of another sphere, the limiting sphere, which is that sphere swept out in reciprocal space by a complete rotation of the Ewald sphere. Problem 4 at the end of this chapter is based on this example. It is sometimes necessary to consider a point just outside the lunes, depending on the probable experimental errors.

The Arndt–Wonnacott camera is a modern version of the oscillation camera, designed for use with crystals having axial spacings greater than about 100 Å (certain protein crystals, for example). Very small oscillations are used, and a complete recording of the diffraction pattern is facilitated by an automatic film-changing device.

3.5.4   Weissenberg Method

The oscillation method suffers from the main disadvantage that it presents three-dimensional information on a two-dimensional film. It is both a distorted and collapsed diagram of the reciprocal lattice. This situation leads, in turn, to overlapping spots, particularly if the reciprocal lattice dimensions are small, and consequent ambiguity in indexing. More advanced X-ray photographic methods permit rapid, unequivocal indexing without graphical construction.

Figure 3.29 is an extension of Figure 3.22. Each cone corresponds to possible reflections of constant h index. In the Weissenberg method, all cones but one are excluded by means of adjustable metal screens, shown in the figure in the position which permits only the $0kl$ reflections to pass through and reach the film. If the film were kept stationary, the exposed record would still look like the zero-layer line in Figure 3.23. However, in the Weissenberg technique, the film is translated parallel to the oscillation axis, synchronously with the oscillatory motion of the crystal. The result of this procedure is a spreading of the spots over the surface of the film on characteristic straight lines and curves (Figure 3.30). Each spot has a particular position on the film, governed, in this example, by the values of k and l. With practice, the film can be indexed by inspection.

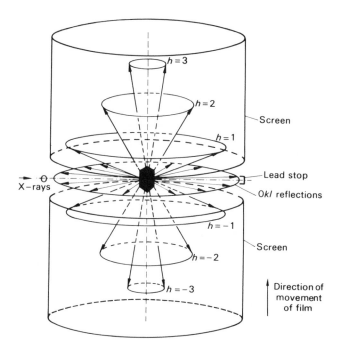

FIGURE 3.29. Cones representing possible directions of diffracted X-rays from a crystal rotating on its *a* axis. The screens are set to exclude all but the zero layer, $0kl$.

If the layer-line screens are moved by the appropriate distance parallel to the oscillation axis, the first layer ($h = 1$) can be recorded on the film. Generally, nonzero layers are recorded by the equi-inclination technique,* which ensures that their interpretation is very similar to that for the corresponding zero layer.

The Weissenberg photograph is clearly much simpler to interpret than the oscillation photograph, but the apparatus required is more complex. Although the photograph of the reciprocal lattice is not collapsed, it is clear from the figure that it is still distorted: the orthorhombic reciprocal net does not appear as rectangular.

Unit-Cell Dimensions from a Weissenberg Photograph

We shall use Figure 3.31, which is a scaled-down reproduction of a Weissenberg photograph, as an example. The principal reciprocal lattice

* See Bibliography.

FIGURE 3.30. Typical Weissenberg X-ray photograph: zero layer, $h0l$ reflections for a monoclinic crystal.

axes, y^* and z^*, are straight lines of spots on the photograph, and have a slope of 2 (tan 63.435°) and a separation of 45.0 mm. The crystal rotates through 2° for every 1 mm of travel of the cassette, so that $\alpha^*(\widehat{y^*z^*})$ is 90°.

Prior to making measurements on the film, the spacing along the rotation axis of the crystal (a in this example) would have been measured from an oscillation photograph taken with the same camera (see Section 3.5.2). We must now obtain the values of b^*, c^*, and α^* from the Weissenberg photograph. Measurement of unit-cell dimensions from the zero-level Weissenberg photograph makes use of the fact that, for a given reflection, 2θ is calculated according to (3.36), where H is the perpendicular distance from a spot to the central line OO', and R is the film radius, as before. A standard Weissenberg camera has a diameter ($2R$) of 180/π, or 57.30 mm.

In practice, the line OO' may be difficult to locate, and the value of a given H may be best obtained by measuring $2H'$ along the actual reciprocal lattice row, across the film between corresponding spots on the top and bottom halves. A zero-layer Weissenberg photograph has inherent twofold symmetry, and each row and curve shown repeats on the other side of OO'. Then we can calculate H from

$$H = \tfrac{1}{2}(2H') \sin(\tan^{-1}2) \tag{3.38}$$

From (3.37) we have, as before,

$$d^* = 2 \sin[180H/2\pi R] = 2 \sin H \tag{3.39}$$

where H is measured in mm. The angle α^* is determined by measuring the separation between y^* and z^* in the OO' direction and converting to degrees by multiplying by the camera constant of 2 deg mm^{-1}.

In Figure 3.31, the value of $H(090)$, for example, is 61.2 mm, using (3.38) with H' measured as 38.0 mm and multiplied by the scale factor of 1.8. hence $d^*(090)$ is 1.753 and b^* is 0.195. Similarly, $H(009)$ is (38.5 × 1.8) sin tan^{-1}(2), or 62.0 mm, whence c^* is 0.196. The distance labeled α^* in Figure 3.31 is 25.0 mm, whence α^* is 25.0 × 1.8 (scale factor) × 2 (camera constant), or 90°, which is correct for the orthorhombic system. Since the photograph was taken with Cu $K\alpha$ radiation ($\lambda = 1.542$ Å), it follows that $b = 7.91$ Å and $c = 7.87$ Å. In other crystal systems, similar relationships apply. A monoclinic crystal is very conveniently studied in this way. If the crystal is mounted along b, an oscillation photograph (in the Weissenberg camera) enables b to be determined. Then, from the corresponding zero-level Weissenberg photograph, we can obtain a, c, and β.

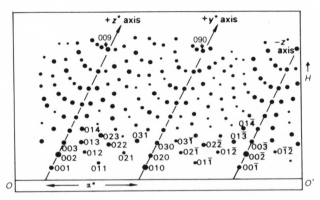

FIGURE 3.31. Sketch of one-half of a partly indexed zero-layer Weissenberg photograph of an orthorhombic crystal mounted about *a*. The horizontal travel of the film is 1 mm per 2° rotation of the crystal, and α^* is 90°, which is correct for an orthorhombic crystal. The reflections shown are confined to a single line (zero layer) on the corresponding oscillation photograph. However, the $0kl$ reciprocal net is still presented in a distorted manner on the Weissenberg photograph, although it is no longer collapsed. The linear scale of the diagram is 1/1.8, λ is 1.542 Å, and $2R$ is 57.30 mm. A sketch of a precession photograph for the same crystal is shown in Figure 3.33.

Further examples of Weissenberg photographs can be found in Chapter 8.

3.5.5 Precession Method

The precession method produces an undistorted picture of the reciprocal lattice. This is achieved, in principle, by ensuring that a crystal axis *t* precesses about the X-ray beam and that the film follows the precession motion in such a way that the film is always perpendicular to the crystal axis (Figure 3.32). This is much easier to say than to carry out, and an apparatus of appreciable mechanical complexity is required.* However, the precession photograph is symmetry-true and readily indexed, as shown by Figure 3.33. It is a sort of "contact print" of a layer of the reciprocal lattice.

Unit-Cell Dimensions from a Precession Photograph

A reduced version of the precession photograph of the same crystal as that used for the Weissenberg photograph is shown in Figure 3.33. It

* See Bibliography.

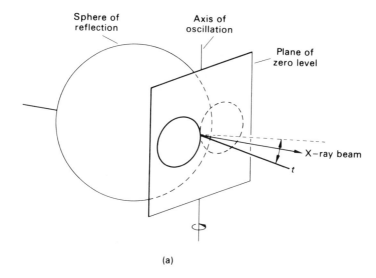

(a)

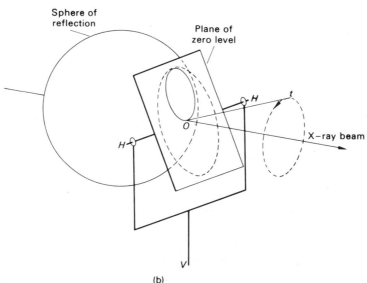

(b)

FIGURE 3.32. Oscillation and precession geometry compared: (a) reciprocal lattice zero-level plane whose normal t is oscillating to equal limits on each side of the X-ray beam. Maximum symmetry information about the direction t is $2mm$, the symmetry of an oscillation movement; (b) reciprocal lattice zero-level plane whose normal t is precessing about the X-ray beam. H and V are horizontal and vertical axes. [Reproduced from *The Precession Method* by M. J. Buerger, with the permission of John Wiley and Sons Inc., New York.]

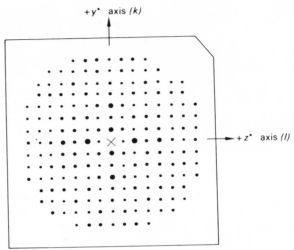

FIGURE 3.33. Sketch of a precession photograph of an orthorhombic crystal precessing about a; an undistorted $0kl$ reciprocal net is represented, permitting b^*, c^*, and α^* to be obtained directly from the film. Remember that $\underline{b^*}$ and $\underline{c^*}$ are $\underline{\text{magnified by a factor equal to the crystal-to-film distance,}}$ $\underline{60.0 \text{ mm in this example.}}$ The linear scale of the diagram is 1/3.75 and λ (Cu $K\alpha$) is 1.542 Å. A sketch of a Weissenberg photograph for the same crystal is shown in Figure 3.31.

was taken in a precession camera with a crystal-to-film distance of 60.0 mm,[†] with the crystal mounted such that a is along the X-ray beam when the camera is in the stationary position, that is, when the precession angle, defined as the angle between t and the X-ray beam in Figure 3.32, is zero.

Since the precession photograph provides a completely undistorted picture of the reciprocal lattice, it is very straightforward to obtain the unit-cell constants. For example, from Figure 3.33 we find that $14b^*$ and $14c^*$ measure 43.7 and 44.1 mm, respectively, which leads to $b^* = 0.195$ and $c^* = 0.197$, in good agreement with the values obtained from the Weissenberg photograph. It is easy to see from Figure 3.33 that $\widehat{y^* \, z^*}$ is 90°, so we may report $b = 7.91$ Å, $c = 7.83$ Å, and $\alpha = 90°$.

In practice, measurements on the film are carried out with specially constructed film holders that are provided with cross-wires and vernier scales, thus facilitating the process of obtaining precise measurements. Nevertheless, the reader is encouraged to make measurements on the films illustrated here, and to obtain the appropriate unit-cell dimensions from them.

[†] Distances in mm measured on the film are divided by 60.0 to convert to dimensionless reciprocal lattice units.

3.6   Recognition of Crystal System

We have mentioned previously several pointers to the system of a crystal under investigation. They include the results of an optical examination under polarized light (Section 3.3) and X-ray photographs that may have been obtained by one or more of the several commonly available techniques (Section 3.5).

Table 3.1 (p. 107) summarizes the main crystallographic features that can be expected from the preliminary optical examination. It indicates that a cubic crystal should be recognized from the optics alone (isotropic). If the crystal is optically uniaxial the optical examination should reveal the direction of the z axis, and the rotational symmetry (three-, four-, or sixfold) could be inferred from a symmetric oscillation photograph (Table 3.2) for a crystal mounted perpendicular to this direction, from a sequence of oscillation photographs taken with the crystal mounted along z (see pp. 131–132), from zero-level and the first-level Weissenberg photographs from a crystal mounted parallel to the z axis, or from zero- and first-level precession photographs from a crystal mounted with z along the X-ray beam when μ is reset to zero. Any zero-level photograph has at least two-fold symmetry, which would modify the actual symmetry about the three-fold axis of a trigonal crystal, giving it the appearance of a six-fold axis; it is therefore necessary to obtain at least one upper-level photograph so as to reveal the true symmetry. For a monoclinic crystal (optically biaxial, Table 3.1), the optical examination should reveal the direction of the unique, y axis. Since this direction could also be one of the three principal axes of an orthorhombic crystal, X-ray photographs would be required to resolve this ambiguity. The symmetry on any oscillation photograph taken with this direction as the oscillation axis would be m perpendicular to that axis). A zero-level Weissenberg photograph would show symmetry 2 for a monoclinic crystal and symmetry $2mm$ for an orthorhombic crystal. Similar results would be obtained from precession photographs of the crystal mounted perpendicular to y with this axis along the beam when $\mu = 0$. For the crystal mounted about y, a zero- and first-level photograph would be required to establish the system as monoclinic (one m plane, parallel to y, disappears for the upper level). If the crystal is orthorhombic and mounted about one axis (say y) it should be possible to take two zero-level precession photographs, separated by dial axis rotation of 90°; each photograph would show $2mm$ symmetry, thus identifying all three crystallographic axes.

TABLE 3.5. Scheme for Recognition of Crystal System and Axes

Classification from optical examination under crosssed polars	Possible crystal system	Axis usually recognizable from optical and/or morphological examination	Type of X-ray photograph and crystal mounting recommended (Minimal features to look for in photograph)[a]	
			Weissenberg	Precession
Isotropic	Cubic only	[100] and/or [010] and/or [001]	0-level: fourfold[b] (for 432, $\bar{4}3m$, or $m3m$); twofold[c] (for 23 or $m3$) [100], [010], or [001], mounting	0-level: fourfold[b] (for 432, $\bar{4}3m$, or $m3m$); twofold[c] (for 23 or $m3$) [100], [010], or [001] precessing, i.e., mounting perpendicular to one of the axes.
Anisotropic/uniaxial	Hexagonal	[0001]	0-level Upper-level } sixfold [01$\bar{1}$0] mounting or equivalent	0-level Upper-level } sixfold [10$\bar{1}$0] mounting or equivalent [0001] precessing
	Trigonal (indexed on hexagonal axes)	[0001]	0-level: sixfold Upper-level: threefold [0001] mounting	0-level: six-fold Upper-level: threefold [10$\bar{1}$0] mounting or equivalent [0001] precessing
	Tetragonal	[001]	0-level Upper-level } fourfold [001] mounting	0-level Upper-level } fourfold [100] mounting [001] precessing

Anisotropic/Biaxial			
Orthorhombic	[100] and/or [010] and/or [001]	0-level Upper-level } mm2 [100], [010] or [001] mounting	0-level Upper-level } mm2 [100], [010], or [001] precessing, i.e., mounting perpendicular to one of the axes recognized.
Monoclinic	[010]	0-level Upper-level } twofold [010] mounting	0-level $mm2$ Upper-level: m ($\perp$[010]) [010] mounting [100] or [001] precessing
	[100] or [001] i.e. $\perp$[010]	0-level: $mm2$ Upper-level: m ($\perp$[010]) [100] or [001] mounting	0-level Upper-level } twofold [100] or [001] mounting [010] precessing
Triclinic	Axes not necessarily revealed by optics or by morphology	0-level: twofold Upper-level: onefold Any axis mounting	0-level: twofold Upper-level: one-fold Any axis precessing

[a] See also Table 1.65.   [b] Full symmetry is actually $4mm$.   [c] Full symmetry is actually $2mm$.

Notes:
1. Optical examination alone should identify a transparent crystal as cubic. For a cubic crystal, the conventional (Bravais) unit cell is either P, I, or F. A nonstandard axial assignment would give rise to a nonstandard unit cell, which should be transformed accordingly.
2. For a triclinic crystal, a unit cell may be defined (see text) with respect to a reciprocal axis parallel to the dial axis of a precession camera. Precession photographs about two real axes perpendicular to this reciprocal axis are required. A nonprimitive triclinic unit cell may be transformed to the standard P form.
3. Preliminary X-ray studies of a single crystal require at least two zero-level and one-upper-level Weissenberg or precession photographs for establishing the unit cell and space group.

Consider finally the two extreme cases, the cubic and triclinic systems. Having established the crystal as cubic (optically isotropic), the direction of *x*, *y*, or *z* may be found from X-ray photographs, which would show the symmetry summarized in Table 3.5. For a triclinic crystal, the choice of axes is arbitrary, but unit-cell definition is best achieved with precession photographs. For this purpose a reciprocal unit-cell axis must be lined up along the dial direction. A series of zone axes along the beam may then be identified for different settings of the dial axis while maintaining the selected reciprocal axis along the dial axis. One such photograph will give two reciprocal axial spacings and the angle between them. Selection of another zone (ideally ≈90° away) would then give the third reciprocal cell spacing and second angle; the third angle would be derived from the difference in dial settings. A mixture of real and reciprocal unit-cell parameters is obtained, requiring use of Table 2.4 to separate them into six real and six reciprocal cell parameters.

Bibliography

Crystal Optics

GAY, P., *An Introduction to Crystal Optics*, London, Longmans (1967).

HARTSHORNE, N. H., and STUART, A., *Crystals and the Polarising Microscope*, London, Arnold (1970).

X-Ray Scattering and Reciprocal Lattice

ARNDT, U. W., and WONACOTT, A. J. (Editors), *The Rotation Method in Crystallography: Data Collection from Macromolecular Crystals*, Amsterdam, North-Holland (1977).

BUERGER, M. J., *X-Ray Crystallography*, New York, Wiley (1942).

JEFFERY, J. W., *Methods in X-Ray Crystallography*, London, Academic Press (1971).

WOOLFSON, M. M., *An Introduction to X-Ray Crystallography*, Cambridge, Cambridge University Press (1970).

Interpretation of X-Ray Diffraction Photographs

Henry, N. F. M., Lipson, H., and Wooster, W. A., *The Interpretation of X-Ray Diffraction Photographs*, London, Macmillan (1960).

Jeffery, J. W., *Methods in X-Ray Crystallography*, London, Academic Press (1971).

Powder Methods

Azaroff, L. V., and Buerger, M. J., *The Powder Method in X-Ray Crystallography*, New York, McGraw-Hill (1958).

Precession Method

Buerger, M. J., *The Precession Method*, New York, Wiley (1964).

Diffractometry

Arndt, U. W., and Willis, B. T. M., *Single Crystal Diffractometry*, Cambridge, Cambridge University Press (1966).

Problems

3.1. Crystals of KH_2PO_4 are needle shaped and show straight extinction parallel to the needle axis. A Laue photograph taken with the X-rays parallel to the needle axis shows symmetry $4mm$.

(a) What is the crystal system and Laue group, and how is the optic axis oriented?
(b) Describe and explain the appearance between crossed Polaroids of a section cut perpendicular to the needle axis.
(c) What minimum symmetry would be observed on both general and symmetric oscillation photographs taken with the crystal mounted on the needle axis?

3.2. Crystals of acetanilide (C_8H_9NO) are brick-shaped parallelepipeda, showing straight extinction for sections cut normal to each of the three edges of the "brick."

(a) What system would you assign to the crystals?

(b) Allocate suitable crystallographic axes.

(c) What minimum symmetry would be shown by general oscillation photographs taken, in turn, about each of the three crystallographic axes?

(d) What symmetry would an oscillation photograph exhibit where the crystal is oscillating about the a axis such that b is parallel to the X-ray beam at the center of the oscillation range?

3.3. Crystals of sucrose show the extinction directions indicated on the crystal drawing of Figure P3.1; the arrows indicate the directions of the cross-wires at extinction.

(a) To what crystal system does sucrose belong?

(b) How are the morphological directions, p, q, and r related to the crystallographic axes?

(c) How would you mount the crystal in order to test your conclusions with (i) Laue photographs, in a single mounting of the crystal, and (ii) oscillation photographs? In each case, indicate the symmetry you would expect the photographs to exhibit in the orientations you have chosen.

3.4. General oscillation photographs of an orthorhombic crystal mounted, in turn, about its a, b, and c axes had layer-line spacings, measured between the zero and first levels, of 5.07, 7.74, and 9.43 mm, respectively. If $\lambda = 1.50$ Å and $R = 30.0$ mm, calculate a, b, and c and a^*, b^*, and c^* for the given wavelength.

Explain why the 146 reflection for this crystal could not be recorded with X-rays of the given wavelength. What symmetry would be

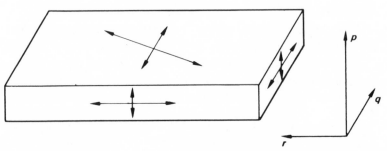

FIGURE P3.1.

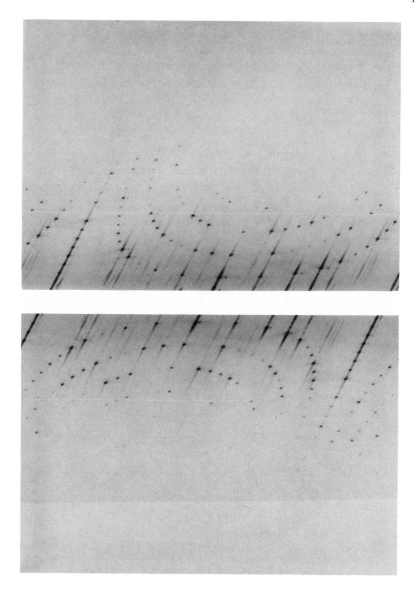

FIGURE P3.2. Weissenberg photograph of the *h0l* layer of euphenyl iodoacetate.

observed on the above oscillation photographs? Is this evidence alone conclusive that the crystal is orthorhombic?

3.5. (a) A tetragonal crystal is oscillated through 15° (i) about [110] and (ii) about the c axis. The layer-line spacings, measured between the $+2$ and -2 layers in each case, were 3.5 and 2.7 cm, respectively. If $\lambda = 1.54$ Å and $R = 3.00$ cm, calculate the a and c dimensions of the unit cell.

(b) How many layers could be recorded in position (ii) if the overall film dimension parallel to the oscillation axis is 8.0 cm?

(c) What is the minimum symmetry obtainable on films such as (i) and (ii)?

(d) How would a third photograph taken after rotating the crystal in (ii) through 90° compare with photograph (ii)?

3.6. Euphenyl iodoacetate is monoclinic, with $a = 7.260$, $b = 11.55$, and $c = 19.22$ Å. Figure P3.2 is the Weissenberg zero level ($h0l$). If the film translation constant is 1 mm per 2° rotation, determine β^* and β.

4

Intensity of Scattering of X-Rays by Crystals

4.1 Introduction

In the previous chapter, we showed how Bragg's equation is used to interpret the geometric features of X-ray photographs. An understanding of the variation in intensity from one diffraction spot to another requires further development of the underlying theory.

The Bragg equation was derived by considering a simple structure in which the atoms were situated at lattice points. Such structures are not unknown, but they are not of a sufficiently general nature for present purposes. Figure 4.1 shows again the reflection of X-rays from any (hkl) family of planes. The plane O passes through the origin of the unit cell, and may be called the zero (hkl) plane. The plane O' is the first (hkl) from the origin and, therefore, is at a perpendicular distance $d(hkl)$ from O. The path

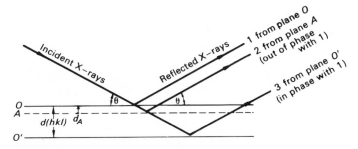

FIGURE 4.1. Construction for Bragg reflection from (hkl) planes, taking into account the variation of atomic distribution in the unit cell. Planes O and O' contain a different sample of atoms from that contained by the parallel conceptual plane A. The ray 2 reflected from plane A is out of phase with those reflected (1 and 3) from planes O and O' in the same (hkl) family.

difference between X-rays reflected from the plane O and those reflected from the adjacent plane O' is, from (3.16), $2d(hkl) \sin \theta(hkl)$.

Consider an atom lying between the planes O and O'; we can imagine a plane A, parallel to (hkl), passing through this atom. For an incident X-ray beam in the correct reflecting position, the planes O and O' will reflect in phase with each other, but, in general, out of phase with the reflection from plane A. We require to determine the path difference between waves reflected by planes O and A in any general situation.

4.2   Path Difference

In Figure 4.2, the plane LMN, at a perpendicular distance p from the origin O, makes intercepts a_p, b_p, and c_p with the general triclinic axes x, y, and z, respectively. The perpendicular OP makes angles χ, ψ, and ω with the same axes. The intercept form of the equation of this plane is, from (1.10),

$$x/a_p + y/b_p + z/c_p = 1 \tag{4.1}$$

where

$$a_p = p/\cos \chi, \qquad b_p = p/\cos \psi, \qquad c_p = p/\cos \omega \tag{4.2}$$

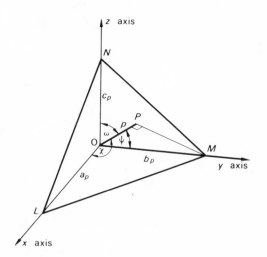

FIGURE 4.2. Plane LMN referred to general (triclinic) axes.

Equation (4.1) may be rewritten as

$$x \cos \chi + y \cos \psi + z \cos \omega = p \tag{4.3}$$

This equation represents a plane parallel to LMN, in terms of the perpendicular distance from the origin, which may be regarded as a variable. Two planes of interest, O and A (Figure 4.1), can be expressed in this way. If plane LMN is parallel to (hkl), then from (4.2)

$$\cos \chi = \frac{d(hkl)}{a/h}, \qquad \cos \psi = \frac{d(hkl)}{b/k}, \qquad \cos \omega = \frac{d(hkl)}{c/l} \tag{4.4}$$

Since plane A is parallel to (hkl) and at a distance d_A from the origin, its equation is

$$x \cos \chi + y \cos \psi + z \cos \omega = d_A \tag{4.5}$$

Substituting for the cosines from (4.4), we have

$$[(hx/a) + (ky/b) + (lz/c)]d(hkl) = d_A \tag{4.6}$$

Let atom A have fractional coordinates $x_A = X_A/a$, $y_A = Y_A/b$, and $z_A = Z_A/c$. Then

$$d_A = (hx_A + ky_A + lz_A)d(hkl) \tag{4.7}$$

By analogy with the Bragg equation (3.16), the path difference δ_A between X-rays reflected from planes O and A is given by

$$\delta_A = 2d_A \sin \theta(hkl) \tag{4.8}$$

or

$$\delta_A = 2d(hkl)[\sin \theta(hkl)](hx_A + ky_A + lz_A) \tag{4.9}$$

which, from Bragg's equation, gives

$$\delta_A = \lambda(hx_A + ky_A + lz_A) \tag{4.10}$$

This relationship is important; it provides a quantitative measure of the path difference between rays reflected by different conceptual planes of atoms, all parallel to any given (*hkl*) plane, with respect to the parallel plane through the origin. It must be realized that in a real situation there need be no atom lying on the (*hkl*) planes; the electron density which is associated with an atom pervades the whole unit cell space.

4.3 Mathematical Representation of a Wave: Amplitude and Phase

Experiments such as those described in Chapter 3 demonstrate clearly both the three-dimensional periodicity of crystals and the wave nature of X-radiation. So-called X-ray diffraction is really one of the best known examples of what may be more aptly called constructive interference. The condition for producing a strong "reflection," that is, a local maximum disturbance, with X-rays scattered by a single crystal is governed by Bragg's equation (3.23). In discussing this effect, we considered the recombination of wavelets which were completely in register, that is, their path difference was $n\lambda$ ($n = 1, 2, 3, \ldots$; Figure 4.3). It has been assumed that two wavelets that are completely out of register, with path difference $n\lambda + \lambda/2$, tend to annul each other (Figure 4.4), and do so completely if both have the same amplitude. These cases are the two extreme situations, and it is important

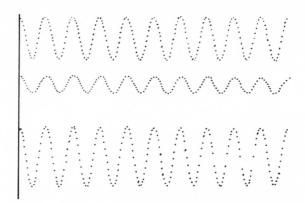

FIGURE 4.3. Combination of two waves that are in exact register (path difference $= \lambda$). The resultant wave has the same wavelength, but with an amplitude equal to the sum of the amplitudes of the two contributing waves.

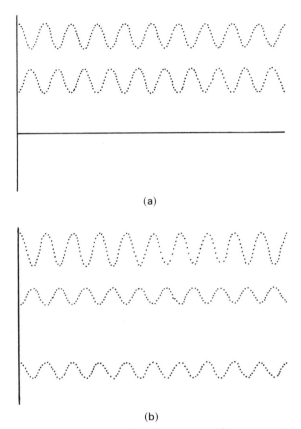

(a)

(b)

FIGURE 4.4. Combination of two waves which are exactly
out of register (path difference $= \lambda/2$). In (a) both waves
have the same amplitude, and the resultant is zero. In (b),
the waves have different amplitudes, and the resultant wave
has an amplitude equal to the difference of the amplitudes
of the contributing waves.

also to consider the interaction between wavelets whose wave disturbances
are neither exactly in register nor exactly out of register (Figure 4.5). This
problem is best treated mathematically, and it introduces a very important
concept in crystal structure analysis, that of phase.

Let W_0 be the transverse displacement at a time t of a wave moving
in the x direction, and let W_0 have a maximum value of f at time $t = 0$
($x = 0$). Then we may write

$$W_0 = f \cos 2\pi x/\lambda \qquad (4.11)$$

FIGURE 4.5. The wave $W = f \cos(\omega t - \phi)$, indicating the phase ϕ relative to the origin (arbitrary). In the illustration, ϕ is $0.7 \times 2\pi$ radian, or $252°$.

W_0 is periodic, with a wavelength λ, since the angle represented by $2\pi x/\lambda$ is equivalent to $2\pi(x + n\lambda)/\lambda$. Since the speed c of the wave is x/t, we can recast (4.11) as

$$W_0 = f \cos 2\pi ct/\lambda \qquad (4.12)$$

which repeats at intervals of λ/c since the angle $2\pi ct/\lambda$ is equivalent to $2\pi c(t + n\lambda/c)/\lambda$. The quantity c/λ represents the number of waves passing a fixed point in unit time, and is known as the frequency, ν. Equation (4.12) may be written as

$$W_0 = f \cos 2\pi\nu t \qquad (4.13)$$

or

$$W_0 = f \cos \omega t \qquad (4.14)$$

where ω is the angular frequency $(2\pi\nu)$ of an object rotating in a circular path of radius f in a time T $(T = 2\pi/\omega = 1/\nu)$. The angular equivalence of one wavelength λ is thus 2π (radians).

Next consider a wave W_1, similar to that represented by W_0 but having its maximum transverse displacement when $t = t_1$ $(x = x_1)$, as shown in Figure 4.5. Let

$$W_1 = f \cos(2\pi x/\lambda - \phi_1) \qquad (4.15)$$

where

$$\phi_1 = 2\pi x_1/\lambda \qquad (4.16)$$

Wave W_1 has the same amplitude and wavelength as does W_0, but it is displaced with respect to W_0 by an amount x_1, which is equivalent to an angular displacement of ϕ_1. In terms of our previous experience, we see that W_0 and W_1 would overlap exactly (Figure 4.3) if $\phi_1 = 0, 2\pi, 4\pi, \ldots,$ that is, $x_1 = n\lambda$, and would be completely out of register if $\phi_1 = \pi, 3\pi, 5\pi, \ldots,$ that is, $x_1 = n\lambda + \lambda/2$. Although conceptually more difficult in crystallographic terms, we tend to retain ϕ to describe the relative positions of waves, rather than use the path difference x. We say that the wave W_1 is *out of phase* with W_0 by the amount ϕ_1, or simply that the phase of W_1 is ϕ_1, that is, with respect to W_0, which has its maximum value f at $x = 0$. Thus, W_0 has a relative phase of zero ($\phi_0 = 0$), and may be thought of as a reference wave.

By analogy with (4.14), we can rewrite (4.15) as

$$W_1 = f \cos(\omega t - \phi_1) \qquad (4.17)$$

For waves of the same wavelength (and frequency), (4.17) is a convenient expression which embodies neatly the relative phase.

4.4 Combination of Two Waves

A given *hkl* reflection consists of the combined scattering by all atoms in the structure. One unit cell is a sufficient representative portion of the structure: Waves scattered by neighboring unit cells will be in phase if Bragg's equation is satisfied. We shall assume that the crystal consists of a geometrically perfect stacking of unit cells; this assumption is considered in Appendix A5.

Scattered X-rays of interest in crystal structure analysis are those scattered without change in wavelength, and so we consider next the combination of two waves (scattered by two atoms) of the same frequency, but having, in general, different amplitudes and phases.

Figure 4.6 illustrates two such waves W_1 and W_2 of angular frequency ω and having amplitudes f_1 and f_2 and corresponding phases ϕ_1 and ϕ_2. These waves may be represented analytically as

$$W_1 = f_1 \cos(\omega t - \phi_1) = f_1[\cos \omega t \cos \phi_1 + \sin \omega t \sin \phi_1] \qquad (4.18)$$

$$W_2 = f_2 \cos(\omega t - \phi_2) = f_2[\cos \omega t \cos \phi_2 + \sin \omega t \sin \phi_2] \qquad (4.19)$$

and their sum $W_1 + W_2$ is

$$W = (\cos \omega t)(f_1 \cos \phi_1 + f_2 \cos \phi_2) + (\sin \omega t)(f_1 \sin \phi_1 + f_2 \sin \phi_2) \tag{4.20}$$

Let W be of the form $F \cos(\omega t - \phi)$. Then

$$W = F[\cos \omega t \cos \phi + \sin \omega t \sin \phi] \tag{4.21}$$

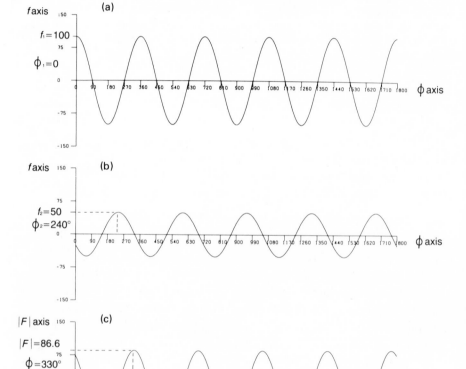

FIGURE 4.6. Combination of two waves of different amplitudes f_1 and f_2 and different phases ϕ_1 and ϕ_2, respectively. The resultant wave has an amplitude $|F|$ and a phase ϕ; no change of wavelength (or frequency) takes place. (a) First wave, (b) second wave, (c) resultant combination of (a) and (b), i.e., $|F| = [(100 + 50 \cos 240°)^2 + (50 \sin 240°)^2]^{\frac{1}{2}}$; $\phi = \tan^{-1}[(50 \sin 240°)/(100 + 50 \cos 240°)]$.

By comparing coefficients in (4.20) and (4.21), we obtain

$$F \cos \phi = f_1 \cos \phi_1 + f_2 \cos \phi_2 \qquad (4.22)$$

and

$$F \sin \phi = f_1 \sin \phi_1 + f_2 \sin \phi_2 \qquad (4.23)$$

Hence

$$F = [(f_1 \cos \phi_1 + f_2 \cos \phi_2)^2 + (f_1 \sin \phi_1 + f_2 \sin \phi_2)^2]^{1/2} \qquad (4.24)$$

and

$$\tan \phi = (f_1 \sin \phi_1 + f_2 \sin \phi_2)/(f_1 \cos \phi_1 + f_2 \cos \phi_2) \qquad (4.25)$$

This analysis shows that the combination of two waves of the same frequency but with different amplitudes and phases results in a third wave of the same frequency but with its own amplitude and phase (Figure 4.6). Continued combination of several such waves gives further similar results.

4.5 Argand Diagram

The combination of waves may be represented in a clear and concise manner by an Argand diagram. The waves are represented as vectors with real and imaginary components. Thus,

$$\mathbf{f}_1 = f_1 \cos \phi_1 + i f_1 \sin \phi_1 \qquad (4.26)$$
$$\mathbf{f}_2 = f_2 \cos \phi_2 + i f_2 \sin \phi_2 \qquad (4.27)$$

These equations are illustrated in Figure 4.7; $\mathbf{F}$ is the resultant vector. De Moivre's theorem states that

$$e^{\pm i\phi} = \cos \phi \pm i \sin \phi \qquad (4.28)$$

and provides an even more convenient summary of these expressions:

$$\mathbf{f}_1 = f_1 e^{i\phi_1}, \qquad \mathbf{f}_2 = f_2 e^{i\phi_2} \qquad (4.29)$$

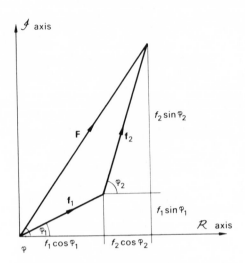

FIGURE 4.7. Combination of two waves as
vectors on an Argand diagram.

Hence,

$$\mathbf{F} = f_1 e^{i\phi_1} + f_2 e^{i\phi_2} \tag{4.30}$$

$\mathbf{F}$, $\mathbf{f}_1$, and $\mathbf{f}_2$ are all vectors,* having both magnitude and direction on the Argand diagram; $e^{i\phi}$ may be regarded as a mathematical operator which rotates a vector anticlockwise in the complex plane through an angle ϕ measured from the real axis. The reader should now use this construction to check the example given in Figure 4.6.

4.6 Combination of N Waves

Extension of the foregoing analysis enables us to combine any number of waves. The resultant of N waves is, from (4.23),

$$\mathbf{F} = f_1 e^{i\phi_1} + f_2 e^{i\phi_2} + f_3 e^{i\phi_3} + \cdots + f_j e^{i\phi_j} + \cdots + f_N e^{i\phi_N} \tag{4.31}$$

or

$$\mathbf{F} = \sum_{j=1}^{N} f_j e^{i\phi_j} \tag{4.32}$$

---

* In many textbooks, $\mathbf{F}$ is not written in bold (vector) type in relation to its representation through the Argand diagram. Our nomenclature has been chosen in order to draw attention to the amplitude and phase characteristics of $\mathbf{F}$.

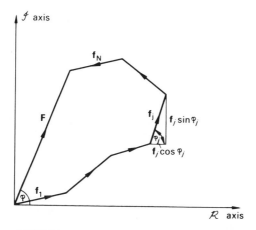

FIGURE 4.8. Combination of N waves ($N = 6$)
on an Argand diagram.

On an Argand diagram, (4.32) expresses a polygon of vectors (Figure 4.8); the resultant $\mathbf{F}$ is given by

$$\mathbf{F} = |F|e^{i\phi} \tag{4.33}$$

The amplitude $|F|$ is given by

$$|F|^2 = \mathbf{F}\mathbf{F}^* \tag{4.34}$$

where $\mathbf{F}^*$ is the complex conjugate of $\mathbf{F}$:

$$\mathbf{F}^* = |F|e^{-i\phi} \tag{4.35}$$

By analogy with (4.24),

$$|F| = (A'^2 + B'^2)^{1/2} \tag{4.36}$$

where

$$A' = \sum_{j=1}^{N} f_j \cos \phi_j \tag{4.37}$$

and

$$B' = \sum_{j=1}^{N} f_j \sin \phi_j \tag{4.38}$$

A' and B' are, respectively, the real and imaginary components of $\mathbf{F}$, and the phase angle ϕ is given by

$$\tan \phi = B'/A' \tag{4.39}$$

4.7 Combined Scattering of X-Rays from the Contents of the Unit Cell

We may now employ the formulae just deduced in order to obtain an equation for the resultant scattering by all the atoms in a unit cell. We shall consider a general structure consisting of N atoms, not necessarily of the same species, occupying fractional coordinates x_j, y_j, z_j ($j = 1, 2, \ldots, N$) in a unit cell.

4.7.1 Phase Difference

The path difference associated with waves scattered by an atom j whose position relative to the origin is specified by the coordinates x_j, y_j, z_j is given, by analogy with (4.17), as

$$\delta_j = \lambda (hx_j + ky_j + lz_j) \tag{4.40}$$

The corresponding phase difference (angular measure) is given by

$$\phi_j = (2\pi/\lambda)\delta_j \tag{4.41}$$

or

$$\phi_j = 2\pi(hx_j + ky_j + lz_j) \tag{4.42}$$

4.7.2 Scattering by Atoms

In order to evaluate the combined scattering from the atoms in the unit cell, we need also the amplitudes of the waves scattered by the atoms, the atomic scattering factors. Although their values are well known in practice, the calculation of atomic scattering factors is complicated; they are tabulated* as functions of $(\sin \theta)/\lambda$, and denoted as $f_{j,\theta}$, or just f_j. The atomic

---

* See Bibliography.

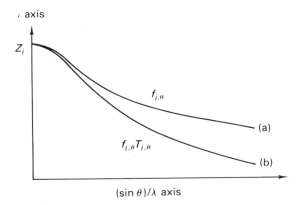

FIGURE 4.9. Atomic scattering factors: (a) stationary atom, $f_{j,\theta}$, (b) atom corrected for thermal vibration, $f_{j,\theta}T_{j,\theta}$.

scattering factor depends upon the nature of the atom, the direction of scattering, the wavelength of X-rays used, and the thermal vibrations of the atom; it is a measure of the efficiency with which an atom scatters with respect to a single electron.

In the first place, f depends upon the number of extranuclear electrons in the atom; its maximum value for a given atom j is Z_j, the atomic number of the jth atomic species. Along the direction of the incident beam [$\sin \theta(hkl) = 0$], f has its maximum value:

$$f_{j,\theta(\theta=0)} = Z_j \qquad (4.43)$$

f is expressed in electrons.

Figure 4.9a shows the general form of the variation of f_j with $(\sin \theta)/\lambda$ for an atom at rest. Since the electrons are distributed over a finite volume in an atom, interference takes place within the atom, and the overall effect of Z_j electrons is diminished. In the forward direction ($\sin \theta = 0$) there is no interference [equation (4.43)]. Interference increases with increasing $\sin \theta$ and decreasing λ, since, in both cases, the path differences within the atom become relatively larger. The decrease of f, being a function of $(\sin \theta)/\lambda$, is also, for a given crystal, a function of hkl.

Thermal Vibration and Temperature Factor

The fourth factor influencing the amplitude of scattering from an atom is the thermal vibration of the particular atom in a given crystal. Each atom in a structure vibrates, in general, in an anisotropic manner, and an exact description of this motion involves several parameters which are dependent upon direction. For simplicity, we shall assume isotropic vibration, in which case the temperature factor correction for the jth atom is

$$T_{j,\theta} = \exp[-B_j(\sin^2\theta)/\lambda^2] \tag{4.44}$$

B_j is the temperature factor of atom j. It is given by

$$B_j = 8\pi^2\overline{U_j^2} \tag{4.45}$$

where $\overline{U_j^2}$ is the mean square amplitude of vibration of the jth atom from its equilibrium position in a direction normal to the reflecting plane, and is a function of temperature. The factor T, like f, is a function of $(\sin\theta)/\lambda$ and, hence, of hkl. We may write for the temperature-corrected atomic scattering factor

$$g_j = f_{j,\theta}T_{j,\theta} \tag{4.46}$$

Thermal vibrations increase the effective volume of the atom, and interference within the atom becomes more noticeable. Consequently, f falls off with increasing $\sin\theta$ more rapidly the higher the temperature (Figure 4.9b).

4.8 Structure Factor

The structure factor $\mathbf{F}(hkl)$ expresses the combined scattering of all atoms in the unit cell compared to that of a single electron; its amplitude $|F(hkl)|$ is measured in electrons. The components required for the combined scattered wave from the (hkl) planes are $g_{j,\theta}$ and $\phi_j(hkl)$, which, *for brevity*, will often be referred to as g_j and ϕ_j, respectively. The individual scattered waves from the various atoms j are vectors of the form given by the general term in (4.31). The resultant wave for the unit cell is therefore

$$\mathbf{F}(hkl) = \sum_{j=1}^{N} g_j \exp(i\phi_j) = \sum_{j=1}^{N} g_j \exp[i2\pi(hx_j + ky_j + lz_j)] \tag{4.47}$$

From Figure 4.10, it may be seen that

$$\mathbf{F}(hkl) = A'(hkl) + iB'(hkl) \tag{4.48}$$

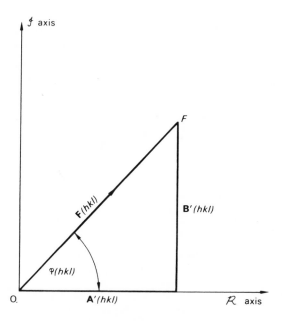

FIGURE 4.10. Structure factor $\mathbf{F}(hkl)$ plotted on an Argand diagram; $\phi(hkl)$ is the resultant phase, and the amplitude $|F(hkl)|$ is represented by OF.

where

$$A'(hkl) = \sum_{j=1}^{N} g_j \cos 2\pi(hx_j + ky_j + lz_j) \qquad (4.49)$$

and

$$B'(hkl) = \sum_{j=1}^{N} g_j \sin 2\pi(hx_j + ky_j + lz_j) \qquad (4.50)$$

By comparison with (4.33), we may write

$$\mathbf{F}(hkl) = |F(hkl)|e^{i\phi(hkl)} \qquad (4.51)$$

where the amplitude is given by

$$|F(hkl)| = [A'^2(hkl) + B'^2(hkl)]^{1/2} \qquad (4.52)$$

and the phase by

$$\tan \phi(hkl) = B'(hkl)/A'(hkl) \qquad (4.53)$$

From Figure 4.10, it may be seen that

$$A'(hkl) = |F(hkl)| \cos \phi(hkl) \qquad (4.54)$$

and

$$B'(hkl) = |F(hkl)| \sin \phi(hkl) \qquad (4.55)$$

4.9 Intensity Expressions

The energy associated with a cosine wave is proportional to the square of the amplitude of the wave. In an X-ray diffraction experiment, it is expressed in terms of the intensity of the scattered wave from the unit cell, $I_o(hkl)$; the subscript o signifies an experimentally observed quantity. Since the amplitude of the combined wave is, from (4.52), $|F(hkl)|$, we use the symbol $I(hkl)$ to represent $|F(hkl)|^2$, sometimes called the ideal intensity. Hence,

$$I_o(hkl) \propto |F_o(hkl)|^2 \qquad (4.56)$$

or

$$I_o(hkl) = K^2 C(hkl)|F_o(hkl)|^2 \qquad (4.57)$$

where $C(hkl)$ combines several geometric and physical factors which depend upon both hkl and the experimental conditions, notably the Lorentz, polarization, and absorption corrections (see Appendix A.5). The factor K is a scale factor associated with $|F_o(hkl)|$.

Equation (4.57) forms the basis of X-ray structure analysis: The experimental quantities $I_o(hkl)$ are directly related to the structure through $|F_o(hkl)|$ [vide (4.47) and (4.51)], and represent the information from which the crystallographer begins to unravel a structure in terms of the atomic coordinates.

4.10 Phase Problem in Structure Analysis

Structure analysis is hampered fundamentally by the inability to determine, in an X-ray diffraction experiment, the complete vectorial structure factor. The modulus $|F(hkl)|$ can be obtained from the intensity data, but the corresponding phase $\phi(hkl)$ is not directly measurable. A simple explanation of a physical limitation on phase measurement is given by the following argument. The practical determination of phase would involve the measurement of a time corresponding, in general, to a fraction of 2π. For Cu $K\alpha$ radiation ($\lambda = 1.542$ Å), the frequency is approximately 2×10^{18} Hz. Hence, for a phase difference of, say, $2\pi/5$ radian, the time involved is approximately 10^{-19} s. Time measurements on this scale have not yet been carried out routinely. Determination of a structure must, however, be carried out in terms of both amplitude and phase.

Image formation in a microscope is obtained mechanically by focusing the lens system. Scattered X-rays cannot be focused by any feasible experimental procedure. We shall see in Chapter 6 that the electron density distribution in a crystal can be recovered mathematically provided that both the amplitude, $|F|$, and the phase, ϕ, are available. Figure 6.5 shows the importance of the phase and emphasizes the fact that the amplitude values alone cannot be used to determine atomic positions directly. From Figure 6.5 it is evident also that incorrect phases will affect the peak (atom) positions in the calculated electron density, thus leading to an incorrect structure. Values for the phase angles must, therefore, be obtained, and this process calls for a solution of the phase problem—the central problem in X-ray structure analysis—which is the main subject of Chapter 6.

4.11 Applications of the Structure Factor Equation

The remainder of this chapter will be concerned with applications of the structure factor equation to preliminary structure analysis, and the development of relationships which have a bearing on the more detailed techniques to be discussed later.

4.11.1 Friedel's Law

In normal circumstances, the X-ray diffraction pattern from a crystal is centrosymmetric, whatever the crystal class (see pages 45 and 127). A diffraction pattern may be thought of as a reciprocal lattice with each point weighted by the corresponding value of $I(hkl)$. Friedel's law states the centrosymmetric property of the diffraction pattern as

$$I(hkl) = I(\bar{h}\bar{k}\bar{l}) \tag{4.58}$$

which may be derived as follows.

Since the atomic scattering factor is a function of $(\sin \theta)/\lambda$, g_j will be the same for both the hkl and the $\bar{h}\bar{k}\bar{l}$ reflections. Thus,

$$g_{j,\theta} = g_{j,-\theta} \tag{4.59}$$

because reflection from opposite sides of any plane will occur at the same Bragg angle θ. This equality may be shown also through (2.16) and (3.32); however, it should be noted that it depends upon the spherically symmetric model of an atom, which is generally used in the calculation of f values.

From (4.47),

$$\mathbf{F}(hkl) = \sum_{j=1}^{N} g_{j,\theta} \exp[i2\pi(hx_j + ky_j + lz_j)] \tag{4.60}$$

and

$$\mathbf{F}(\bar{h}\bar{k}\bar{l}) = \sum_{j=1}^{N} g_{j,-\theta} \exp[-i2\pi(hx_j + ky_j + lz_j)] \tag{4.61}$$

From (4.48),

$$\mathbf{F}(\bar{h}\bar{k}\bar{l}) = A'(\bar{h}\bar{k}\bar{l}) + iB'(\bar{h}\bar{k}\bar{l}) \tag{4.62}$$

where $A'(\bar{h}\bar{k}\bar{l})$ and $B'(\bar{h}\bar{k}\bar{l})$ are given by (4.49) and (4.50), respectively.

From (4.28), and (4.59)–(4.62), we can now derive

$$\mathbf{F}(\bar{h}\bar{k}\bar{l}) = A'(hkl) - iB'(hkl) \tag{4.63}$$

The vectorial representations of $\mathbf{F}(hkl)$ and $\mathbf{F}(\bar{h}\bar{k}\bar{l})$ are shown on an Argand diagram in Figure 4.11. Several important relationships now follow:

$$\phi(\bar{h}\bar{k}\bar{l}) = -\phi(hkl) \tag{4.64}$$

$$|F(hkl)| = |F(\bar{h}\bar{k}\bar{l})| = [A'^2(hkl) + B'^2(hkl)]^{1/2} \tag{4.65}$$

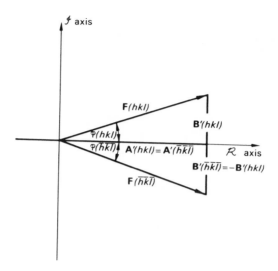

FIGURE 4.11. Relationship between $F(hkl)$ and $F(\bar{h}\bar{k}\bar{l})$ leading to Friedel's law, from which $|F(hkl)| = |F(\bar{h}\bar{k}\bar{l})|$.

Since $I(hkl) \propto |F(hkl)|^2$, from (4.56),

$$I(hkl) = I(\bar{h}\bar{k}\bar{l}) \qquad (4.66)$$

which is Friedel's law. Apart from minor variations connected with anomalous dispersion (page 282), this equality is observable within the limits of experimental error.

4.11.2 Structure Factor for a Centrosymmetric Crystal

One of the questions arising at the outset of a crystal structure analysis is whether the space group is centrosymmetric or not. In the following discussion of the determination of space groups from systematic absences, it turns out that often an ambiguity arises which is connected with centrosymmetry.

Consider a centrosymmetric structure with N atoms per unit cell, the origin of coordinates being at a center of symmetry in the unit cell. There will be $N/2$ atoms in the unit cell with positions independent of the center of symmetry, assuming that no atoms lie on a center of symmetry. Thus, for any atom at x_j, y_j, z_j, there is a centrosymmetrically related atom of the same type at $\bar{x}_j$, $\bar{y}_j$, $\bar{z}_j$, and the real and imaginary components of the structure factor (4.49) and (4.50) become

$$A'(hkl) = \sum_{j=1}^{N/2} g_j[\cos 2\pi(hx_j + ky_j + lz_j)$$
$$+ \cos 2\pi(-hx_j - ky_j - lz_j)] \tag{4.67}$$

and

$$B'(hkl) = \sum_{j=1}^{N/2} g_j[\sin 2\pi(hx_j + ky_j + lz_j)$$
$$+ \sin 2\pi(-hx_j - ky_j - lz_j)] \tag{4.68}$$

Since, for any angle ϕ, $\cos(-\phi) = \cos \phi$ and $\sin(-\phi) = -\sin \phi$.

$$A'(hkl) = 2 \sum_{j=1}^{N/2} g_j \cos 2\pi(hx_j + ky_j + lz_j) \tag{4.69}$$

$$B'(hkl) = 0 \tag{4.70}$$

The summation in (4.69) is taken over $N/2$ noncentrosymmetrically related atoms, and $A'(hkl)$ is now equivalent to $F(hkl)$.

This important result has the further consequence that the phase angle has only two possible values. From (4.53), if $A'(hkl)$ is positive,

$$\phi(hkl) = 0 \tag{4.71}$$

whereas if $A'(hkl)$ is negative,

$$\phi(hkl) = \pi \tag{4.72}$$

From (4.54) and (4.55), we see that the phase angle then attaches itself to $|F(hkl)|$ as a positive or negative sign, and we often speak of the signs of the reflections in centrosymmetric crystals instead of the phases. If we use the symbol $s(hkl)$ as the sign of a centrosymmetric reflection, then

$$F(hkl) = s(hkl)|F(hkl)| \tag{4.73}$$

These results are illustrated in Figure 4.12. By taking the origin at a center of

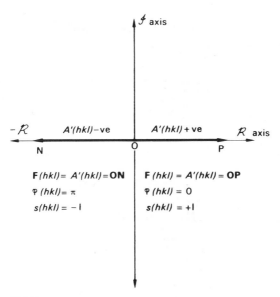

FIGURE 4.12. Structure factor for a centrosymmetric crystal (origin on $\bar{1}$); $\mathbf{F}(hkl) = A'(hkl)$ and can have one of two possible phases, 0 (right-hand side) or π (left-hand side).

symmetry, the value of the phase angle is restricted to 0 or π. The phase problem is much simplified, and centrosymmetric crystals usually present fewer difficulties to the structure analyst than do noncentrosymmetric crystals, in which the phase angles range from 0 to 2π.

4.11.3 Limiting Conditions and Systematic Absences

The X-ray diffraction pattern can always be used to determine the unit-cell type corresponding to the chosen axial system. From the form of (4.47), it would be an unexpected coincidence for many intensities to be zero. With primitive unit cells, the intensity of reflection is not, in general, zero for any particular combination of h, k, and l. Very weak intensities may arise for certain reflections because of the particular structure under investigation; we call these unobservable reflections "accidental absences."

Body-Centered Unit Cell

In any body-centered unit cell an atom at x_j, y_j, z_j is related by translation to another atom of the same type at $\frac{1}{2} + x_j$, $\frac{1}{2} + y_j$, $\frac{1}{2} + z_j$. From

(4.47), assuming N atoms in the unit cell, we can write a summation for the $N/2$ atoms *not* related by translation:

$$\mathbf{F}(hkl) = \sum_{j=1}^{N/2} g_j \{\exp[i2\pi(hx_j + ky_j + lz_j)] \cdot$$

$$+ \exp[i2\pi(hx_j + ky_j + lz_j + \tfrac{1}{2}(h + k + l))]\} \qquad (4.74)$$

The term in curly brackets may be expressed as

$$\{\exp[i2\pi(hx_j + ky_j + lz_j)]\}\{1 + \exp[i\pi(h + k + l)]\} \qquad (4.75)$$

From (4.28), since $(h + k + l)$ is an integer, say m,

$$1 + e^{i\pi(h+k+l)} = 1 + \cos(m\pi) \qquad (4.76)$$

and

$$1 + \cos(m\pi) = 2\cos^2(m\pi/2) \qquad (4.77)$$

or

$$2\cos^2(m\pi/2) = 2\cos^2[2\pi(h + k + l)/4] = G(hkl), \quad \text{say} \qquad (4.78)$$

From (4.74)–(4.78) we can state some useful results. The reduced structure factor equation for a body-centered unit cell may be written as

$$\mathbf{F}(hkl) = 2\cos^2[2\pi(h + k + l)/4] \sum_{j=1}^{N/2} g_j \exp[i2\pi(hx_j + ky_j + lz_j)]$$

$$(4.79)$$

and this equation can be broken down into its components $A'(hkl)$ and $B'(hkl)$ in the usual way. Further simplification is possible since $G(hkl)$, or $2\cos^2[2\pi(h + k + l)/4]$, can take only two values: If $h + k + l$ is an even number,

$$G(hkl) = 2 \qquad (4.80)$$

whereas if it is odd,

$$G(hkl) = 0 \qquad (4.81)$$

Hence, we may write the limiting conditions for a body-centered unit cell, the conditions under which reflection can occur, as

$$hkl: \quad h + k + l = 2n \qquad (n = 0, \pm 1, \pm 2, \ldots) \qquad (4.82)$$

The same situation expressed in terms of systematic absences, the conditions under which reflection cannot occur, is

$$hkl: \quad h + k + l = 2n + 1 \qquad (4.83)$$

Both terms are in common use, and the reader should distinguish between them carefully.

4.11.4 Determination of Unit-Cell Type

Expressions analogous to (4.79) may be derived for any unit-cell type. The reader should attempt the derivations for C and F, and check the results against Table 4.1.

We may summarize these results by saying that if a reflection arises from a centered unit cell, the structure factor has the same form as that for the primitive unit cell, but multiplied by a factor $G(hkl)$, which is the number of lattice points associated with the particular unit-cell type. The summation in the reduced structure factor equation is taken over the number of atoms in the unit cell *not* related by the unit-cell translational symmetry.

In practice, the diffraction pattern is recorded and indexed, and a scrutiny of the reflections present leads to a deduction of the unit-cell type, following Table 4.1.

TABLE 4.1. Limiting Conditions for Unit-Cell Type

Unit-cell type	Limiting conditions		Translations associated with the unit-cell type	$G(hkl)$
P	None		None	1
A	$hkl:$	$k + l = 2n$	$b/2 + c/2$	2
B	$hkl:$	$h + l = 2n$	$a/2 + c/2$	2
C	$hkl:$	$h + k = 2n$	$a/2 + b/2$	2
I	$hkl:$	$h + k + l = 2n$	$a/2 + b/2 + c/2$	2
F	$\begin{cases} hkl: \\ hkl: \\ hkl: \end{cases}$	$\begin{cases} h + k = 2n \\ k + l = 2n \\ (h + l = 2n)^a \end{cases}$	$\begin{cases} a/2 + b/2 \\ b/2 + c/2 \\ a/2 + c/2 \end{cases}$	4
$R_{hex}{}^b$	$hkl:$ $\quad -h + k + l = 3n$ (obv) or $hkl:$ $\quad h - k + l = 3n$ (rev)		$\begin{cases} a/3 + 2b/3 + 2c/3 \\ 2a/3 + b/3 + c/3 \end{cases}$ $\begin{cases} a/3 + 2b/3 + c/3 \\ 2a/3 + b/3 + 2c/3 \end{cases}$	3 3

[a] This condition is not independent of the other two, as may be shown easily.
[b] See page 66 and Table 2.3.

4.11.5   Structure Factors and Symmetry Elements

In the structure factor equation (4.47), some of the N atoms in the unit cell may be equivalent under symmetry operations other than unit-cell centering translations. Let N' be the number of atoms in the asymmetric unit (not related by symmetry) and Z the number of asymmetric units in one unit cell. Thus,

$$N = N'Z \qquad (4.84)$$

Symbolically, we may write

$$\sum_{j=1}^{N} \equiv \sum_{r=1}^{N'} \times \sum_{s=1}^{Z} \qquad (4.85)$$

where the sum over r refers to the symmetry-independent atoms, and that over s to the symmetry-related atoms. The structure factor equation thus contains two parts, which may be considered separately.

The summation over Z symmetry-related atoms of any one type is expressed in terms of the coordinates of the general equivalent positions. Following (4.49) and (4.50),

$$A(hkl) = \sum_{s=1}^{Z} \cos 2\pi(hx_s + ky_s + lz_s) \qquad (4.86)$$

and

$$B(hkl) = \sum_{s=1}^{Z} \sin 2\pi(hx_s + ky_z + lz_s) \qquad (4.87)$$

Extending to the N' atoms in the asymmetric unit, with one such term for each atom r,

$$A'(hkl) = \sum_{r=1}^{N'} g_r A_r(hkl) \qquad (4.88)$$

and

$$B'(hkl) = \sum_{r=1}^{N'} g_r B_r(hkl) \qquad (4.89)$$

The terms $A(hkl)$ and $B(hkl)$ are independent of the structural arrangement of atoms in the asymmetric unit; they are a property of the space group and are called geometric structure factors. Much can be learnt about the diffraction pattern of a crystal from its geometric structure factors. We shall consider several examples, starting in the monoclinic system. Generally, for convenience, the subscript s in (4.86) and (4.87) will not be retained, as all Z positions are related to x, y, z by symmetry. It may be noted, in passing, that these and similar results assume that the atoms in the unit cell are vibrating isotropically. However, where anisotropic vibration is postulated, similar relationships can be derived, but the treatment is correspondingly complex and too detailed for inclusion here.

Space Group $P2_1$

General equivalent positions: x, y, z; $\bar{x}$, $\frac{1}{2}+y$, $\bar{z}$ (see Figure 2.30).
Geometric structure factors:

$$A(hkl) = \cos 2\pi(hx + ky + lz) + \cos 2\pi(-hx + ky - lz + k/2) \tag{4.90}$$

Using

$$\cos P + \cos Q = 2\cos[(P+Q)/2]\cos[(P-Q)/2] \tag{4.91}$$

we obtain

$$A(hkl) = 2\cos 2\pi(hx + lz - k/4)\cos 2\pi(ky + k/4) \tag{4.92}$$

$$B(hkl) = \sin 2\pi(hx + ky + lz) + \sin 2\pi(-hx + ky - lz + k/2) \tag{4.93}$$

Using

$$\sin P + \sin Q = 2\sin[(P+Q)/2]\cos[(P-Q)/2] \tag{4.94}$$

we find

$$B(hkl) = 2\cos 2\pi(hx + lz - k/4)\sin 2\pi(ky + k/4) \tag{4.95}$$

The reader should have no difficulty in evaluating the corresponding expressions for space group $P2$.

Systematic Absences in P2₁. Geometric structure factors enable one to determine limiting conditions, or to predict which reflections will be systematically absent in the X-ray diffraction pattern. If we can show, for given values of h, k, and l, that both $A(hkl)$ and $B(hkl)$ are zero, then $I(hkl)$ is zero, regardless of the structure. For $P2_1$, we can cast (4.92) and (4.95) in the following forms, according to the parity (oddness or evenness) of k. Expanding (4.92), using

$$\cos(P \pm Q) = \cos P \cos Q \mp \sin P \sin Q \qquad (4.96)$$

we may write

$$A(hkl)/2 = [\cos 2\pi(hx + lz) \cos 2\pi(k/4) + \sin 2\pi(hx + lz) \sin 2\pi(k/4)]$$
$$\times [\cos 2\pi ky \cos 2\pi(k/4) - \sin 2\pi ky \sin 2\pi(k/4)] \qquad (4.97)$$

In multiplying the right-hand side of (4.97), terms such as

$$\cos 2\pi(hx + lz) \cos 2\pi(k/4) \sin 2\pi ky \sin 2\pi(k/4) \qquad (4.98)$$

occur. Using the identity

$$\sin 2P = 2 \sin P \cos P \qquad (4.99)$$

we find that (4.98) becomes

$$\tfrac{1}{2} \cos 2\pi(hx + lz) \sin 2\pi ky \sin 4\pi(k/4) \qquad (4.100)$$

which is zero, since k is an integer. Hence, (4.97) becomes

$$A(hkl)/2 = \cos 2\pi(hx + lz) \cos 2\pi ky \cos^2 2\pi(k/4)$$
$$- \sin 2\pi(hx + lz) \sin 2\pi ky \sin^2 2\pi(k/4) \qquad (4.101)$$

In a similar manner, and using

$$\sin(P + Q) = \sin P \cos Q + \cos P \sin Q \qquad (4.102)$$

we can derive

$$B(hkl)/2 = \cos 2\pi(hx + lz) \sin 2\pi ky \cos^2 2\pi(k/4)$$
$$+ \sin 2\pi(hx + lz) \cos 2\pi ky \sin^2 2\pi(k/4) \qquad (4.103)$$

Separating for k even and odd, we obtain

$$k = 2n: \qquad A(hkl) = 2 \cos 2\pi(hx + lz) \cos 2\pi ky \qquad (4.104)$$

$$B(hkl) = 2 \cos 2\pi(hx + lz) \sin 2\pi ky \qquad (4.105)$$

$$k = 2n + 1: \qquad A(hkl) = -2 \sin 2\pi(hx + lz) \sin 2\pi ky \qquad (4.106)$$

$$B(hkl) = 2 \sin 2\pi(hx + lz) \cos 2\pi ky \qquad (4.107)$$

Only one systematic condition can be extracted from these equations: If both h and l are zero, then from (4.106) and (4.107)

$$A(hkl) = B(hkl) = 0 \qquad (4.108)$$

In other words, the limiting condition, associated with the 2_1 axis, is

$$0k0: \quad k = 2n$$

4.11.6 Limiting Conditions from Screw-Axis Symmetry

The example of the 2_1 axis has been treated in detail; it shows again the ability of the diffraction pattern to reveal translational symmetry elements (cf. the I unit cell on page 173). We can show how the limiting conditions for a 2_1 axis arise from a consideration of the Bragg construction. Figure 4.13 is a schematic illustration of a 2_1 symmetry pattern; the motif ♥ represents a structure at a height z, and ○ the structure at a height $\bar{z}$ after operating on it with the 2_1 axis. MM' represents a family of $(0k0)$ planes and NN' a family of $(02k,0)$ planes.

Reflections of the type $0k0$ from MM' planes are cancelled by the reflections from the NN' planes because their phase change relative to MM' is 180°. Clearly, this result is not obtained with the $02k,0$ reflections. Although the figure illustrates the situation for $k = 1$, the same argument can be applied with any value of k.

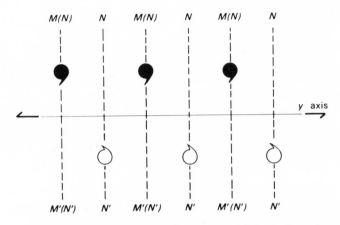

FIGURE 4.13. Pattern of a structure containing a 2_1 axis: $d(NN') = d(MM')/2$; the MM' planes are "halved" by the NN' family.

Limiting conditions for other screw axes, and in other orientations, can be written down by analogy. Try to decide what the conditions for the following screw axes will be, and then check your results from Table 4.2:

2_1 parallel to a
3_2 parallel to c
4_3 parallel to b
6_5 parallel to c

Notice that pure rotation axes R, as in $P2$, do not introduce any limiting conditions.

TABLE 4.2. Limiting Conditions for Screw Axes

Screw axis	Orientation	Limiting condition		Translation component
2_1	$\parallel a$	$h00$:	$h = 2n$	$a/2$
2_1	$\parallel b$	$0k0$:	$k = 2n$	$b/2$
2_1	$\parallel c$	$00l$:	$l = 2n$	$c/2$
3_1 or 3_2	$\parallel c$	$000l$:	$l = 3n$	$c/3, 2c/3$
4_1 or 4_3	$\parallel c$	$00l$:	$l = 4n$	$c/4$
4_2	$\parallel c$	$00l$:	$l = 2n$	$2c/4\ (c/2)$
6_1 or 6_5	$\parallel c$	$000l$:	$l = 6n$	$c/6, 5c/6$
6_2 or 6_4	$\parallel c$	$000l$:	$l = 3n$	$2c/6, 4c/6\ (c/3, 2c/3)$
6_3	$\parallel c$	$000l$:	$l = 2n$	$3c/6\ (c/2)$

4.11.7 Centrosymmetric Zones

In space group $P2_1$ and other space groups of crystal class 2, the $h0l$ reflections are of special interest. Among equations (4.104)–(4.107), only (4.104) is relevant, because zero is an even number, and $\sin(2\pi 0y) = 0$. Hence,

$$A(h0l) = 2\cos 2\pi(hx + lz) \qquad (4.109)$$

and

$$B(h0l) = 0 \qquad (4.110)$$

From (4.53), $\phi(h0l)$ is either 0 or π; in other words, the $[h0l]$ zone is centrosymmetric, which is of importance in a structure analysis in this space group. Centrosymmetric zones occur in noncentrosymmetric space groups which have 2 as a subgroup of their class (page 45).

Space Group Pc

General equivalent positions: x, y, z; $x, \bar{y}, \frac{1}{2}+z$ (Figure 4.14). Geometric structure factors: proceeding as before, we obtain

$$A(hkl) = 2\cos 2\pi(hx + lz + l/4)\cos 2\pi(ky - l/4) \qquad (4.111)$$

$$B(hkl) = 2\sin 2\pi(hx + lz + l/4)\cos 2\pi(ky - l/4) \qquad (4.112)$$

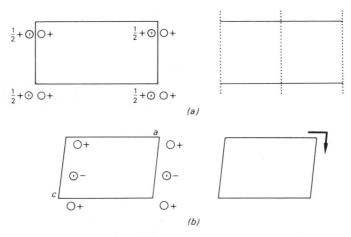

FIGURE 4.14. Space group Pc: (a) viewed along c, (b) viewed along b.

If we separate these equations for l even and odd, we find systematic absences occur only for the $h0l$ reflections:

$$h0l: \quad l = 2n + 1$$

Again, there is a relationship between the index (l) involved in the condition and the symmetry translation ($c/2$).

4.11.8 Limiting Conditions from Glide-Plane Symmetry

Table 4.3 indicates the conditions relating to glide planes of interest to us. They may be deduced, or written down by analogy with Pc.

Space Group $P2_1/c$

This space group, which is often encountered in practice, exhibits a combination of the two translational symmetry elements already considered in detail, a 2_1 axis along b and a c-glide plane perpendicular to b (Figure 2.32 and Problem 2.7a).

$P2_1/c$ is a centrosymmetric space group, and the general equivalent positions may be summarized as

$$\pm\{x, y, z; \ x, \tfrac{1}{2} - y, \tfrac{1}{2} + z\}$$

Geometric structure factors: From the discussion on pages 171–172, we may write

$$A(hkl) = 2\{\cos 2\pi(hx + ky + lz) + \cos 2\pi[hx - ky + lz + \tfrac{1}{2}(k + l)]\} \quad (4.113)$$

$$B(hkl) = 0 \quad\quad\quad\quad\quad\quad\quad\quad\quad\quad\quad\quad\quad\quad\quad\quad\quad\quad (4.114)$$

TABLE 4.3. Limiting Conditions for Glide Planes

Glide plane	Orientation	Limiting condition	Translation component
a	$\perp b$	$h0l:$ $h = 2n$	$a/2$
a	$\perp c$	$hk0:$ $h = 2n$	$a/2$
b	$\perp a$	$0kl:$ $k = 2n$	$b/2$
b	$\perp c$	$hk0:$ $k = 2n$	$b/2$
c	$\perp a$	$0kl:$ $l = 2n$	$c/2$
c	$\perp b$	$h0l:$ $l = 2n$	$c/2$
n	$\perp a$	$0kl:$ $k + l = 2n$	$b/2 + c/2$
n	$\perp b$	$h0l:$ $h + l = 2n$	$a/2 + c/2$
n	$\perp c$	$hk0:$ $h + k = 2n$	$a/2 + b/2$

Carrying out the analysis as before leads to

$$A(hkl) = 4 \cos 2\pi[hx + lz + \tfrac{1}{4}(k+l)] \cos 2\pi[ky - \tfrac{1}{4}(k+l)] \quad (4.115)$$

Separating for $k+l$ even and odd, we obtain

$$k+l = 2n: \qquad A(hkl) = 4 \cos 2\pi(hx+lz) \cos 2\pi ky \quad (4.116)$$

$$k+l = 2n+1: \qquad A(hkl) = -4 \sin 2\pi(hx+lz) \sin 2\pi ky \quad (4.117)$$

We can now *deduce* the limiting conditions:

hkl:	None	—	P unit cell
$h0l$:	$l = 2n$	—	c-glide plane $\perp b$
$0k0$:	$k = 2n$	—	2_1 axis $\parallel b$

These regions are important in monoclinic reciprocal space, because only here can we find the characteristic systematic absences. Despite Friedel's law, the diffraction symmetry determines the true space group in this case:

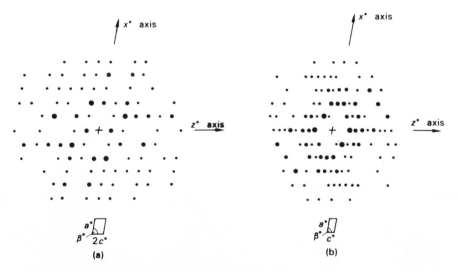

FIGURE 4.15. x^*z^* reciprocal nets for space group Pc, $P2/c$, or $P2_1/c$: (a) $k = 0$, (b) $k > 0$. The c-glide plane, which is perpendicular to b, causes a halving of the rows parallel to x^* when $k = 0$, so that only the rows with $l = 2n$ are present. Hence, the true c^* spacing is not observed on the level $k = 0$, but can be detected on higher reciprocal lattice levels. The symmetry on both levels is 2, in keeping with the diffraction symmetry $2/m$, so that $|F(hkl)|^2 = |F(\bar{h}k\bar{l})|^2$. The reciprocal lattice points in the diagram are weighted according to $|F(hkl)|^2$.

the center of symmetry is revealed through the interaction of the c and 2_1 symmetry elements. Figure 4.15 illustrates weighted reciprocal lattice levels for a possible monoclinic crystal with space group Pc, $P2/c$, or $P2_1/c$.

We shall complete this symmetry study with two examples from the orthorhombic system.

Space Group Pma2

From the data in Figure 4.16, we can write down the expression for the geometric structure factors:

$$A(hkl) = \cos 2\pi(hx + ky + lz) + \cos 2\pi(-hx - ky + lz)$$
$$+ \cos 2\pi(-hx + ky + lz + h/2) + \cos 2\pi(hx - ky + lz + h/2)$$

$$(4.118)$$

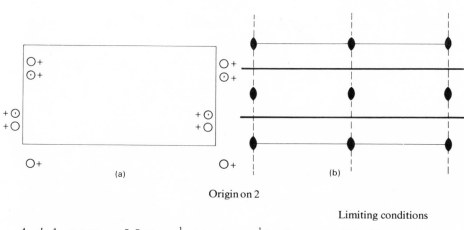

Origin on 2

							Limiting conditions
4	d	1	$x, y, z;$	$\bar{x}, \bar{y}, z;$	$\frac{1}{2}-x, y, z;$	$\frac{1}{2}+x, \bar{y}, z.$	hkl: None
							$0kl$: None
							$h0l$: $h = 2n$
							$hk0$: None
							$h00$: $(h = 2n)$
							$0k0$: None
							$00l$: None
2	c	m	$\frac{1}{4}, y, z;$	$\frac{3}{4}, \bar{y}, z.$			As above
2	b	2	$0, \frac{1}{2}, z;$	$\frac{1}{2}, \frac{1}{2}, z.$			As above +
2	a	2	$0, 0, z;$	$\frac{1}{2}, 0, z.$			hkl: $h = 2n$

Symmetry of special projections

$(001) p2mg$	$(100) pm1(p1m1)$	$(010) p1m (p11m)a' = a/2$

FIGURE 4.16. Space group $Pma2$.

Combining the first and third, and second and fourth terms,

$$A(hkl) = 2 \cos 2\pi(ky + lz + h/4) \cos 2\pi(hx - h/4)$$
$$+ 2 \cos 2\pi(-ky + lz + h/4) \cos 2\pi(hx + h/4) \quad (4.119)$$

Further simplification of this expression requires the separate parts to contain a common factor. We return to (4.118) and make a minor alteration to the term $\cos 2\pi(hx - ky + lz + h/2)$. Since h is an integer, we may write this term as $\cos(\phi + h\pi)$, where $\phi = 2\pi(hx + ky + lz)$. But $\cos(\phi + h\pi) = \cos(\phi - h\pi)$; hence, we change the term $\cos 2\pi(hx - ky + lz + h/2)$ to $\cos 2\pi(hx - ky + lz - h/2)$ in (4.118). Another way of looking at this process is that the fourth general equivalent position in the list has been converted to $-\frac{1}{2} + x$, $\bar{y}$, z, or moved through one unit-cell repeat a in the negative x direction to a crystallographically equivalent position, a perfectly valid and generally applicable tactic.

Returning to $Pma2$, (4.119) now becomes

$$A(hkl) = 2 \cos 2\pi(ky + lz + h/4) \cos 2\pi(hx - h/4)$$
$$+ 2 \cos 2\pi(-ky + lz - h/4) \cos 2\pi(hx - h/4) \quad (4.120)$$

which simplifies to

$$A(hkl) = 2[\cos 2\pi(hx - h/4)][\cos 2\pi(ky + lz + h/4)$$
$$+ \cos 2\pi(-ky + lz - h/4)] \quad (4.121)$$

Combining again,

$$A(hkl) = 4[\cos 2\pi(hx - h/4)] \cos 2\pi(ky + h/4) \cos 2\pi lz \quad (4.122)$$

Similarly,

$$B(hkl) = 4[\cos 2\pi(hx - h/4)] \cos 2\pi(ky + h/4) \sin 2\pi lz \quad (4.123)$$

In the orthorhombic system, the regions of reciprocal space of particular interest are listed under Figure 4.16. Separating (4.122) and (4.123) for even and odd h, we obtain

$$h = 2n: \qquad A(hkl) = 4 \cos 2\pi hx \cos 2\pi ky \cos 2\pi lz \quad (4.124)$$
$$B(hkl) = 4 \cos 2\pi hx \cos 2\pi ky \sin 2\pi lz \quad (4.125)$$

$$h = 2n + 1: \quad A(hkl) = -4 \sin 2\pi hx \sin 2\pi ky \cos 2\pi lz \quad (4.126)$$

$$B(hkl) = -4 \sin 2\pi hx \sin 2\pi ky \sin 2\pi lz \quad (4.127)$$

from which we find the limiting condition, listed in Figure 4.16:

$$h0l: \quad h = 2n$$

The condition

$$h00: \quad (h = 2n)$$

should be considered carefully. One might be excused for thinking that it implies the existence of a 2_1 axis parallel to a, but for the knowledge that there are no symmetry axes parallel to a in class $mm2$. This condition is dependent upon the previous one. We must emphasize the danger of considering limiting conditions out of the following hierarchal order:

hkl	for the unit cell type	
$0kl$		
$h0l$	for glide planes	order of inspection
$hk0$		
$h00$		
$0k0$	for screw axes	
$00l$		

One should proceed to a lower level in this list only after considering the full implications of a condition arising from a higher level. Conditions such as that for $h00$ in $Pma2$ are called redundant, or dependent, and are placed in parentheses. Reflections involved in such conditions are certainly absent from the diffraction record, but do not, necessarily, contribute to a determination of the space-group symmetry.

Space Group $Pman$

This space group may be derived from $Pma2$ by the inclusion of an n-glide plane perpendicular to the c axis [with a translational component of $(a+b)/2$]. We have now seen on several occasions that it is advantageous to take the origin at $\bar{1}$ wherever possible; Figure 4.17 shows $Pman$ drawn in this orientation. It is left to the reader to show that the geometric structure

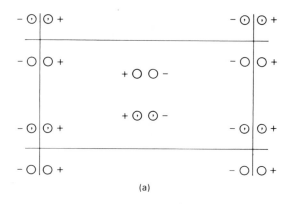

(a)

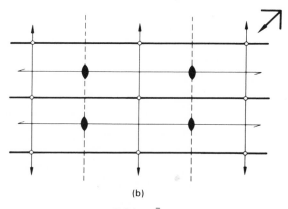

(b)

Origin at $\bar{1}$

Limiting conditions

$\pm\{x, y, z; \quad \bar{x}, y, z; \quad \frac{1}{2}+x, \frac{1}{2}-y, z; \quad \frac{1}{2}-x, \frac{1}{2}-y, z\}$

hkl:	None
$0kl$:	None
$h0l$:	$h = 2n$
$hk0$:	$h + k = 2n$
$h00$:	$(h = 2n)$
$0k0$:	$(k = 2n)$
$00l$:	None

FIGURE 4.17. Space group *Pman*: (a) general equivalent positions, (b) symmetry elements, coordinates of general equivalent positions and limiting conditions.

factors are

$$A(hkl) = 8 \cos 2\pi hx \cos 2\pi[ky - (h + k)/4] \cos 2\pi[lz + (h + k)/4]$$

$$(4.128)$$

$$B(hkl) = 0$$

$$(4.129)$$

TABLE 4.4 Preliminary Stages in a Structure Analysis

Stage	Procedures
1	Optical examination and selection of crystals
2	Determination of crystal system and unit-cell dimensions
3	Measurement of crystal density
4	Calculation of unit-cell contents and enumeration of atom types present
5	Recording of $I(hkl)$ for the practicable ranges of h, k, and l
6	Determination of space group

and, thence, to derive the limiting conditions. Notice that redundant conditions sometimes do and sometimes do not relate to the presence of the corresponding symmetry element.

4.12   Preliminary Structure Analysis

We may now make a brief survey of the various aspects of structure analysis which we have introduced so far. Table 4.4 summarizes the steps which are usually taken in a preliminary examination of a crystal.

Stages 1 and 2 have been discussed in Chapter 3. Stages 3 and 4 would normally be undertaken before 5 and 6, but for the moment we will bypass 3 and 4, and assume that stage 5 has been taken far enough for the determination of the space group of the crystal.

4.12.1   Practical Determination of Space Groups

Crystal structure analysis is always aided by a knowledge of the space group at the outset of the detailed investigation. We shall examine samples of several diffraction patterns of monoclinic and orthorhombic crystals in order to discuss the process of space-group determination.

It is important to bear in mind that X-ray techniques can reveal directly the presence of four types of translational symmetry:

1. Translations relating to the unit cell (a, b, and c).
2. Translations relating to the centering of the unit cell (integer fractions of a, b, and c).
3. Translations relating to glide planes.
4. Translations relating to screw axes.

Categories 2–4 are concerned with systematic absences.

TABLE 4.5. Some Reflection Data for
Monoclinic Crystal (I)

hkl	hkl	hkl	hkl
200	401	112	510
201	402	113	020
202	600	114	040
203	110	310	060
400	111	311	080

Monoclinic Space Groups

Single-crystal X-ray photographs taken with a monoclinic crystal showed typically the reflections listed in Table 4.5. From the important reflection types, hkl, $h0l$, and $0k0$, we can deduce the conditions

hkl: $h + k = 2n$
$h0l$: $(h = 2n)$
$0k0$: $(k = 2n)$

from which we must conclude, using Table 4.7, that the space group could be $C2$, Cm, or $C2/m$. At this stage in the structure analysis of this crystal, there is no way of distinguishing among these three possibilities.

Table 4.6 provides the next list of reflections for inspection. Now there is no limiting condition on hkl, but $h0l$ are restricted by l being even, and $0k0$ are restricted by k being even. This space group is determined uniquely as $P2_1/c$. The ambiguity which arose for monoclinic crystal (I) is not unusual, and we shall consider in Chapter 6 how it might be resolved.

The limiting conditions for the 13 monoclinic space groups are listed in Table 4.7 in their standard orientations. In practice, it is possible, by an inadvertent choice of axes, to find oneself working with a nonstandard

TABLE 4.6. Some Reflection Data for
Monoclinic Crystal (II)

hkl	hkl	hkl	hkl
100	204	111	322
200	402	122	020
300	502	113	040
400	110	311	060
202	310	322	080

TABLE 4.7. Limiting Conditions for Monoclinic Space Groups

Conditions limiting possible X-ray reflections	Possible space groups
hkl: none $h0l$: none $0k0$: none	$P2, Pm, P2/m$
hkl: none $h0l$: none $0k0$: $k=2n$	$P2_1, P2_1/m$
hkl: none $h0l$: $l=2n$ $0k0$: none	$Pc, P2/c$
hkl: none $h0l$: $l=2n$ $0k0$: $k=2n$	$P2_1/c$
hkl: $h+k=2n$ $h0l$: none $0k0$: none	$C2, Cm, C2/m$
hkl: $h+k=2n$ $h0l$: $l=2n, (h=2n)$ $0k0$: $(k=2n)$	$Cc, C2/c$

space-group symbol. Generally, a fairly straightforward transformation of axes will provide the standard setting (see Problems 2.11 and 4.6b and c).

Orthorhombic Space Groups

We begin with sample data in Table 4.8.

TABLE 4.8. Some Reflection Data for Orthorhombic Crystal (I)

hkl	hkl	hkl	hkl
111	011	110	020
112	021	120	040
212	012	310	060
312	101	200	002
322	203	400	004
332	303	600	006

From these data, we deduce the following conditions.

hkl: none $h00$: $h = 2n$
$0kl$: none $0k0$: $k = 2n$
$h0l$: none $00l$: $l = 2n$
$hk0$: none

All of these reflection classes must be examined in the orthorhombic system, in the correct hierarchy. We deduce 2_1 axes parallel to a, b, and c: The space group is uniquely determined as $P2_12_12_1$ (see pages 98 and 99).

In the next two examples we shall state only the conclusions obtained from the inspection of reflection data. Consider the following list:

hkl: none $h00$: none
$0kl$: $k = 2n$ $0k0$: $(k = 2n)$
$h0l$: $l = 2n$ $00l$: $(l = 2n)$
$hk0$: none

These conditions apply to space groups $Pbc2_1$ and $Pbcm$; the difference between them involves a center of symmetry. In the list

hkl: none $h00$: $(h = 2n)$
$0kl$: $k = 2n$ $0k0$: $(k = 2n)$
$h0l$: $l = 2n$ $00l$: $(l = 2n)$
$hk0$: $h = 2n$

space group $Pbca$ is uniquely determined.

These results seem quite reasonable and straightforward, but, nevertheless, the beginner might be tempted to question their validity. For example, in orthorhombic crystal (I), is there a space group in class mmm which would give the same systematic absences as those in Table 4.8? Experience tells us that there is not. Since no glide planes are indicated by the systematic absences, the three symmetry planes would have to be mirror planes. Three mirror planes could not be involved with three screw axes except in a centered space group, such as $Immm$, and so our original conclusion was correct.

The practicing crystallographer is assisted by the space group information which is tabulated fully in Volume I of the *International Tables for X-Ray Crystallography*.* Combined with a working knowledge of sym-

* See Bibliography.

metry, these tables enable most situations arising in the course of a structure analysis to be treated correctly.

Bibliography

Structure Factor and Intensity

STOUT, G. H., and JENSEN, L. H., *X-Ray Structure Determination—A Practical Guide*, New York, Macmillan (1968).

WOOLFSON, M. M., *An Introduction to X-Ray Crystallography*, Cambridge, Cambridge University Press (1970).

Determination of Space Groups

HAHN, T. (Editor), *International Tables for Crystallography*, Vol. A, Dordrecht, D. Reidel (1983).

HENRY, N. F. M., and LONSDALE, K. (Editors), *International Tables for X-Ray Crystallography*, Vol. I, Birmingham, Kynoch Press (1965).

STOUT, G. H., and JENSEN, L. H., *X-Ray Structure Determination—A Practical Guide*, New York, Macmillan (1968).

Atomic Scattering Factor Data

IBERS, J. A., and HAMILTON, W. C. (Editors), *International Tables for X-Ray Crystallography*, Vol. IV, Birmingham, Kynoch Press (1974).

Problems

4.1. A two-dimensional structure has four atoms per unit cell, two of type P and two of type Q, with the following fractional coordinates:

	x	y
P_1	0.1	0.2
P_2	0.9	0.8
Q_1	0.2	0.7
Q_2	0.8	0.3

Calculate $\mathbf{F}(hkl)$ for the reflections 5, 0; 0, 5; 5, 5; and 5, 10 in terms of the scattering factors for the two species, g_P and g_Q. If g_P is equal to $2g_Q$, what are the phase angles for these reflections?

4.2. α-Uranium (U) crystallizes in the orthorhombic system with four U atoms in the special positions

$$\pm\{0, y, \tfrac{1}{4}; \ \tfrac{1}{2}, \tfrac{1}{2}+y, \tfrac{1}{4}\}$$

Use the data below to decide whether y is best chosen as 0.10 or 0.15.

hkl	$|F_o(hkl)|$	$g_U(hkl)$
020	88.5	70.0
110	268.9	80.0

4.3. The unit-cell dimensions of α-U are $a = 2.85$, $b = 5.87$, and $c = 5.00$ Å. Use the value of y_U from problem 4.2 to determine the shortest U—U distance in the structure. It may be helpful to plot the U atom positions in a few neighboring unit cells.

4.4. In the examples listed below, conditions limiting possible X-ray reflections are given for crystals in the monoclinic system. In each case, write down the symbols of the possible space groups corresponding to this information.

(a) hkl: no conditions
     $h0l$:   no conditions
     $0k0$: $k = 2n$.

(b) hkl: no conditions
     $h0l$: $h = 2n$
     $0k0$:   no conditions.

(c) hkl: $h + k = 2n$
     $h0l$: $l = 2n; (h = 2n)$
     $0k0$: $(k = 2n)$.

(d) hkl: no conditions
     $h0l$:   no conditions
     $0k0$:   no conditions.

4.5. Repeat question 4.4 for the conditions given below relating to orthorhombic space groups.

(a) hkl:
     $0kl$:   no conditions
     $h0l$:
     $hk0$:

    $h00$: $h = 2n$
    $0k0$: $k = 2n$
    $00l$:   no conditions

(b) hkl :   no conditions $h00$:   no conditions
     $0kl$: $k = 2n$ $0k0$: $k = 2n$
     $h0l$:   no conditions $00l$:   no conditions
     $hk0$:   no conditions
(c) hkl : $h + k + l = 2n$ $h00$: $h = 2n$
     $0kl$: $k = 2n, l = 2n$ $0k0$: $k = 2n$
     $h0l$: $h + l = 2n$ $00l$: $l = 2n$
     $hk0$: $h + k = 2n$

Rewrite (c) with parentheses to indicate the redundant conditions.

4.6. (a) Write the conditions limiting possible X-ray reflections for the following space groups: (i) $P2_1/a$; (ii) Pc; (iii) $C2$; (iv) $P2_122$; (v) $Pcc2$; (vi) $Imam$. In each case write the symbols of the space groups, if any, in the same crystal system with the same conditions.

(b) Write the conditions limiting possible reflections in the monoclinic space group $P2_1/n$ (nonstandard symbol).

(c) Give the conventional symbols for space groups $A2/a$ and $B2_122_1$.

5

Methods in X-Ray Structure Analysis. I

5.1 Introduction

We have reached the stage where we begin to consider the contents of the unit cell and attempt to assign possible types of locations to atoms or molecules in the asymmetric unit.

5.2 Analysis of the Unit-Cell Contents

The density D_m of the crystals under examination may be measured by suspending them in a liquid or liquid mixture.* The composition of the liquid is altered until the crystals neither rise nor fall; then the density of the liquid, equal to D_m, is measured. The flotation procedure is best carried out in a thermostat. It may still happen, however, that the demarcation between sinking and floating is a little ill defined. Generally, inclusion of air or solvent in the crystal will lead to a smaller apparent density, and a position corresponding to maximum measured density should be appropriate.

If the crystal contains the number Z of chemical species, each of relative molar mass M_r, in the unit cell, then the following relationship holds:

$$D_m = ZM_r u / V_c \qquad (5.1)$$

where D_m is in kg m^{-3}, u is the atomic mass unit $(1.6606 \times 10^{-27}$ kg), and V_c is the unit cell volume in m^3. In practice, V_c is expressed in nm^3 and u is replaced by 1.6606. An alternative formulation is

$$D_m = ZM / V_c N_A \qquad (5.2)$$

---

* If a large quantity of crystals is available, a simple displacement method may be used.

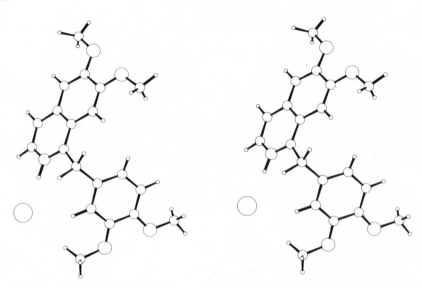

FIGURE 5.1. Molecular conformation of papaverine hydrochloride; the circles, in
decreasing order of size, represent Cl, N, C, and H.

where N_A is the Avogadro constant $(6.0221 \times 10^{23} \text{ mol}^{-1})$ and M is numerically the same as M_r but in the units of g mol^{-1}, so that with D_m in g cm^{-3}
and V_c in Å^3, N_A may be replaced by 0.60221.

5.2.1 Papaverine Hydrochloride, $C_{20}H_{21}NO_4 \cdot HCl$

Crystal Data

> System: monoclinic.
> Unit-cell dimensions: $a = 13.059, b = 15.620, c = 9.130 \text{ Å}, \beta = 92.13°$.
> V_c: 1861.0 Å^3.
> D_m: 1.33 g cm^{-3}.
> M: 375.8.
> Z: 4 to the nearest integer [3.98 from (5.2)].
> Unit-cell contents: 80 C, 88 H, 4 N, 16 O, 4 Cl atoms.
> Absent spectra: $h0l$: l odd; $0k0$: k odd.
> Space group: $P2_1/c$ (Figure 2.33 and page 94).

All atoms are in general equivalent positions. The molecular conformation, obtained by a complete structural analysis,* is shown in Figure 5.1.

* C. D. Reynolds *et al., Journal of Crystal and Molecular Structure* **4**, 213 (1974).

5.2.2 Naphthalene, $C_{10}H_8$

Crystal Data

System: monoclinic.

Unit-cell dimensions: $a = 8.658$, $b = 6.003$, $c = 8.235$ Å, $\beta = 122.92°$.

V_c: 359.2 Å^3.

D_m: 1.152 g cm^{-3}.

M: 128.2.

Z: 2 to the nearest integer [1.94 from (5.2)].

Unit-cell contents: 20 C, 16 H atoms.

Absent spectra: $h0l$: $l = 2n + 1$; $0k0$: $l = 2n + 1$.

Space group: $P2_1/c$.

5.2.3 Molecular Symmetry

In papaverine hydrochloride, the four molecules in the unit cell are, as complete entities, able to satisfy the requisite general positions in the space group. The molecules are said to be in general positions, and each atom at coordinates x_j, y_j, z_j ($j = 1, 2, \ldots, 48$) is repeated by the space-group symmetry mechanism to build up the crystal structure. There are, therefore, 48 atoms, including hydrogen, in the asymmetric unit to be located by the structure analysis.

Naphthalene is not quite so straightforward. With two molecules per unit cell, there are 20 carbon atoms and 16 hydrogen atoms, which may be distributed in four equivalent-position sets of five and four atoms, respectively, in the unit cell. This means that in order to solve the structure, we have to locate five carbon atoms and four hydrogen atoms. This number is only half that expected: since Z is 2, each atom is related by one of the symmetry elements of the space group to a second atom of the same type in the same molecule, so as to generate $C_{10}H_8$ from C_5H_4. There are three different symmetry elements to consider: the 2_1 axis, the c-glide plane, and the center of symmetry. The screw axis and glide plane are eliminated because they involve translational symmetry, which would generate an infinite molecule with translational repeats. We must, therefore, conclude that the atom pairs are related by a center of symmetry, which means that the molecule of naphthalene is centrosymmetric.

The symmetry analysis for naphthalene has served two very useful purposes: It has halved the work of the subsequent structure analysis, and shown that the molecules in the crystal exhibit a certain minimum symmetry

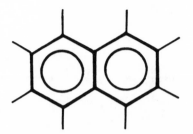

FIGURE 5.2. Naphthalene molecule.

($\bar{1}$). This result is, of course, in agreement with chemical knowledge, which ordinarily we are quite entitled to use. The conventional notion that naphthalene should have *mmm* symmetry (Figure 5.2) is not supported directly, although the crystal structure analysis shows that this symmetry holds within experimental error.

5.2.4 Special Positions

The molecules of naphthalene are said to lie on special positions in $P2_1/c$ (Figure 5.3). Special position sites correspond in symmetry to one of the 32 crystallographic point groups, and in subsequent examples, we shall see that both atoms and molecules can occupy special positions.

Glide planes and screw axes do not usually accommodate atoms or molecules; an atom lying exactly on a translational symmetry element would introduce a pseudo-half-axial translation, thus creating special conditions which, depending on the atomic number, may be observable among the X-ray reflections (see Problem 5.1 at the end of this chapter).

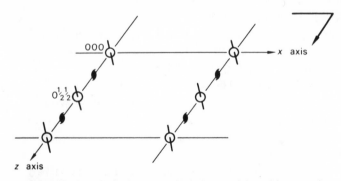

FIGURE 5.3. Grouping of one of the special position sets in $P2_1/c$; the arrangement of molecules (symmetry $\bar{1}$) with their centers at 0, 0, 0 and 0, $\frac{1}{2}$, $\frac{1}{2}$ is shown.

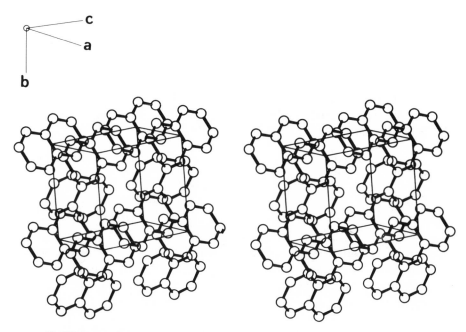

FIGURE 5.4. Stereoview of the naphthalene structure (H atoms are not shown).

Although they are in special positions, the molecules of naphthalene are subject to the space-group mechanism inherent in the general positions: If one molecule is located at 0, 0, 0, then the second molecule is at $0, \frac{1}{2}, \frac{1}{2}$. This set may be determined by substituting $x = y = z = 0$ into the set of general positions. The structure of naphthalene* is shown in Figure 5.4. The reader may like to consider the three other possible sets of special positions that could be used to represent this structure, and then show from the structure factor equation that $|F(hkl)|$ is invariant with respect to each set of special positions.

5.2.5   Nickel Tungstate, NiWO₄

Crystal Data

System: monoclinic.
Unit-cell dimensions: $a = 4.60$, $b = 5.66$, $c = 4.91$ Å, $\beta = 90.1°$.

---

* D. W. J. Cruickshank, *Acta Crystallographica* **10**, 504 (1957), who used the nonstandard space group $P2_1/a$, equivalent to $P2_1/c$, with the a and c axes quoted here interchanged.

V_c: 127.8 Å^3.
D_m: 7.964 g cm^{-3}.
M: 306.5.
Z: 2 to the nearest integer [2.00 from (5.2)].
Unit-cell contents: 2 Ni, 2 W, 8 O atoms.
Absent spectra: $h0l$: $l = 2n + 1$.
Possible space groups: Pc or $P2/c$.

We shall use space group $P2/c$, since the structure was determined successfully only with this space group.*

The general equivalent positions in $P2/c$ are

$$\pm\{x, y, z;\ x, \bar{y}, \tfrac{1}{2} + z\}$$

but in order to study NiWO$_4$ further, we must consider the possible special positions for this space group; they are located on either the twofold axes or the centers of symmetry. The reader should make a drawing for space group $P2/c$, using the coordinates listed above.

Special Positions on Twofold Axes

The twofold axes lie along the lines $[0, y, \tfrac{1}{4}]$, $[\tfrac{1}{2}, y, \tfrac{1}{4}]$, $[0, y, \tfrac{3}{4}]$ and $[\tfrac{1}{2}, y, \tfrac{3}{4}]$. The equivalent positions generated by the space-group symmetry show that the special position sets are

$$\pm\{0, y, \tfrac{1}{4}\} \quad \text{or} \quad \pm\{\tfrac{1}{2}, y, \tfrac{1}{4}\}$$

and each set satisfies $P2/c$ symmetry by accommodating two structural entities with symmetry 2 in the unit cell.

Special Positions on Centers of Symmetry

If we repeat the above analysis for the eight centers of symmetry in the space group, we will develop four special position sets:

$$0, 0, 0;\quad 0, 0, \tfrac{1}{2}$$

$$\tfrac{1}{2}, 0, 0;\quad \tfrac{1}{2}, 0, \tfrac{1}{2}$$

$$0, \tfrac{1}{2}, 0;\quad 0, \tfrac{1}{2}, \tfrac{1}{2}$$

$$\tfrac{1}{2}, \tfrac{1}{2}, 0;\quad \tfrac{1}{2}, \tfrac{1}{2}, \tfrac{1}{2}$$

*R. O. Keeling, *Acta Crystallographica* **10**, 209 (1957).

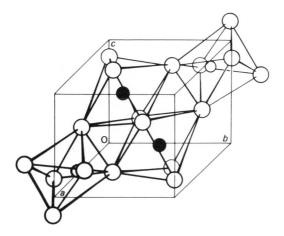

FIGURE 5.5. WO_6 and NiO_6 octahedra in $NiWO_4$: large open circles O, small open circles W, small black circles Ni.

The Ni and W atoms must lie on special positions, with either 2 or $\bar{1}$ symmetry. Nothing can be said about the position of the oxygen atoms, and without further detailed analysis we cannot define this structure further. However, to complete the picture, we list the atomic parameters for this structure, and illustrate it in Figure 5.5:

2 Ni $\pm\{\frac{1}{2}, 0.653, \frac{1}{4}\}$

2 W $\pm\{0, 0.180, \frac{1}{4}\}$

4 O $\pm\{0.22, 0.11, 0.96; 0.22, 0.89, 0.46\}$

4 O' $\pm\{0.26, 0.38, 0.39; 0.26, 0.62, 0.89\}$

The heavy atoms were found to occupy the four twofold axes in pairs. This conclusion, although not uniquely derivable from the symmetry analysis alone, was at least partially indicated by it. Once again, a pencil and paper operation saved considerable effort in the subsequent detailed structure analysis by pointing to the proper course of action.

In these few examples, we have shown the value of the symmetry analysis in the early stages of a structure determination. This procedure may be regarded as a routine to be carried out before the more detailed calculations required in the elucidation of the atomic parameters.

5.3 Two Early Structure Analyses Revisited

The structures of NaCl and FeS_2 (pyrite) are largely of historical interest. Both structures were solved by Sir Lawrence Bragg, and although we are not considering entirely the methods which he used in 1913, we are reminded through these structures that he laid the firm foundations for the subject of this book. We shall show that they can provide both further examples of procedures already discussed and introduce new features in our study of crystal structure analysis.

5.3.1 Sodium Chloride, NaCl

Crystal Data

> System: cubic.
> Unit-cell dimensions: $a = 5.638$ Å.
> V_c: 179.2 Å^3.
> D_m: 2.165 g cm^{-3}.
> M: 58.44.
> Z: 4 to the nearest integer [4.00 from (5.2)].
> Unit-cell contents: 4 Na and 4 Cl atoms.
> Conditions limiting possible X-ray reflections: hkl: $h + k = 2n$, $k + l = 2n$, $(l + h = 2n)$.
> Possible space groups: $F23$, $Fm3$, $F432$, $F\bar{4}3m$, $Fm3m$.

Symmetry Analysis

Since $Z = 4$, we can define an origin in the unit cell by placing one atom, say Na, at 0, 0, 0. An atom can be used conveniently to define an origin in this manner if it occurs N_L times per unit cell, where N_L is the number of lattice points per unit cell.

The symmetry-equivalent Na atoms occupy the positions

$$0, 0, 0; \quad 0, \tfrac{1}{2}, \tfrac{1}{2}; \quad \tfrac{1}{2}, 0, \tfrac{1}{2}; \quad \tfrac{1}{2}, \tfrac{1}{2}, 0$$

A survey of cubic F space groups* shows that there are only two possible situations for the Cl atoms:

(a) $\tfrac{1}{2}, 0, 0; \quad 0, \tfrac{1}{2}, 0; \quad 0, 0, \tfrac{1}{2}; \quad \tfrac{1}{2}, \tfrac{1}{2}, \tfrac{1}{2}$

(b) $\tfrac{1}{4}, \tfrac{1}{4}, \tfrac{1}{4}; \quad \tfrac{3}{4}, \tfrac{3}{4}, \tfrac{1}{4}; \quad \tfrac{1}{4}, \tfrac{3}{4}, \tfrac{3}{4}; \quad \tfrac{3}{4}, \tfrac{1}{4}, \tfrac{3}{4}$

* See Bibliography.

TABLE 5.1. Observed and Calculated Structure Factor Amplitudes
for NaCl

hkl	$|F_o|$	$|F_c|$	$K|F_o|$	$\Delta F(=K|F_o|-|F_c|)$
200	209.0	81.6	86.5	4.9
400	115.6	45.6	47.9	2.3
600	53.2	25.3	22.0	-3.3
800	26.7	12.7	11.1	-1.6
220	162.9	64.3	67.4	3.1
440	61.5	28.1	25.5	-2.6
660	24.5	10.6	10.1	-0.5
880	4.2	3.1	1.7	-1.4
111	49.7	19.0	20.6	1.6
222	127.7	53.3	52.9	-0.4
333	19.6	7.3	8.1	0.8
444	41.7	18.6	17.3	-1.3
555	4.9	3.7	2.0	-1.7

$\sum|F_o|=901.2$ $\sum|F_c|=373.2$ $\sum K|F_o|=373.1$ $\sum|\Delta F|=25.5$

Although there are five possible space groups, there are no positional
parameters to determine, and it is of no consequence to know the true space
group at this stage.*

Structure Factor Calculations

Distinction between structural arrangements (a) and (b) can be effected
by comparing the *observed* structure factor amplitudes $|F_o(hkl)|$ listed in
Table 5.1 with the corresponding calculated values $|F_c(hkl)|$ based on each
of the models in turn.

In any F unit cell, we may write, from (4.49),(4.50), and Table 4.1, for
hkl all odd or all even,

$$A'(hkl) = 4 \sum_{j=1}^{N/4} g_j \cos 2\pi(hx_j + ky_j + lz_j) \tag{5.3}$$

$$B'(hkl) = 4 \sum_{j=1}^{N/4} g_j \sin 2\pi(hx_j + ky_j + lz_j) \tag{5.4}$$

For all other hkl, $A'(hkl) = B'(hkl) = 0$.

---

* This information is necessary in order to effect true refinement in the final stages of the
analysis. Alternative (b) corresponds only to $F23$ and $Fm3$. The Laue symmetry (Table 1.6)
$m3m$ provides a partial answer, because $F23$ and $Fm3$ correspond to Laue group $m3$,
whereas $F432$, $F\bar{4}3m$, and $Fm3m$ belong to Laue group $m3m$.

Structure (a) is centrosymmetric, with the origin on $\bar{1}$. Hence, from (5.3) and (5.4),

$$A'(hkl) = 4\{g_{Na^+} + g_{Cl^-}\cos 2\pi[(h+k+l)/2]\} \qquad (5.5)$$

$$B'(hkl) = 0 \qquad (5.6)$$

A similar analysis for structure (b), keeping the same positions for Na, leads to the following equations:

$$A'(hkl) = 4\{g_{Na^+} + g_{Cl^-}\cos 2\pi[(h+k+l)/4]\} \qquad (5.7)$$

$$B'(hkl) = 4g_{Cl^-}\sin 2\pi[(h+k+l)/4] \qquad (5.8)$$

Determination of the Structure

It is not difficult to decide at this stage which structure model, (a) or (b), is correct. Table 5.2 lists expressions for $|F_c(hkl)|$ for a few selected reflections for comparison with Table 5.1, and it is easy to see that only model (a) will produce the correct *pattern* of calculated structure factor amplitudes ($Z_{Na^+} = 10$, $Z_{Cl^-} = 18$). Table 5.1 lists also the results of a quantitative comparison of $|F_o|$ and $|F_c|$ over all the experimental data. Values of g_{Na^+} and g_{Cl^-} have been derived by correcting the calculated values of f_{Na^+} and f_{Cl^-} for isotropic thermal vibration, taking $\overline{U_{Na^+}^2} = \overline{U_{Cl^-}^2} = 0.025 \text{ Å}^2$. The $|F_o|$ data have been scaled by a factor K of 0.414, since the experimental data are obtained on a true relative but arbitrary scale. The crystal structure is illustrated in Figure 1.1.

Scale Factor

Assuming that all the atoms in the unit cell are included in the model, the $|F_o|$ values may be scaled by making their sum over all values of hkl equal

TABLE 5.2. Structure Factors Calculated for Models (a) and (b)

hkl	$|F_c(hkl)|$, model (a)	$|F_c(hkl)|$, model (b)
200	$4(g_{Na^+} + g_{Cl^-})$	$4(g_{Na^+} - g_{Cl^-})$
220	$4(g_{Na^+} + g_{Cl^-})$	$4(g_{Na^+} + g_{Cl^-})$
111	$4(g_{Na^+} - g_{Cl^-})$	$4(g_{Na^+}^2 + g_{Cl^-}^2)^{1/2}$
222	$4(g_{Na^+} + g_{Cl^-})$	$4(g_{Na^+} - g_{Cl^-})$

to the same sum with respect to $|F_c|$. The scale factor K is thus given by

$$K = \sum_{hkl} |F_c(hkl)| \Big/ \sum_{hkl} |F_o(hkl)| \qquad (5.9)$$

where the sum extends over all symmetry-independent reflections. If only a fraction of the atoms in the unit cell are present in a trial structure, a realistic value of K is given by

$$K \stackrel{\sim}{\times} \sum_{j=1}^{T} Z_j \Big/ \sum_{j=1}^{N} Z_j \qquad (5.10)$$

where N is the number of atoms in the unit cell, T is the number of atoms in the trial structure, and Z_j is the atomic number of the jth atom.

Reliability Factor

The differences between the scaled-observed and the calculated structure-factor amplitudes are a measure of the quality of the trial structure. Large differences correspond to poor reliability, and vice versa. An overall reliability factor (R-factor) is defined as

$$R = \sum_{hkl} |K|F_o| - |F_c|| \Big/ \sum_{hkl} K|F_o| \qquad (5.11)$$

For a well-refined structure model, the value of R approaches a small value, corresponding to the errors in both the experimental data and the model. In the early stages of the analysis, however, it may be between 0.4 and 0.5. For the NaCl data in Table 5.1, the bottom line shows the components of R (0.068) and K (0.414). In completing the structure analysis of NaCl, we have introduced the first criterion of correctness, namely, good agreement between $|F_o|$ and $|F_c|$, expressed through the R-factor. It should be noted that trial structures with an R-factor of more than 50% have been known to be capable of refinement—it is only a rough guide at that stage of the analysis. A better basis for judgment is a comparison of the *pattern* of $|F_o|$ and $|F_c|$, which requires care and experience.

5.3.2 Pyrite, FeS$_2$

This problem is a little more complicated than that of NaCl, because a positional parameter has to be determined to specify the crystal structure.

Crystal Data

> System: cubic.
> Unit-cell dimensions: $a = 5.407$ Å.
> V_c: 158.1 Å^3.
> D_m: 4.87 g cm^{-3}.
> M: 120.0.
> Z: 4 to the nearest integer [3.87 from (5.2)].
> Unit-cell contents: 4 Fe and 8 S atoms.
> Absent spectra: $0kl$: $k = 2n + 1$.
> Space group: $Pa3$.

Table 5.3 lists $|F_o|^2$ values for pyrite, relative to $|F_o(200)|^2 = 100.0$.

Structure Analysis

A symmetry analysis of space group $Pa3$ shows that the Fe and S atoms must lie in special equivalent positions:

4Fe: $0, 0, 0;\quad 0, \frac{1}{2}, \frac{1}{2};\quad \frac{1}{2}, 0, \frac{1}{2};\quad \frac{1}{2}, \frac{1}{2}, 0$

8 S: $\pm\{x, x, x;\quad \frac{1}{2}+x, \frac{1}{2}-x, \bar{x};\quad \bar{x}, \frac{1}{2}+x, \frac{1}{2}-x;\quad \frac{1}{2}-x, \bar{x}, \frac{1}{2}+x\}$

The iron atoms occupy sites which correspond to an F unit cell. These atoms dominate the X-ray diffraction pattern, since the atomic numbers of Fe and S are 26 and 16, respectively. For this reason, only reflections with $h + k$, $k + l$, and $(l + h)$ even are listed (see Table 4.1), the others being comparatively weak.

TABLE 5.3. $|F_o(hkl)|^2$ Data for Pyrite, Relative to $|F_o(200)|^2 = 100.0$

hkl	$|F_o|^2$	hkl	$|F_o|^2$	hkl	$|F_o|^2$
200	100.0	220	55.0	111	44.0
400	0.4	440	37.0	222	36.0
600	3.2	660	4.0	333	19.0
800	14.4	880	4.0	444	1.0
10,00	4.0			555	10.0

Using methods developed already, we can show that the structure factor equations assume the following forms* for pyrite:

$$A'(hkl) = 4\{g_{Fe} + 2g_S \cos 2\pi[hx_S + (h+k)/4]$$

$$\times \cos 2\pi[kx_S - (k+l)/4] \cos 2\pi[lx_S - (h-l)/4]\}$$

$$(5.12)$$

$$B'(hkl) = 0 \qquad\qquad (5.13)$$

We can solve this structure from the $h00$ and hhh reflections, for which cases (5.12) becomes

$$A'(h00) = 4(g_{Fe} + 2g_S \cos 2\pi hx_S) \qquad (5.14)$$

with h even, and with $h+k$ and $k+l$ both even

$$A'(hhh) = 4(g_{Fe} + 2g_S \cos^3 2\pi hx_S) \qquad (5.15)$$

From an inspection of Table 5.3, we see that among the $h00$ reflections, the intensities of 400 and 600 are comparatively weak, whereas that of 800, bearing in mind that it is a high-order reflection, for which the temperature-corrected f factors (g) will be fairly small, is quite strong. From (5.14), it is to be expected that intensities will tend to be weakest where the Fe and S contributions are in opposition ($\cos 2\pi hx_S \approx -1$).

The implications of these observations are summarized in Table 5.4, which also contains information for two hhh reflections. The most consistent values of x_S are those marked with asterisks under "Possible values," which, in fact, represent a unique value for x_S since they are, in each case, related by the center of symmetry at the origin. Selecting the fourth column under "Possible values" and finding the mean value, we have $x_S = 0.39$. A few further structure factor calculations confirm this result. The comparison of $|F_o|$ and $|F_c|$ and the calculation of the R-factor are left as an exercise for the reader. The crystal structure is illustrated in Figure 5.6.

---

* This is easily proved, particularly if the last S position is changed to $\frac{1}{2}+x, x, -\frac{1}{2}-x$.

TABLE 5.4. Solution of the x Parameter of the S atom in FeS_2

hkl	$|F_o(hkl)|^2$	Interpretation	Solution in terms of possible values for x_S in the unit cell	Possible values of x_S in 120ths[a]
400	0.4	$A'(400) = 4(f_{Fe} + 2f_S \cos 8\pi x_S)$ If $|A'| \to$ min, $\cos 8\pi x_S \to -1$	$(2n+1)/8 = \frac{1}{8}, \frac{3}{8}, \frac{5}{8}, \frac{7}{8}$	— 15 — 45* — 75* — 105
600	3.2	$A'(600) = 4(f_{Fe} + 2f_S \cos 12\pi x_S)$ If $|A'| \to$ min, $\cos 12\pi x_S \to -1$	$(2n+1)/12 = \frac{1}{12}, \frac{3}{12}, \frac{5}{12}, \frac{7}{12}, \frac{9}{12}, \frac{11}{12}$	— 10 30 50* — 70* 90 110
800	14.4	$A'(800) = 4(f_{Fe} + 2f_S \cos 16\pi x_S)$ If $|A'| \to$ max, $\cos 16\pi x_S \to +1$	$n/8 = 0, \frac{1}{8}, \frac{2}{8}, \frac{3}{8}, \frac{4}{8}, \frac{5}{8}, \frac{6}{8}, \frac{7}{8}$	0 15 30 45* 60 75* 90 105
444	1.0	$A'(444) = 4(f_{Fe} + 2f_S \cos^3 8\pi x_S)$ If $|A'| \to$ min, $\cos^3 8\pi x_S \to -1$	$(2n+1)/8 = \frac{1}{8}, \frac{3}{8}, \frac{5}{8}, \frac{7}{8}$	— 15 — 45* — 75* — 105
555	10.0	$A'(555) = 4(f_{Fe} + 2f_S \cos^3 10\pi x_S)$ If $|A'| \to$ max, $\cos^3 10\pi x_S \to +1$	$n/5 = 0, \frac{1}{5}, \frac{2}{5}, \frac{3}{5}, \frac{4}{5}$	0 — 24 48* — 72* 96 —

[a] The most consistent values of x are marked with asterisks. Other agreements among the first four rows show the inability to differentiate between x_S and $\frac{1}{2} - x_S$ when h is an even number.

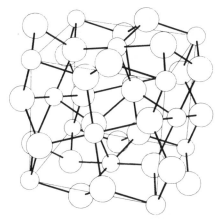

FIGURE 5.6. Unit cell of the pyrite structure: large circles S, small circles Fe.

Bibliography

Introductory Structure Analysis

BRAGG, W. L., *A General Survey* (*The Crystalline State*, Vol. I), London, Bell (1949).

HAHN, T. (Editor), *International Tables for Crystallography*, Vol. A, Dordrecht, D. Reidel (1983).

HENRY, N. F. M., and LONSDALE, K. (Editors), *International Tables for X-Ray Crystallography*, Vol. I, Birmingham, Kynoch Press (1965).

Problems

5.1. A structure with the apparent space group $P2_1/c$ consists of atoms at $0.2, \frac{1}{4}, 0.1$ and the symmetry-related positions; the center of symmetry is at the origin. Evaluate the geometric structure factor for the four given positions in the unit cell, and, hence, determine the systematic absences among the *hkl* reflections. What are the consequences of these absences as far as the true structure is concerned? Sketch the structure in projection along *b*. What is the true space group?

5.2. Rh_2B crystallizes in space group *Pnma* with $a = 5.42$, $b = 3.98$, $c = 7.44$ Å, and $Z = 4$. Consider Figure 2.36. Show that if no two Rh atoms may approach within 2.5 Å of each other, they cannot lie in general positions. Where might the Rh atoms be placed? Illustrate your answer with a sketch showing possible positions for these atoms in projection on (010).

5.3. Trimethylammonium chloride,

$$\left[\begin{array}{c} H_3C \\ H_3C-N-H \\ H_3C \end{array}\right]^+ Cl^-,$$

crystallizes in a monoclinic, centrosymmetric space group, with $a = 6.09$, $b = 7.03$, $c = 7.03$ Å, $\beta = 95.73°$, and $Z = 2$. The only limiting condition is $0k0$: $k = 2n$. What is the space group? Comment on the probable positions of (a) Cl, (b) C, (c) N, and (d) H atoms.

5.4. Potassium hexachloroplatinate(IV), $K_2[PtCl_6]$, is cubic, with $a = 9.755$ Å. The atomic positions are as follows ($Z = 4$):

$$(0,0,0; \quad 0,\tfrac{1}{2},\tfrac{1}{2}; \quad \tfrac{1}{2},0,\tfrac{1}{2}; \quad \tfrac{1}{2},\tfrac{1}{2},0)+$$

4   Pt: 0, 0, 0

8   K: $\tfrac{1}{4},\tfrac{1}{4},\tfrac{1}{4}; \quad \tfrac{3}{4},\tfrac{3}{4},\tfrac{3}{4}$

24   Cl: $\pm\{x, 0, 0; \quad 0, x, 0; \quad 0, 0, x\}$

Show that $F_c(hhh) = A'(hhh)$, where

$$A'(hhh) = 4g_{Pt} + 8g_K \cos(3\pi h/2) + 24g_{Cl} \cos 2\pi h x_{Cl}$$

Calculate $|F_c(hhh)|$ for the values of h tabulated below, with $x_{Cl} = 0.23$ and 0.24. Obtain R-factors for the scaled $|F_o|$ data for the two values of x_{Cl}, and indicate which value of x_{Cl} is the more acceptable. Calculate the Pt—Cl distance, and sketch the $[PtCl_6]^{2-}$ ion. What is the point group of this species?

hkl	111	222	333
$|F_o|$	491	223	281
g_{Pt}	73.5	66.5	59.5
g_K	17.5	14.5	12.0
g_{Cl}	15.5	13.0	10.5

Atomic scattering factors g may be taken to be temperature-corrected values.

5.5. USi crystallizes in space group *Pbnm*, with $a = 5.65$, $b = 7.65$, $c = 3.90$ Å, and $Z = 4$. The U atoms lie at the positions

$$\pm\{x, y, \tfrac{1}{4};\ \tfrac{1}{2}-x, \tfrac{1}{2}+y, \tfrac{1}{4}\}$$

Obtain a reduced expression for the geometric structure factor ($\bar{1}$ at 0, 0, 0) for the U atoms. From the data below, determine approximate values for x_U and y_U; the Si contributions may be neglected.

hkl	200	111	210	231	040	101	021	310
$I_o(hkl)$	0	236	251	200	0	170	177	0

Proceed by using 200 to find a probable value for x_U. Then find y_U from 111, 231, and 040.

5.6. Methylamine forms a complex with boron trifluoride of composition $CH_3NH_2BF_3$.

Crystal data

System: monoclinic.
Unit-cell dimensions: $a = 5.06$, $b = 7.28$, $c = 5.81$ Å, $\beta = 101.5°$.
V_c: 209.7 Å^3.
D_m: 1.54 g cm^{-3}.
M: 98.86.
Z: 1.97, or 2 to the nearest integer.
Unit-cell contents: 2 C, 10 H, 2 N, 2 B, and 6 F atoms.
Absent spectra: $0k0$: $k = 2n + 1$.
Possible space groups: $P2_1$ or $P2_1/m$ ($P2_1/m$ may be assumed).

Determine what you can about the crystal structure.

Methods in X-Ray Structure Analysis. II

6.1 Introduction

In this chapter, we shall introduce Fourier series and show how they are used in structure analysis. This discussion leads naturally to the Patterson function and the very important heavy-atom and isomorphous replacement techniques for solving the phase problem.

6.2 Fourier Series

According to Fourier's theorem, a continuous, single-valued, periodic function can be represented by a series composed of sine and cosine terms. A typical periodic function $\psi(x)$ is shown in Figure 6.1, in which the repeat period along the x axis is a. The corresponding Fourier series may be written as

$$\psi(x) = \frac{1}{a} \sum_{h=-\infty}^{\infty} \left[C(h) \cos 2\pi \frac{hx}{a} + S(h) \sin 2\pi \frac{hx}{a} \right] \qquad (6.1)$$

The index h of the hth term in this equation represents its frequency or wavenumber, which is the number of times its own wavelength fits into the

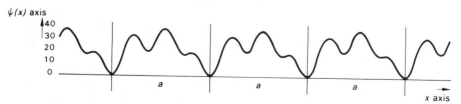

FIGURE 6.1. Periodic function.

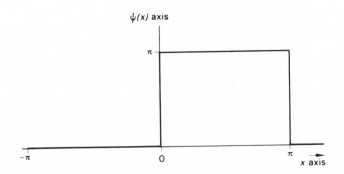

FIGURE 6.2. Square wave.

repeat period. The coefficients $C(h)$ and $S(h)$ may be expressed by the equations

$$C(h) = \int_{-a/2}^{a/2} \psi(x) \cos(2\pi hx/a)\, dx \qquad (6.2)$$

and

$$S(h) = \int_{-a/2}^{a/2} \psi(x) \sin(2\pi hx/a)\, dx \qquad (6.3)$$

If the form of the function $\psi(x)$ is known, $C(h)$ and $S(h)$ can be evaluated for values of h. We shall carry out this process for the square wave in Figure 6.2.

6.2.1 Computation of $\psi(x)$ for a Square Wave

This function is defined in the range $-\pi$ to π, with a repeat of 2π. Hence,

$$\psi(x) = \psi(x \pm m2\pi) \qquad (6.4)$$

where $m = 0, 1, \ldots, \infty$. In the range $-\pi \leq x \leq 0$,

$$\psi(x) = 0 \qquad (6.5)$$

and in the range $0 \leq x \leq \pi$,

$$\psi(x) = \pi \qquad (6.6)$$

From (6.2), with $a/2$ replaced by π and using (6.5) and (6.6),

$$C(h) = \int_{-\pi}^{0} 0 \cos hx \, dx + \int_{0}^{\pi} \pi \cos hx \, dx \qquad (6.7)$$

Integrating with respect to x, and since the first term on the right-hand side is zero,

$$C(h) = \frac{\pi \sin hx}{h} \Bigg]_{0}^{\pi} \qquad (6.8)$$

The limit $h = 0$ must be considered separately from (6.7), since h occurs in both the numerator and the denominator. Thus,

$$C(0) = \int_{0}^{\pi} \pi \, dx \qquad (6.9)$$

Integrating,

$$C(0) = \pi^2 \qquad (6.10)$$

Similarly for $S(h)$, from (6.3),

$$S(h) = \int_{-\pi}^{0} 0 \sin hx \, dx + \int_{0}^{\pi} \pi \sin hx \, dx \qquad (6.11)$$

Integrating as before,

$$S(h) = -\frac{\pi}{h} \cos hx \Bigg]_{0}^{\pi} = \frac{\pi}{h}(1 - \cos h\pi) \qquad (6.12)$$

From (6.11), it is clear that

$$S(0) = 0 \qquad (6.13)$$

Substituting for $C(h)$ and $S(h)$ in (6.1),

$$\psi(x) = \frac{1}{2\pi} \left[\pi^2 + \sum_{\substack{h=-\infty \\ h \neq 0}}^{\infty} \frac{\pi}{h}(1 - \cos h\pi) \sin hx \right] \qquad (6.14)$$

The term corresponding to $h = 0$ is taken outside the summation. From (6.12),

$$S(-h) = -S(h) \qquad (6.15)$$

Now,

$$(1/h) \sin hx = (1/-h) \sin(-hx) \qquad (6.16)$$

and since $1 - \cos h\pi$ is equal to $1 - (-1)^h$, $\psi(x)$ is zero for *even values* of h. Thus

$$\psi(x) = \frac{\pi}{2} + 2 \sum_{h=1}^{\infty} \frac{1}{h} \sin hx, \qquad (h = 2n+1) \qquad (6.17)$$

Range of X

The variable x defines a sampling point in any repeat interval of the function. We can describe the function $\psi(x)$ by choosing values of x from any point x_0 to $x_0 + 2\pi$. For convenience, we shall choose x_0 as 0, and define a sampling interval of $2\pi/50$. In this example, we shall calculate $\psi(x)$ at $0, 2\pi/50, 4\pi/50, \ldots, 98\pi/50, 100\pi/50$, but we shall note from the results that we could, and, in general, would, have quite properly made use of the reflection symmetry of the function at $x = (2m+1)\pi/2$, $m = 0, 1, 2, \ldots, \infty$.

Range of h

The summations in Fourier series extend, theoretically, from $-\infty$ to ∞. In practice, however, this range is $h_{\min} \leq h \leq h_{\max}$, where $h_{\min}$ and $h_{\max}$ are some preset limits of the frequency variable. In this example, $h_{\min}$ is unity, and the effect of two different values for $h_{\max}$ is illustrated in Figures 6.3a and 6.3b, drawn from the results in Table 6.1. Even at this level, the value of $h_{\max}$ has a dramatic effect on the series. By increasing the number of terms, the series (6.17) approaches more closely a representation of the square-wave function. In general, the more independent terms that can be included in a Fourier series, the better it represents the periodic function under investigation, from which the terms are derived.

The process of determining the coefficients of a Fourier series is called Fourier analysis, and the process of reconstructing the function by summa-

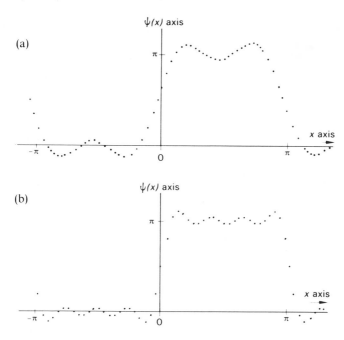

FIGURE 6.3. Calculated square waves: (a) $h_{max} = 3$, (b) $h_{max} = 5$. The positive and negative fluctuations across $\psi(x) = 0$ and π are known as series termination errors; they arise because there are insufficient terms to provide good convergence of the series.

tion of a series such as (6.17) is called Fourier synthesis. A microscope, in forming an image of an object, effectively performs its own Fourier synthesis of the light scattered by the object (see also page 169).

6.2.2 Exponential Form of Fourier Expressions

Using de Moivre's theorem, (4.28), we obtain

$$\cos(2\pi hx/a) = (e^{i2\pi hx/a} + e^{-i2\pi hx/a})/2 \tag{6.18}$$

$$\sin(2\pi hx/a) = (e^{i2\pi hx/a} - e^{-i2\pi hx/a})/2i \tag{6.19}$$

and substituting (6.18) and (6.19) in (6.1),

$$\psi(x) = \frac{1}{2a} \sum_{h=-\infty}^{\infty} [C(h)(e^{i2\pi hx/a} + e^{-i2\pi hx/a}) - iS(h)(e^{i2\pi hx/a} - e^{-i2\pi hx/a})]$$

$$\tag{6.20}$$

TABLE 6.1. Values of the Function $\psi(x) = \frac{1}{2}\pi + 2 \sum_{h=1}^{h_{max}} (1/h) \sin hx$ (h odd) for $h_{max} = 3$ and $= 5$. Compared with the True Value of $\psi(x) = \pi(0 < x < \pi)$ or $\psi(x) = 0(\pi < x < 2\pi)^a$

$x/2\pi$	$\psi(x) \simeq \frac{1}{2}\pi + 2 \sin x$ $+\frac{2}{3}\sin 3x$ $h_{max} = 3$	$\psi(x) \simeq \frac{1}{2}\pi + 2 \sin x$ $+\frac{2}{3}\sin 3x$ $+\frac{2}{5}\sin 5x$ $h_{max} = 5$	$\psi(x)$
0/50	1.571	1.571	3.142
1/50	2.067	2.522	3.142
2/50	2.525	3.186	3.142
3/50	2.910	3.428	3.142
4/50	3.200	3.330	3.142
5/50	3.380	3.109	3.142
6/50	3.454	2.977	3.142
7/50	3.433	3.017	3.142
8/50	3.343	3.158	3.142
9/50	3.215	3.265	3.142
10/50	3.081	3.249	3.142
11/50	2.972	3.137	3.142
12/50	2.912	3.034	3.142
m———			———m
13/50	2.912	3.034	3.142
14/50	2.972	3.137	3.142
15/50	3.081	3.249	3.142
16/50	3.215	3.265	3.142
17/50	3.343	3.158	3.142
18/50	3.433	3.017	3.142
19/50	3.454	2.977	3.142
20/50	3.380	3.109	3.142
21/50	3.200	3.330	3.142
22/50	2.910	3.428	3.142
23/50	2.525	3.186	3.142
24/50	2.067	2.522	3.142
25/50	1.571	1.571	3.142
26/50	1.075	0.619	0
27/50	0.617	−0.044	0
28/50	0.231	−0.287	0
29/50	−0.058	−0.188	0
30/50	−0.239	0.033	0
31/50	−0.312	0.164	0
32/50	−0.291	0.125	0
33/50	−0.201	−0.017	0
34/50	−0.073	−0.123	0
35/50	0.061	−0.107	0
36/50	0.169	0.005	0
37/50	0.230	0.108	0
m———			———m
38/50	0.230	0.108	0

TABLE 6.1—*cont.*

$x/2\pi$	$\psi(x) \simeq \frac{1}{2}\pi + 2\sin x$ $+\frac{2}{3}\sin 3x$ $h_{max} = 3$	$\psi(x) \simeq \frac{1}{2}\pi + 2\sin x$ $+\frac{2}{3}\sin 3x$ $+\frac{2}{5}\sin 5x$ $h_{max} = 5$	$\psi(x)$
39/50	0.169	0.005	0
40/50	0.061	−0.107	0
41/50	−0.073	−0.123	0
42/50	−0.201	−0.017	0
43/50	−0.291	0.125	0
44/50	−0.312	0.164	0
45/50	−0.239	0.033	0
46/50	−0.058	−0.188	0
47/50	0.231	−0.287	0
48/50	0.617	−0.044	0
49/50	1.075	0.619	0
50/50	1.571	1.571	3.142

[a] The corresponding curves are shown in Figure 6.3.

Collecting terms, and following the notation of (4.48),

$$\psi(x) = \frac{1}{2a} \sum_{h=-\infty}^{\infty} [\mathbf{G}(h)e^{-i2\pi hx/a} + \mathbf{G}(-h)e^{i2\pi hx/a}] \qquad (6.21)$$

where

$$\mathbf{G}(h) = C(h) + iS(h) \qquad (6.22)$$

and

$$\mathbf{G}(-h) = C(h) - iS(h) \qquad (6.23)$$

Since h ranges from $-\infty$ to ∞, both expressions under the summation in (6.21) take a sequence of identical values within the range of the variable h. Hence, (6.21) may be written

$$\psi(x) = \frac{1}{a} \sum_{h=-\infty}^{\infty} \mathbf{G}(h)e^{-i2\pi hx/a} \qquad (6.24)$$

From (6.2), (6.3), and (6.22),

$$\mathbf{G}(h) = \int_{-a/2}^{a/2} \psi(x)[\cos(2\pi hx/a) + i\sin(2\pi hx/a)]\,\mathrm{d}x \qquad (6.25)$$

and, using (4.28),

$$\mathbf{G}(h) = \int_{-a/2}^{a/2} \psi(x)e^{i2\pi hx/a}\,\mathrm{d}x \qquad (6.26)$$

Although not of great practical use, the complex forms of (6.24) and (6.26) can provide a useful starting point for further manipulation of Fourier equations. The functions $\psi(x)$ and $\mathbf{G}(h)$, as defined in these equations, are said to be Fourier transforms of each other. If $\mathbf{G}(h)$ is known for all values of h, we can calculate $\psi(x)$. Similarly, if $\psi(x)$ is known over the periodic range $-a/2$ to $a/2$, we can calculate $\mathbf{G}(h)$. Equation (6.24) represents the Fourier synthesis and (6.26) the Fourier analysis of the function $\psi(x)$.

6.3 Representation of Crystal Structures by Fourier Series

Because of the underlying lattice structure of crystals, the contents of the unit cell are repeated periodically by the translations a, b, and c. We may consider an analogy between the square wave and a conceptual one-dimensional crystal, or a projection of a crystal structure onto an axis. The first applications of Fourier series in crystallography used this restricted geometric form, and for good reason. The mathematical treatment for three-dimensional periodicity is correspondingly more complicated, but it follows the form of the one-dimensional example, and we shall obtain the necessary equations by analogy with (6.24) and (6.26). It is important, however, to grasp the significance of the functions in crystallography that correspond to $\psi(x)$ and $\mathbf{G}(h)$.

6.3.1 Electron Density and Structure Factors

We have shown in Chapter 4 how X-rays are scattered by the electrons associated with the atoms in a crystal. Atoms with high atomic numbers provide a greater concentration of electrons than do atoms of low atomic numbers. This concentration of electrons and its distribution around the atom is called the electron density ρ, and is usually measured in electrons per Å^3. Since it is, in general, a function of position, we specify the electron density at the point x, y, z as $\rho(x, y, z)$. The periodicities of the lattice are impressed on the electron density in the crystal, and, therefore, we become concerned with a three-dimensional, periodic electron density function: we may identify $\rho(x)$ with $\psi(x)$ of the previous example [see (6.24)], but we must determine the meaning of $\mathbf{G}(h)$.

In Chapter 4, we considered the electrons in an atom as though they were concentrated at a point, and their distribution was specified by a shape factor, the atomic scattering factor f. The exponential term modifying f represented the phase at the atomic position with respect to the origin of the unit cell.

Consider the one-dimensional electron density function in Figure 6.4. In a small interval dx along the x axis, the electron density $\rho(x)$ may be regarded as constant. The electron count in this strip is, therefore, $\rho(x)\,dx$. Following (4.32) and (4.42), its contribution to the hth structure factor is given by

$$\rho(x)e^{i2\pi hx/a}\,dx \qquad (6.27)$$

Hence, $\mathbf{F}(h)$ is given by the integration of this expression over the unit cell:

$$\mathbf{F}(h) = \int_0^a \rho(x)e^{i2\pi hx/a}\,dx \qquad (6.28)$$

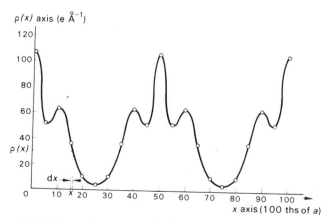

FIGURE 6.4. One-dimensional electron density projection for pyrite (FeS$_2$).

Equation (6.28) is a generalized form for the structure factor; similar representations for two- and three-dimensional structure factors can be written down by analogy with (6.28). We have already identified $\psi(x)$ with $\rho(x)$, and substituting (6.24) in (6.28) gives

$$\mathbf{F}(h) = \int_0^a \frac{1}{a} \sum_{h'=-\infty}^{\infty} \mathbf{G}(h') e^{-i2\pi h' x/a} e^{i2\pi h x/a} \, dx \qquad (6.29)$$

where h' indicates a range of values of h under the summation sign. Since the integral of the sum in (6.29) may be thought of as a sum of the integrals of the separate terms, we write

$$\mathbf{F}(h) = \frac{1}{a} \sum_{h'=-\infty}^{\infty} \mathbf{G}(h') \int_0^a e^{i2\pi x(h-h')/a} \, dx \qquad (6.30)$$

Performing the integration in (6.30) leads to

$$\frac{\exp[i2\pi x(h-h')/a]}{i2\pi(h-h')/a} \Bigg]_0^a \qquad (6.31)$$

which, on substituting the limits, becomes

$$\frac{\exp[i2\pi(h-h')]-1}{i2\pi(h-h')/a} \qquad (6.32)$$

Since h and h' are both integers, the numerator of (6.32) is zero, from (4.28), unless $h = h'$. In this special case (6.32) is indeterminate, but from (6.30) we see that the integral is now

$$\int_0^a dx \qquad (6.33)$$

which has the value a. Since only a *single* value of h' (namely h) leads to a nonzero value for the integral in (6.30), this equation becomes

$$\mathbf{F}(h) = \mathbf{G}(h) \qquad (6.34)$$

and from (6.24)

$$\rho(x) = \frac{1}{a} \sum_{h=-\infty}^{\infty} \mathbf{F}(h)e^{-i2\pi hx/a} \tag{6.35}$$

Since $\rho(x)$ is periodic, we could use the limits $-a/2$ to $a/2$ instead of 0 to a, and, by analogy with (6.24) and (6.26), we see that (6.28) is the Fourier transform of the electron density (6.35). In other words, the structure factors provide the coefficients for the Fourier synthesis of the electron density, which returns us to the phase problem (page 169). We have, in fact, considered already simple examples of the solution of the phase problem in studying NaCl and FeS$_2$ (pages 202–209). Once the atomic positions are known, even approximately, phases can be calculated for each value of $|F_o(hkl)|$ and a Fourier synthesis performed. This has been done for FeS$_2$ in Figure 6.4, from which we can deduce that x_S is 0.11 (compare page 207).

6.3.2 Electron Density Equations

The general electron density function is expressed as a three-dimensional Fourier series. By analogy with (6.24), and introducing fractional coordinates $(x = x/a$, etc.), we may write

$$\rho(x, y, z) = \frac{1}{V_c} \sum_{h}\sum_{k}\sum_{l}^{\infty} \mathbf{F}(hkl)e^{-i2\pi(hx+ky+lz)} \tag{6.36}$$

where V_c is the unit-cell volume, (2.19). Distinguish carefully between the points x, y, z at which the electron density is calculated, and the points x_j, y_j, z_j (in the structure factor equation), which represent *actual* atomic positions; both sets of points cover the same field, the unit cell. The summations, in practice, extend over a finite set of hkl values, which are limited by both the nature of the crystal and the particular experimental arrangement.

The exponential form of (6.36) implies that the electron density is a complex function, whereas, in fact, it is real throughout the unit cell. The derivations of practicable expressions depend upon the use of (4.28), (4.48), (4.51), and (4.63), leading to

$$\rho(x, y, z) = \frac{1}{V_c} \sum_{h}\sum_{k}\sum_{l}^{\infty} |F(hkl)|e^{i\phi(hkl)}e^{-i2\pi(hx+ky+lz)} \tag{6.37}$$

or

$$\rho(x, y, z) = \frac{1}{V_c} \sum_h \sum_{k-\infty}^{\infty} \sum_l [A'(hkl) \cos 2\pi(hx + ky + lz)$$

$$+ B'(hkl) \sin 2\pi(hx + ky + lz)] \qquad (6.38)$$

where the summations are taken over the appropriate practical values of h, k, and l. Using (4.54) and (4.55), (6.38) becomes

$$\rho(x, y, z) = \frac{1}{V_c} \sum_h \sum_{k-\infty}^{\infty} \sum_l |F(hkl)| \cos[2\pi(hx + ky + lz) - \phi(hkl)] \quad (6.39)$$

Further simplification of (6.38) and (6.39) depends on the use of Friedel's law (4.58). Hence, (6.38) becomes

$$\rho(x, y, z) = \frac{2}{V_c} \sum_{h=0} \sum_{k-\infty}^{\infty} \sum_l [A'(hkl) \cos 2\pi(hx + ky + lz)$$

$$+ B'(hkl) \sin 2\pi(hx + ky + lz)] \qquad (6.40)$$

and (6.39),

$$\rho(x, y, z) = \frac{2}{V_c} \sum_{h=0} \sum_{k-\infty}^{\infty} \sum_l |F(hkl)| \cos[2\pi(hx + ky + lz) - \phi(hkl)] \quad (6.41)$$

since the maximum necessary summations now take place over a hemisphere in reciprocal space. In these and similar equations, the term $F(000)$, which represents the total number of electrons in the unit cell, is multiplied by $1/V_c$ and, in practice, the term $F(000)/V_c$ is added separately to the result of the summations. It does not matter which of h, k, or l has zero as its lower summation limit.

Equation (6.40) is the most useful for electron density calculations, but (6.41) shows clearly the dependence of $\rho(x, y, z)$ on $\phi(hkl)$. The one-dimensional function in Figure 6.1 is shown again, now called $\rho(x)$, in Figure 6.5, together with its four components corresponding to $h = 0, 1, 2$, and 3. This function may be written, following (6.41), as

$$\rho(x) = \frac{20}{a} + \frac{2}{a} \{5 \cos[2\pi(x) - 3.019] + 2.5 \cos[2\pi(2x) - 2.397]$$

$$+ 4 \cos[2\pi(3x) - 4.062]\} \qquad (6.42)$$

where, for convenience, a may be taken as unity. The process of X-ray diffraction corresponds to a Fourier analysis, or breakdown of the object $\rho(x)$ into its constituent terms, with the exception that, in recording the *intensities*, the phase information is lost. Fourier synthesis is equivalent to a summation of the terms in order to reconstruct the object, but it cannot be achieved without regaining the phase information.

We will look at the synthesis (6.42) in terms of Figure 6.5. The zeroth-order term $F(0)$ impresses a constant, positive electron density on to the distribution of $\rho(x)$. It is a term of zero phase, and tells us nothing about the structure, only about the contents of the repeat unit a. Note that the zeroth-order term does not lie within the multiplying factor $2/a$.

$F(1)$ shows the magnitude of the period a and introduces a little information about the structure. $F(2)$ and $F(3)$ add successively finer detail about the structure, and $\rho(x)$ is the algebraic sum of these waves, as we can see from Figure 6.5. In general, the more orders of diffracted spectra that are included in the sum, the better is the resolution of the image. However, this sum is correct only because the correct phases are used in the equation. The reader may like to show (very easily with tracing paper) that while a change in phase corresponds only to a movement of the wave along the x axis, an entirely different form of $\rho(x)$ may ensue. Try altering the phase of $F(1)$ from 0.47 to 0.17 and then resumming (6.42).

The methods of structure analysis* seek to extract phase information starting from the $|F_o|$ data. A considerable amount of computation is involved in this process. Fortunately for the modern crystallographer, programs for the calculation of structure factors and electron density fields are now standard "equipment" and become relatively easy on high-speed, large-capacity computers. However, it is very important that the student should appreciate the nature of these calculations and the underlying theory of their use in crystallography.

6.3.3 Interpretation of Electron Density Distributions

Electron density is concentrated in the vicinity of atoms, rising to peaks at the electron density maxima (atomic "positions") and falling to relatively low values between them. The wavelengths of X-rays used in crystal structure analysis are too long to reveal the intimate electronic structure of atoms themselves, which are seen, therefore, somewhat blurred in the

---

* An apparent paradox, since it refers to Fourier synthesis.

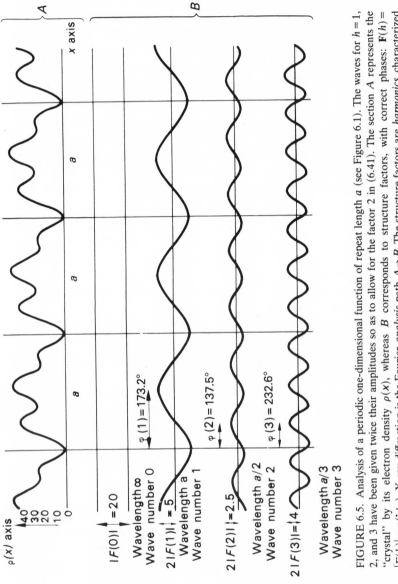

FIGURE 6.5. Analysis of a periodic one-dimensional function of repeat length a (see Figure 6.1). The waves for $h = 1$, 2, and 3 have been given twice their amplitudes so as to allow for the factor 2 in (6.41). The section A represents the "crystal" by its electron density $\rho(x)$, whereas B corresponds to structure factors, with correct phases: $\mathbf{F}(h) = |F(h)| \exp(i\phi_h)$. X-ray diffraction is the Fourier analysis path $A \rightarrow B$. The structure factors are *harmonics*, characterized by both amplitude and phase. Crystal structure determination, exemplified by the path $B \rightarrow A$, cannot be carried out directly because only the amplitudes $|F(h)|$, and not the phases ϕ_h, are recorded experimentally. The values shown for the phases are based on fractions of their own wavelengths a, $a/2$, and $a/3$, in order to comply with the definition of phase (page 158); they influence greatly the positions of the peaks in the $\rho(x)$ summation.

calculated electron density function. Atoms appear as peaks in this function, and the peak position of a given atom is assumed to correspond to its atomic center, within the limit of experimental errors. In general, the more complete and accurate the experimental $|F|$ data, the better will be the atomic resolution and the more precise the final structure model.

Peak Heights and Weights

To a first approximation, the heights of the peaks in an electron density distribution of a crystal are proportional to the corresponding atomic numbers. The hydrogen atom, at the extreme low end of the atomic numbers, is not resolved in electron density maps; its small electron density merges into the background density that arises from errors in both the data and the structure model. However, hydrogen atoms can be detected by a difference-Fourier technique, as discussed in a later section (page 264).

A better measure of the electron content of a given atom may be obtained from an integrated peak weight, in which the values of $\rho(x, y, z)$ are summed over the volume occupied by the atom. This technique makes some allowance for the variation of individual atomic temperature factors, high values of which tend to decrease peak heights for a given electron content.

Computation and Display of Electron Density Distributions

Assuming for the moment that phases are available, the electron density function may be calculated over a grid of chosen values of x, y, and z. For this purpose, the unit cell is divided into a selected number of equal divisions, in a manner similar to that employed in the synthesis of the square-wave function (page 216). Intervals corresponding to about 0.3 Å are satisfactory for most electron density maps. The symmetry of $\rho(x, y, z)$ corresponds to the space group of the crystal under investigation. Consequently, a summation over a volume equal to, or just greater than, that of the asymmetric unit is adequate.

In order to facilitate the interpretation of $\rho(x, y, z)$, it is essential to present the distribution of the numerical values in such a way that the geometric relationships between the peaks are easily inspected. This feature is afforded by first calculating the electron density in sections, each corresponding to a constant value of x, y, or z using (6.40). Each section consists of a field of figures arranged on a grid, closely true to scale for preference, which may be contoured by lines passing through points of equal electron

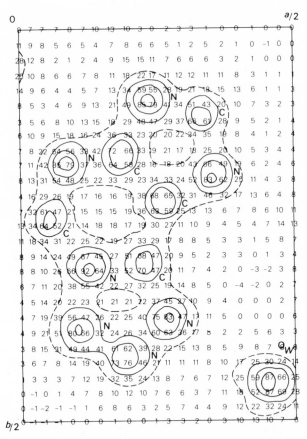

FIGURE 6.6. Two-dimensional electron density projection $\rho(x, y)$ for azidopurine monohydrate, $C_5H_3N_7 \cdot H_2O$ [calculated from the data of Glusker *et al.*, *Acta Crystallographica B* **24**, 359 (1968)]. The isolated peak (O_w) in the lower right-hand region of the map represents the oxygen atom of the water molecule. Hydrogen atom positions are not obtained in a direct electron density synthesis (see page 264). The field figures are $10\rho(x, y)$ in electrons per Å^2 contoured at intervals of 20 units.

density, interpolating as necessary (Figure 6.6). The grading of the contour intervals is selected to produce a reasonable number of contours around the higher density areas. The contouring should be carried out with care; this exercise leads to precise peak positions and a desirable familiarity with the problem. Sophisticated map-plotting and peak-searching facilities are available, but they should be treated with caution by the beginner.

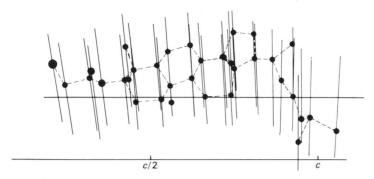

FIGURE 6.7. Three-dimensional model of euphenyl iodoacetate (see Figures 1.2 and 1.3).

The contoured sections are finally transferred to a transparent medium, such as thin perspex or clear acetate sheets, which are then stacked at the requisite spatial intervals and viewed over a diffuse light source. Figure 1.2 is a photograph of such a display, extending through 17 sections.

An alternative method of displaying the results of an electron density calculation is by means of a ball-and-stick model. An example of this form of representation is shown in Figure 6.7.

Projections

The use of two-dimensional studies in crystallography is fairly restrictive, but, nevertheless, worthy of mention because of the relative ease of calculation and preparation of Fourier maps. For example, the function

$$\rho(x, z) = \frac{2}{A_b} \sum_h \sum_l |F(h0l)| \cos[2\pi(hx + lz) - \phi(h0l)] \qquad (6.43)$$

is calculated with the data from only one level of the reciprocal lattice, the zero level, perpendicular to b, and plotted over the area A_b of the ac plane, or the asymmetric portion thereof. The simplification in the calculations is offset, however, by a corresponding complexity in the interpretation of the maps, arising from the superposition of peaks in projection onto the given plane, although this effect is not as severe as in one dimension. Equation (6.43) corresponds to the projection of the electron density along the b axis: it is essential to appreciate the difference between the meaning of $\rho(xz)$ and $\rho(x0z)$, for example; the latter represents the section of the three-dimensional electron density function at $y = 0$. Equations for projections along other principal axes may be written by analogy with (6.43).

Even simple atomic arrangements may appear distorted in projection, with individual molecules overlapping to some degree, but we would not wish to discourage their consideration. We shall restrict their use to examples illustrating various aspects of structure analysis. Practice in the calculation and interpretation of Fourier series is afforded by Problems 6.5 and 6.6 at the end of this chapter (see also Chapter 8).

6.4   Methods of Solving the Phase Problem

The set of $|F_o(hkl)|$ data constitute the starting point of all X-ray structure determinations. The approximate number of symmetry-independent reflections measurable may be calculated in the following manner.

6.4.1   Number of Reflections in the Data Set

The radius of the limiting sphere (page 138) is 2 RU, and its volume is therefore 33.51 RU^3. The number of reciprocal lattice points within the limiting sphere is approximately equal to the number of times the reciprocal unit cell, volume V^*, will fit into 33.51; using (2.20), this number is 33.51 V_c/λ^3. The number of symmetry-independent reflections observable, N_{max}, in a given experiment in which θ_{max} represents the practical upper limit is given by

$$N_{max} = 33.51 V_c \sin^3\theta_{max}/\lambda^3 Gm \qquad (6.44)$$

where G is the unit-cell translation constant (Table 4.1) for nonzero reflections and m is the number of symmetry-equivalent reflections (the number of general equivalent points in the appropriate Laue group). For zones and rows, m may take different values from that for hkl, and a number of systematic absences within the sphere of radius 2 sin θ_{max} may have to be subtracted.

As an example, consider an orthorhombic crystal of space group $Cmm2$, with unit cell dimensions $a = 9.00$, $b = 10.00$, and $c = 11.00$ Å. For Cu $K\alpha$ radiation ($\bar{\lambda} = 1.542$ Å) and θ_{max} of 85°, N_{max} is (33.51×9×10× 11×sin^{3}85°)/(1.542^3×2×8) = 559. If Mo $K\alpha$ radiation ($\bar{\lambda} = 0.7107$ Å) had been used instead of Cu $K\alpha$, the number would have been 5710. Such a structure might contain, say, 15 atoms in the asymmetric unit. In the structure analysis, each atom would be determined by three positional parameters (x_j, y_j, z_j) and, say, one isotropic thermal vibration parameter, which, with an overall scale factor, totals 61 variables. Even with Cu $K\alpha$ radiation, there are nine reflections per variable, a situation which, from a mathematical point of view, is considerably overdetermined. This feature is important, since the experimental intensity measurements contain random errors which cannot be eliminated, and the preponderance of data is needed to ensure precision in the structural parameters. We shall consider this situation again in Chapter 7.

6.4.2 The Patterson Function

It is interesting to note that although the connection between Fourier theory and X-ray diffraction was recorded first in 1915, it was not until about 1930 that practical use was made of it. Before the advent of computing facilities, the calculation of even a Fourier projection involved considerable time and effort. Add to this the phase problem, which necessitated many such calculations, and it is easy to understand that X-ray analysts were not anxious to become involved with extensive Fourier calculations; many early structure analyses were based on two projections.

In 1934, Patterson reported a new Fourier series which could be calculated directly from the experimental intensity data. However, because phase information is not used in the Patterson series, the result cannot be interpreted as a set of atomic positions, but rather as a collection of interatomic vectors all taken to a common origin. Patterson was led to the formulation of his series from considerations of an earlier theory of Debye on the scattering of X-rays by liquids—a much more difficult problem.

Patterson functions are of considerable importance in X-ray structure analysis, and their application will be considered in some detail. We will study first a one-dimensional function.

One-Dimensional Patterson Function

The electron density at any fractional coordinate x is $\rho(x)$, and that at the point $(x+u)$ is $\rho(x+u)$. The average product of these two electron densities in a repeat of length a, for a given value of u, is

$$A(u) = \int_0^1 \rho(x)\rho(x+u)\,\mathrm{d}x \qquad (6.45)$$

where the upper limit of integration corresponds to the use of fractional coordinates. Using (6.36) in a form appropriate to a one-dimensional unit repeat, we obtain

$$A(u) = \int_0^1 \frac{1}{a^2} \sum_h \mathbf{F}(h) e^{-i2\pi hx} \sum_{h'} \mathbf{F}(h') e^{-i2\pi h'(x+u)} \, dx \qquad (6.46)$$

The index h' lies within the same range as h, but is used to effect distinction between the Fourier series for $\rho(x)$ and $\rho(x+u)$. Separating the parts dependent upon x, and remembering that the integral of a sum is the sum of the integrals of the separate terms, we may write

$$A(u) = \frac{1}{a^2} \sum_h \sum_{h'} \mathbf{F}(h)\mathbf{F}(h') e^{-i2\pi h'u} \int_0^1 e^{-i2\pi(h+h')x} \, dx \qquad (6.47)$$

Considering the integral

$$\int_0^1 e^{-i2\pi(h+h')x} \, dx = \frac{e^{-i2\pi(h+h')x}}{-i2\pi(h+h')} \Bigg]_0^1 \qquad (6.48)$$

$e^{-i2\pi(h+h')}$ is unity, since h and h' are integral, from (4.28), and the integral is, in general, zero. However, for the particular value of h' equal to $-h$, it becomes indeterminate and we must consider making this substitution before integration. Thus,

$$\int_0^1 dx = 1 \qquad (6.49)$$

Hence, from (6.47), for nonzero value of $A(u)$, where $h' = -h$,

$$A(u) = \frac{1}{a^2} \sum_h \sum_{-h} \mathbf{F}(h)\mathbf{F}(-h) e^{i2\pi hu} \qquad (6.50)$$

Equation (6.50) is not really a double summation, since h and $-h$ cover the same field of the function. Furthermore, $\mathbf{F}(-h)$ is really the conjugate $\mathbf{F}^*(h)$, and using (4.34), we obtain

$$A(u) = \frac{1}{a^2} \sum_h |F(h)|^2 e^{i2\pi hu} \qquad (6.51)$$

where the index h ranges from $-\infty$ to ∞. Now using Friedel's law (4.58), we find

$$A(u) = \frac{1}{a^2} \sum_h (|F(h)|^2 e^{i2\pi hu} + |F(h)|^2 e^{-i2\pi hu}) \tag{6.52}$$

and from de Moivre's theorem (4.28),

$$A(u) = \frac{2}{a^2} \sum_h |F(h)|^2 \cos 2\pi hu \tag{6.53}$$

where h now ranges from 0 to ∞. The corresponding Patterson function $P(u)$ is usually defined as

$$P(u) = \frac{2}{a} \sum_h |F(h)|^2 \cos 2\pi hu \tag{6.54}$$

a trivial difference from the averaging function $A(u)$.

The practical evaluation of $P(u)$ proceeds through (6.54), but its physical interpretation is best considered in terms of (6.45), neglecting the small difference between $P(u)$ and $A(u)$.

Figure 6.8a shows one unit cell of a one-dimensional structure containing two different atoms A and B situated at fractional coordinates x_A and x_B, respectively. Equation (6.45) represents the value of the electron density product $\rho(x)\rho(x+u)$, for any constant value of u, averaged over the repeat period of the unit cell. The average will be zero if one end of the vector u always lies in zero regions of electron density, small if both ends of the vector encounter low electron densities, large if the electron density products are large, and a *maximum* where u is of such a length that it spans two atomic positions in the unit cell.

For values of u less than u_{min} in Figure 6.8a, no peak will arise from the pair of atoms. As u is increased, however, both ends of the vector will come simultaneously under the electron density peaks, and from (6.45) a finite value of $A(u)$, or $P(u)$, will be obtained. The integration can be simulated by sliding a vector of a given magnitude u along the x axis, evaluating the product $\rho(x)\rho(x+u)$ for all sampling intervals between zero and unit fractional repeat; this process is carried out for all fractional values of u between zero and one. The graph of $P(u)$ as a function of u is similar in appearance to an electron density function, but we must be careful not to interpret it in this way.

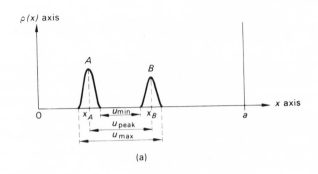

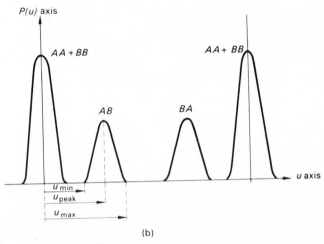

FIGURE 6.8. Development of a one-dimensional Patterson function
for a two-atom structure.

As we proceed through the values of u, we encounter u_{peak}, the interatomic vector $A-B$, which gives rise to the maximum value of $P(u)$, labeled AB in Figure 6.8b. As u increases to u_{max}, the electron density product falls to zero and $P(u)$ decreases correspondingly. Since we are concerned with interatomic *vectors*, negative values of u are equally important; $-AB$ is marked off on the negative side of the origin, or at BA within the given unit cell.

If we consider next very small values of u, both ends of such vectors will lie inside one and the same electron density peak, and $P(u)$ will be large. In the limit as $u \to 0$, the product involves that of the electron density maximum with itself, which is a local maximum for each atom, and a very large peak at the origin ($u = 0$) is to be expected. Thus the Patterson function is represented as a map of interatomic vectors, including null vectors, all taken to the origin.

The reader should confirm from Figure 6.8, using tracing paper, that the positions of the peaks in Patterson space can be plotted graphically by placing each atom of the structure $\rho(x)$ in turn at the origin of the Patterson map, in parallel orientation, and marking the positions of the other atoms onto the Patterson unit cell. Because of the centrosymmetry of the Patterson function (page 225), it is not strictly necessary to plot vectors lying outside one-half of the unit cell.

Three-Dimensional Patterson Function

If we replace $\rho(x)$ and $\rho(x+u)$ in (6.45) by the three-dimensional analogs $\rho(x, y, z)$ and $\rho(x+u, y+v, z+w)$ and integrate over a unit fractional volume, we can derive the three-dimensional Patterson function:

$$P(u, v, w) = \frac{2}{V_c} \sum_h \sum_k \sum_l |F(hkl)|^2 \cos 2\pi(hu + kv + lw) \qquad (6.55)$$

where the summations range, in the most general case, over one-half of experimental reciprocal space. This equation should be compared with (6.41): It is a Fourier series with zero phases and $|F|^2$ as coefficients. In practice, it may be handled like the corresponding electron density equation, with u, v, w replacing x, y, z, but it should be remembered that both functions explore the same field, the unit cell. The roving vector is now specified by three coordinates, u, v, and w, and $P(u, v, w)$ is a maximum where the corresponding vector spans two atoms in the crystal.

Positions and Weights of Peaks in the Patterson Function

The positions of the peaks in $P(u, v, w)$ may be plotted in three dimensions by placing each atom of the unit cell of a structure in turn at the origin of Patterson space, in parallel orientation, and mapping the positions of all other atoms onto the Patterson unit cell. Examples of this process are illustrated graphically in Figure 6.9; for simplicity the origin peak is not shown in Figure 6.9d. In Figure 6.9a, all atoms and their translation equivalents produce vector peaks lying on the points of a lattice that is identical in shape and size to the crystal lattice. For example, atom 1 at x, y, z and its translation equivalent, $1'$, at $x, 1+y, z$ give rise to a vector ending at $0, 1, 0$ on the Patterson map. Peaks of this nature accumulate at the corners

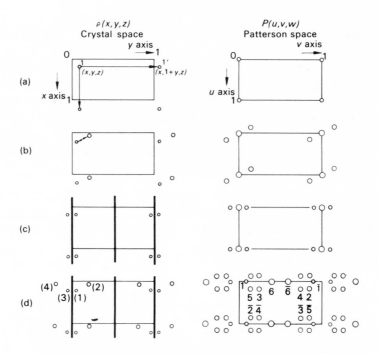

FIGURE 6.9. Effects of symmetry-related and symmetry-independent atoms on the Patterson function. The weights of the peaks are approximately proportional to the diameters of the circles: (a) $P1$ $(N = 1)$; (b) $P1$ $(N = 2)$—two atoms per unit cell produce $(2^2 - 2)$ nonorigin peaks; (c) Pm $(N = 2)$—two nonorigin peaks, but with coordinates $\pm\{0, 2y, 0\}$; (d) Pm $(N = 4)$—twelve nonorigin peaks per unit cell; for clarity the origin peak has not been drawn. The *Patterson* space group is $P\bar{1}$ in (a) and (b) and $P2/m$ in (c) and (d). Figure 6.9d is discussed again on pages 238–239.

of the Patterson unit cell in exactly the same way as those of the origin peak, $P(0, 0, 0)$. From (6.55), we can derive the height of the origin peak:

$$P(0, 0, 0) = \frac{2}{V_c} \sum_{h=0}^{\infty} \sum_{k, l=-\infty}^{} |F_o(hkl)|^2 \tag{6.56}$$

In general, (6.56) is equivalent to a superposition at the origin of all N products like $\rho(x_j, y_j, z_j)\rho(x_j, y_j, z_j)$, where N is the number of atoms in the unit cell. Since $\rho(x_j, y_j, z_j)$ is proportional to the atomic number Z_j of the jth atom (page 227), we have

$$P(0, 0, 0) \propto \sum_{j=1}^{N} Z_j^2 \qquad (6.57)$$

A single vector interaction between two atoms j and k (Figure 6.9b) will have a Patterson peak of height proportional to $Z_j Z_k$. Hence, the height $H(j, k)$ of this peak will be given by

$$H(j, k) \approx P(0, 0, 0) Z_j Z_k \bigg/ \sum_{j=1}^{N'} Z_j^2 \qquad (6.58)$$

where $P(0, 0, 0)$ is calculated from (6.56). This equation can serve as a useful guide, but overlapping vectors may give rise to misleading indications. The reservations on peak heights already mentioned (page 227) apply also to Patterson peaks. It should be remembered that the correct geometrical interpretation of Patterson peaks is of far greater significance than is an adherence to (6.58).

In a structure with N atoms per unit cell, each atom forms a vector with the remaining $N-1$ atoms. There are, thus, $N(N-1)$ nonorigin peaks. From (6.55), substitution of $-u$, $-v$, $-w$ for u, v, w, respectively, leaves $P(u, v, w)$ unaltered, which is a statement of the centrosymmetry of the Patterson function.

The Patterson unit cell is the same size and shape as the crystal unit cell, but it has to accommodate N^2 rather than N "peaks" and is, therefore, correspondingly overcrowded. Thus, peaks in Patterson space tend to overlap when there are many atoms in the unit cell, a feature which introduces difficulties into the process of unraveling the function in terms of the correct distribution of atoms in the crystal.

Sharpened Patterson Function

In a conceptual point atom, the electrons would be concentrated at a point. The atomic scattering factor curves (Figure 4.9) would be parallel to the abscissa and f would be equal to the atomic number for all values of $(\sin \theta)/\lambda$ and at all temperatures. The electron density for a crystal composed of point atoms would show a much higher degree of resolution than does that for a real crystal. Put another way, the broad peaks representing real atoms (Figure 6.4) would be replaced by peaks of very narrow breadth in the point-atom crystal.

A plot of the mean value of $|F_o|^2$ against $(\sin \theta)/\lambda$ for a typical set of data is shown in Figure 6.10. The radial decrease in $\overline{|F_o|^2}$ can be reduced by modifying $|F_o|^2$ by a function which increases as $(\sin\theta)/\lambda$ increases. The coefficients for a sharpened Patterson synthesis may be calculated by the following equation (see also page 296):

$$|F_{\mathrm{mod}}(hkl)|^2 = \frac{|F_o(hkl)|^2}{\exp[-2B(\sin^2\theta)/\lambda^2]\{\sum_{j=1}^{N} f_j\}^2} \qquad (6.59)$$

N is the number of atoms in the unit cell and B is an overall isotropic temperature factor (pages 166 and 257).

The effect of sharpening on a Patterson synthesis is illustrated in Figure 6.17d, the Harker section $(u, \frac{1}{2}, w)$ for papaverine hydrochloride. It should be compared with Figure 6.17b; the increased resolution is very apparent.

Oversharpening of Patterson coefficients may lead to spurious peaks, particularly where heavy atoms are present, and the technique should not be applied without care. Sometimes the coefficients can be further modified to advantage by multiplication by a function such as $\exp(-m\sin^3\theta)$, where m is chosen by trial, but might be about 5. This function has the effect of decreasing the magnitude of the $\overline{|F_o|^2}$ curve at the high θ values. Many other sharpening functions have been proposed, but we shall not dwell on this

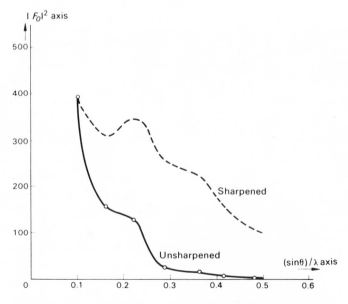

FIGURE 6.10. Effect of sharpening on the radial decrease of the local average intensity $\overline{|F_o|^2}$.

subject. It is sometimes helpful to calculate both the normal and sharpened Patterson functions for comparison.

Symmetry of the Patterson Function for a Crystal of Space Group Pm

An inspection of Figures 6.9c and 6.9d shows that the peaks on the line $[0, v, 0]$ arise from atom pairs related by the m planes. The vector interactions for case (d) are listed in Table 6.2, and may be easily verified by the reader; the values $z_1 = z_2 = 0.0$ were chosen for convenience only.

The m planes in Pm are carried over into Patterson space, and relate the following pairs of peaks in the vector set:

$$1, \bar{1}; \quad 2, 5; \quad \bar{2}, \bar{5}; \quad 3, 4; \quad \bar{3}, \bar{4}; \quad 6, \bar{6} \qquad (6.60)$$

Furthermore, the introduction of a center of symmetry generates a pattern of $2/m$ symmetry in the Patterson map, which corresponds to the Laue

TABLE 6.2. Vectors Generated by Two Independent Atoms and Their Symmetry Equivalents in Space Group Pm [a]

Atom pair	Analytical form of vector	Subtraction of coordinates		Reduced to one unit cell		Point in Figure 6.9d
		u	v	u	v	
(1), (3)	$\pm\{0, 2y_1, 0\}$	0	0.10	0	0.10	1
		0	−0.10	0	0.90	$\bar{1}$
(1), (2)	$\pm\{x_1 - x_2, y_1 - y_2,$	0.15	−0.15	0.15	0.85	2
	$z_1 - z_2\}$	−0.15	0.15	0.85	0.15	$\bar{2}$
(1), (4)	$\pm\{x_1 - x_2, y_1 + y_2,$	0.15	0.25	0.15	0.25	3
	$z_1 - z_2\}$	−0.15	−0.25	0.85	0.75	$\bar{3}$
(2), (3)	$\pm\{x_1 - x_2, -y_1 - y_2,$	0.15	−0.25	0.15	0.75	4
	$z_1 - z_2\}$	−0.15	0.25	0.85	0.25	$\bar{4}$
(3), (4)	$\pm\{x_1 - x_2, -y_1 + y_2,$	0.15	0.15	0.15	0.15	5
	$z_1 - z_2\}$	−0.15	−0.15	0.85	0.85	$\bar{5}$
(2), (4)	$\pm\{0, 2y_2, 0\}$	0	0.40	0	0.40	6
		0	−0.40	0	0.60	$\bar{6}$

[a] The coordinates of the four atoms in two sets of general positions are x, y, z; $x, \bar{y}, z$ with $x_1 = 0.20$, $y_1 = 0.05$, $x_2 = 0.05$, $y_2 = 0.20$, and $z_1 = z_2 = 0.00$.

symmetry of all monoclinic crystals. Evidently, the symmetry of the diffraction pattern is impressed onto the Patterson function by the use of $|F|^2$ coefficients in the Patterson–Fourier series. As a consequence, the Patterson synthesis is computed in the primitive space group corresponding to the Laue symmetry of a crystal, and this situation is similar for all space groups.

We can detect the presence of the twofold axis parallel to b in Figure 6.9d through vector peaks such as $5,\bar{2}$ and $3,\bar{4}$. Finally, the symmetry-related pairs of atoms in the crystal, 1,3 and 2,4, give rise to vectors along the line $[0, v, 0]$—the peaks $1,6,\bar{6},\bar{1}$ in Patterson space. The presence of a large number of peaks along an axis in a three-dimensional Patterson map may be used as evidence for a mirror plane perpendicular to that axis in the crystal. This feature is important because an m plane does not give rise to systematic absences in the diffraction pattern (pages 81 and 190). The existence of peaks, arising from symmetry-related atoms, in certain regions of Patterson space was noted first by Harker in 1936. The line $[0, v, 0]$ for Pm is called a

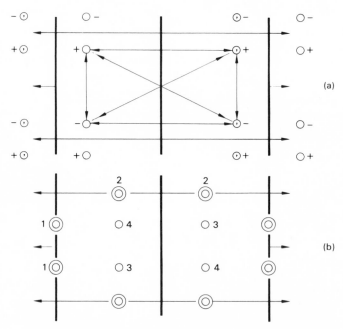

FIGURE 6.11. (a) Vectors between symmetry-related atoms in general equivalent positions in space group $P2/m$. Coordinates like $\bar{x}$ have been treated as $(1 - x)$ in drawing the vectors. (b) One unit cell of the twofold axes intersect the m planes in centers of symmetry. Note the single-weight and double-weight peaks, and their relation to the space-group symmetry.

Harker line; *planes* containing peaks arising from pairs of symmetry-related atoms are called Harker sections. We shall consider some examples below.

Vector Interactions in Other Space Groups

We shall consider atoms in general positions in a number of space groups which should be now familiar.

Space Group P$\bar{1}$.
General positions: x, y, z; $\bar{x}, \bar{y}, \bar{z}$.
Vectors: $\pm\{2x, 2y, 2z\}$.
Harker peaks lie in general positions in Patterson space.

Space Group P2.
General positions: x, y, z; $\bar{x}, y, \bar{z}$.
Vectors: $\pm\{2x, 0, 2z\}$.
Harker section: $(u, 0, w)$.

It may be noted that for complex structures, not all of the peaks on Harker sections are necessarily true Harker peaks. If in this structure there are two atoms not related by symmetry, which, by chance, have the same or nearly the same y coordinates, the vector between them will produce a peak on the Harker section.

Space Group P2/m.
General positions: x, y, z; $\bar{x}, \bar{y}, \bar{z}$; $x, \bar{y}, z$; $\bar{x}, y, \bar{z}$.
Vectors: $\pm\{2x, 0, 2z\}$ double weight   type 1
  $\pm\{0, 2y, 0\}$ double weight   type 2
  $\pm\{2x, 2y, 2z\}$ single weight   type 3
  $\pm\{2x, 2\bar{y}, 2z\}$ single weight   type 4.
Harker section: $(u, 0, w)$.
Harker line: $[0, v, 0]$.

Vector type 1 arises in two ways, once from the pair x, y, z; $\bar{x}, y, \bar{z}$ and once from the pair $x, \bar{y}, z$; $\bar{x}, \bar{y}, \bar{z}$. These two interactions give rise to identical vectors, which therefore superimpose in Patterson space and form a double-weight peak. Similar comments apply to type 2, but the centrosymmetrically related atoms give rise to single-weight peaks, types 3 and 4. Figure 6.11 illustrates these vectors, as seen along the z axis. The reader may now consider how the Patterson function might be used to differentiate among space groups *P2*, *Pm*, and *P2/m*; a clue has already been given on page 240. Statistical methods, outlined briefly in Chapter 7, are often employed to verify the results obtained from a study of the vector distribution.

6.4.3 Examples of the Use of the Patterson Function in Solving the Phase Problem

In this section, we shall consider how the Patterson function was used in the solution of three quite different structures.

Bisdiphenylmethyldiselenide, $(C_6H_5)_2CHSe_2CH(C_6H_5)_2$

Crystals of this compound form yellow needles, with straight extinction under crossed Polaroids for all directions parallel to the needle axis, and oblique extinction on the section normal to the needle axis. Photographs taken with the crystal oscillating about its needle axis show only a horizontal m line, while zero- and upper-layer Weissenberg photographs show only symmetry 2. The crystals are therefore monoclinic, with b along the needle direction.

Crystal Data
System: monoclinic.
Unit-cell dimensions: $a = 18.72, b = 5.773, c = 12.594$ Å, $\beta = 125.47°$.
V_c: 1107.1 Å^3.
D_m: 1.49 g cm^{-3}.
M: 492.4.
Z: 2.02, or 2 to the nearest integer.
Unit-cell contents: 4 Se, 52 C, and 44 H atoms.
Absent spectra: hkl: $h + k = 2n$.
Possible space groups: $C2$, Cm, $C2/m$.

Symmetry Analysis. Where the space group is not determined uniquely by the X-ray diffraction pattern, it may be possible to eliminate certain alternatives at the outset of the structure determination by other means. Space groups $C2$ and Cm each require four general positions:

$$C2: \quad (0, 0, 0; \; \tfrac{1}{2}, \tfrac{1}{2}, 0) + \{x, y, z; \; \bar{x}, y, \bar{z}\}$$

$$Cm: \quad (0, 0, 0; \; \tfrac{1}{2}, \tfrac{1}{2}, 0) + \{x, y, z; \; x, \bar{y}, z\}$$

Since Z is 2, the molecular symmetry is either 2, in $C2$, or m, in Cm. In both $C2$ and Cm, all atoms could satisfy general position requirements, and neither arrangement would be stereochemically unreasonable.

Space group $C2/m$ requires eight general equivalent positions per unit cell. Only special position sets, such as 0, 0, 0 and $\tfrac{1}{2}$, $\tfrac{1}{2}$, 0 correspond with $Z = 2$. These positions have symmetry $2/m$, but it is not possible to construct the molecule in this symmetry without contradicting known chemical facts. Consequently, we shall regard this space group as highly improbable for the compound under investigation.

Patterson Studies. Whatever the answer to the questions remaining from this symmetry analysis, we expect, from chemical knowledge,* that the two selenium atoms will be covalently bonded at a distance of about 2.3 Å. This Se—Se interaction will produce a strong peak in the Patterson function at about 2.3 Å from the origin.

The atomic numbers of Se, C, and H are 34, 6, and 1, respectively. Hence, the important vectors in the Patterson function would have single-weight peak heights, from (6.58), as follows:

(a)   Se—Se: 1156.
(b)   Se—C: 204.
(c)   C—C: 36.

Because of the presence of identical vectors arising from the C unit cell, all vectors will be double these values.

Figure 6.12 is the Patterson section $P(u, 0, w)$, calculated with 1053 data of $|F_o(hkl)|^2$, with grid intervals of 50ths along u, v, and w. The origin peak $P(0, 0, 0)$ was scaled to 100 and, from (6.57), $\sum_{j=1}^{N} Z_j^2 = 6540$. Hence, the vector interactions (a), (b), and (c) should have the approximate peak heights of 35, 6, and 1, respectively.

The section is dominated by a large peak of height 39 at a distance of about 2.3 Å from the origin. Making the reasonable assumption that it represents the Se—Se vector, and since there are no significant peaks on the v axis, it follows that the space group cannot be Cm, thus leaving $C2$ as the most logical choice.

By measurement on the section, the Patterson coordinates are $u = 6.7/50$ and $w = 2.2/50$, and from the study of space group $P2$ (pages 88 and 241), it follows that $x_{Se} + 0.067$ and $z_{Se} = 0.022$.

In space group $C2$, the unit cell origin is fixed in the xz plane by the twofold axis. There is no symmetry element that defines the origin in the y direction, which must be fixed by specifying the y coordinate for a selected atom. For convenience, we may set $y_{Se} = 0$, and our analysis so far may be given as the positions

$$\text{Se:} \quad 0.067, 0, \quad 0.022$$

$$\text{Se}': \quad -0.067, 0, -0.022$$

* See Bibliography.

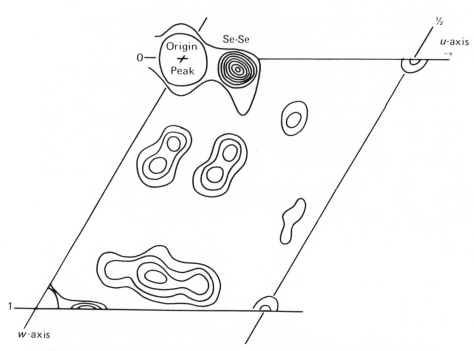

FIGURE 6.12. Patterson section, $P(u, 0, w)$; the origin peak (height = 100) has not been contoured. Contours around the Se—Se peaks are at intervals of 4; elsewhere at intervals of 2.

A space-group ambiguity is not always resolved in this manner. Sometimes it is necessary to proceed further with the structure analysis before confirmation is obtained.

What of the atoms other than selenium? Is it possible to determine the positions of the carbon and hydrogen atoms? We shall find that we can locate the carbon atoms in this structure from the Patterson synthesis. To explain the procedure, we consider first only part of the structure, including one phenyl ring of the asymmetric unit (Figure 6.13a), and neglect all but the C—Se vectors. The vector set generated by the two Se atoms and 14 C atoms in this hypothetical arrangement contains two images of the structure fragment (one per Se atom), which are displaced from each other by the Se—Se vector. The idealized vector set is shown in Figure 6.13b. By shifting one of these images by a *reverse** Se—Se vector displacement, it is possible

---

* If the forward direction of this vector is used, the structure obtained would, in general, be inverted through the origin. This does not happen with the example under study because the molecule possesses twofold symmetry.

to bring the two images into coincidence. Verify this statement by making a transparent copy of Figure 6.13b and placing its origin over an Se—Se vector position in the same figure, keeping the pairs of u and w axes parallel. Certain peaks overlap, producing a single, displaced image of the structure. Shade the peaks that overlap. This image is displaced with respect to the true space-group origin, which we know to be midway between the two Se atoms. A correctly placed image of the structure can be recovered by inserting the true origin position onto the tracing and neglecting any peaks that are not shaded.

The partial vector set was formed from the image of all atoms of the fragment in each Se atom; each image is weighted by Z_j, the atomic number of the jth atom (carbon, in this example) imaged in Se. The displacement arises because the Patterson synthesis transfers all vectors to a common origin.

Patterson Superposition. The technique just described depends upon the recognition of the vector interaction from a given pair of atoms, the two Se atoms in this example. At least a partial unscrambling of the structure images in the Patterson function was effected by correctly displacing two copies of the Patterson map and noting the positions of overlap.

To illustrate the method further and to derive a systematic procedure for its implementation, we return to the Patterson section in Figure 6.12. The two Se atoms have the same y coordinate, which means that the vector shift takes place in this section. Now make two copies on tracing paper of the half unit-cell outline, $x = 0$ to $\frac{1}{2}$ and $z = 0$ to 1, and label them copy 1 and copy 2.

On copy 1 mark in the position S of the point, $-(x_{Se}, z_{Se})$, which is at -0.067, -0.022, and on copy 2 mark in the position S' of the point $-(x_{Se'}, z_{Se'})$, which is at $0.067, 0.022$. Think of these two unit cells as existing in crystal space, not Patterson space. Place copy 1 over the Patterson $(u, 0, w)$ section, maintaining a parallel orientation, with S over the origin, and trace out the Patterson map (Figure 6.14a). Repeat this procedure with copy 2, placing S' over the Patterson section origin (Figure 6.14b).

Finally, superimpose copy 1 and copy 2. As in the exercise with Figures 6.13a and 6.13b, some peaks overlap and some lie over blank regions in one or the other map. The overlaps correspond to regions of high electron density in the crystal. They are best mapped out by compiling a new diagram which contains the *minimum* value of the vector density between copy 1 and copy 2 for each point, thus eliminating or decreasing in height those regions where one copy has no or only slight overlap. A map prepared in this way is shown in Figure 6.14c.

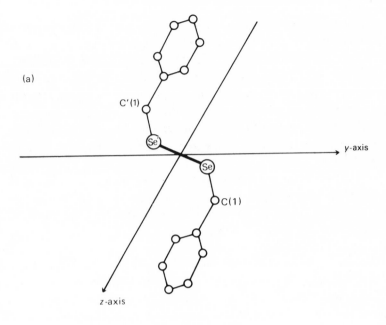

(a)

C'(1)

Se'

y-axis

Se

C(1)

z-axis

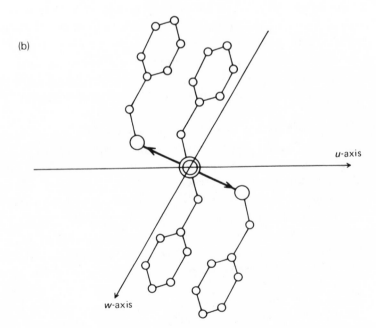

(b)

u-axis

w-axis

FIGURE 6.13. (a) Hypothetical structure fragment $C_6H_5CHSe_2CHC_6H_5$;
(b) idealized set of Se—Se and Se—C vectors.

Minimum Function

The technique outlined above follows the method of Buerger.* An analytical expression for the minimum function $M_n(x, y, z)$ is given by (6.61); it may be regarded as an approximation to the electron density $\rho(x, y, z)$.

$$M_n(x, y, z) = \text{Min}[P(u - x_1, v - y_1, w - z_1), P(u - x_2, v - y_2, w - z_2),$$
$$\ldots, P(u - x_n, v - y_n, w - z_n)] \qquad (6.61)$$

where $\text{Min}(P_1, P_2, \ldots, P_n)$ is the lowest value at the point x, y, z in the set of superpositions $P_1, P_2, \ldots, P_n$; n corresponds with the number of known or trial atomic positions. The following general comments on the application of the minimum function procedure should be noted:

(a) The n trial atoms should form within themselves a set or sets of points related by the appropriate space-group symmetry.

(b) In a noncentrosymmetric space group, n should be three or more if it is necessary to remove the Patterson center of symmetry.

(c) If the various n trial atoms have different atomic numbers, the corresponding Patterson copies should be weighted accordingly in order to even out the different image strengths.

(d) Incorrectly placed atoms in the trial set tend to confuse the structure image. New atom sites therefore should be added to the model with caution.

Figure 6.15 shows a composite electron density map of the atoms in the asymmetric unit that were revealed by a three-dimensional minimum function M_2. This result is quite satisfactory; only C(9), C(10), and C(11) are not yet located. The composite map of the complete structure† and the packing of the molecules in the unit cell are shown in Figure 6.16. In favorable circumstances, the Patterson function can be solved for the majority of the heavier atoms in the crystal structure. The atoms not located by M_2 in this example were obtained from an electron density map phased on those atoms that were found, a standard method for attempting to complete a partial structure (see page 260).

* See Bibliography.
† H. T. Palmer and R. A. Palmer, *Acta Crystallographica*, **B25**, 1090 (1969). The $|F_o|$ data for this compound may be obtained from one of the authors (R.A.P.).

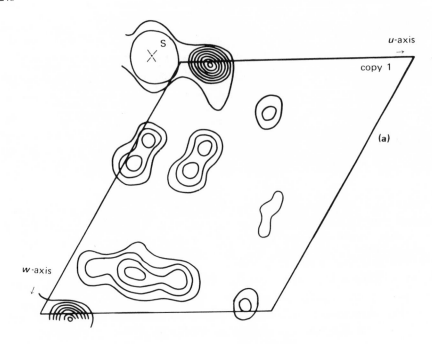

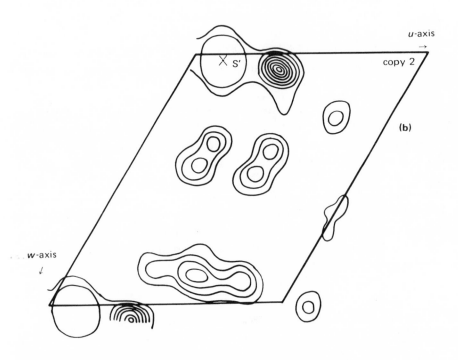

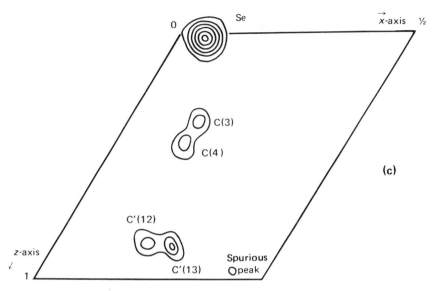

FIGURE 6.14. (a, b) Shifted copies 1 and 2 prepared from the $(u, 0, w)$ section; (c) minimum-function M_2 section at $y = 0.0$; $C'(12)$ and $C'(13)$ are symmetry-related to $C(12)$ and $C(13)$ in Figure 6.15.

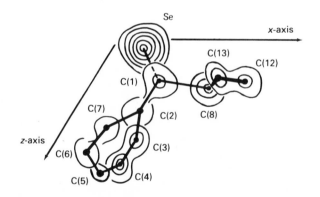

FIGURE 6.15. Composite map of the three-dimensional minimum function $M_2(x, y, z)$.

Determination of the Chlorine Atom Positions in Papaverine Hydrochloride

The crystal data for this compound have been given in Chapter 5 (page 196). The calculated origin peak height is approximately 4700, and a single-weight Cl—Cl vector would have a height of about 6% of that of the origin peak. The Cl—Cl vector may not be located as easily as that of Se—Se in the previous example. The general equivalent positions in $P2_1/c$ (see pages 95–96 and problems 2P.7 and 2P.8) give rise to the vectors shown in Table 6.3. The assignment of coordinates to the chlorine atoms follows the recognition of peaks A, B, and C as Cl—Cl vectors on the Patterson maps (Figures 6.17a–c). Figure 6.17d is the sharpened section, $(u, \frac{1}{2}, w)$. The steps in the solution of the problem are set out in Table 6.4.

The results are completely self-consistent, and we may list the Cl coordinates in the unit cell:

4 Cl: 0.025,   0.169,   0.038;   0.025,   0.331,   0.538

         −0.025, −0.169, −0.038;   −0.025, −0.331, −0.538

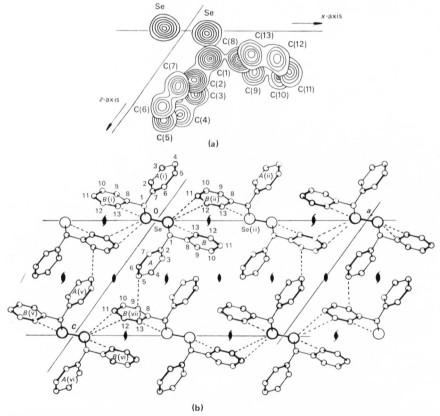

FIGURE 6.16. (a) Composite electron density map as seen along b; (b) crystal structure as seen along b; the dashed lines indicate the closest intermolecular contacts.

TABLE 6.3

Label	Vector	Peak strength	Harker region
A	$\pm\{0, \frac{1}{2}+2y, \frac{1}{2}\}$	Double weight	Line: $[0, v, \frac{1}{2}]$
B	$\pm\{2x, \frac{1}{2}, \frac{1}{2}+2z\}$	Double weight	Section: $(u, \frac{1}{2}, w)$
C	$\pm\{2x, 2y, 2z\}$	Single weight	General region
D	$\pm\{2x, 2\bar{y}, 2z\}$	Single weight	General region

TABLE 6.4

Patterson map	Label	Vector coordinates[a]	Cl coordinates
Figure 6.17a, level $v = 8.4/52$	A	$\frac{1}{2} - 2y = 8.4/52$	$y = 0.169$
Figure 6.17b, level $v = \frac{1}{2}$	B	$2x = 2.2/44$, $\frac{1}{2} + 2z = 17.3/30$	$x = 0.025$ $z = 0.038$
Figure 6.17c, level $v = 17.6/52$	C	$2x = 2.2/44$ $2y = 17.6/52$ $2z = 2.3/30$	$x = 0.025$ $y = 0.169$ $z = 0.038$

[a] The Patterson synthesis was computed with the intervals of subdivision 44, 52, and 30 along u, v, and w, respectively.

For simplicity, peak A was assigned as $-(\frac{1}{2} + 2y)$, which is crystallographically the same as $\frac{1}{2} - 2y$, in order to obtain $y \leq \frac{1}{2}$. For a similar reason, B was retained as $\frac{1}{2} + 2z$.

The specification of the peak parameters in this manner is, to some extent, dependent on the observer. A different choice, for example, $\frac{1}{2} + 2y$ in A, merely results in a set of atomic positions located with respect to one of the other centers of symmetry as origin. In space groups where the origin might be defined with respect to other symmetry elements, similar arbitrary peak specification may be possible.

Determination of the Mercury Atom Positions in KHg$_2$

This example illustrates the application of the Patterson function to the determination of the coordinates of atoms in special positions of space group *Imma*.

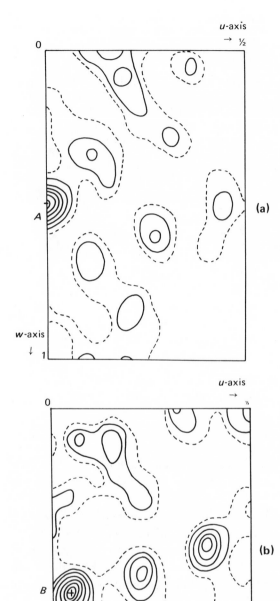

FIGURE 6.17. Three-dimensional Patterson sections for papaverine hydrochloride; the Cl—Cl vectors are labeled A, B, and C: (a) $v = 8.4/52$, (b) $v = \frac{1}{2}$, (c) $v = 17.6/52$, (d) $v = \frac{1}{2}$ (sharpened section).

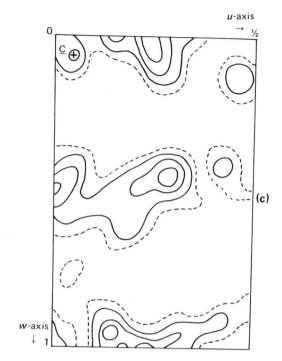

(c)

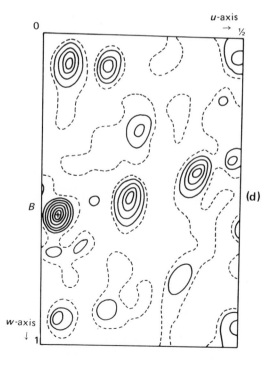

(d)

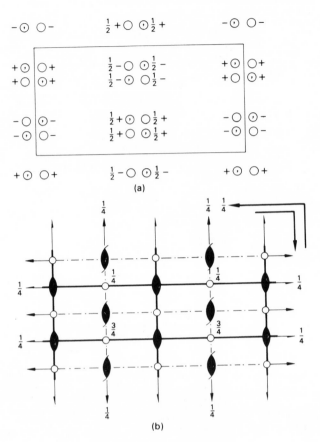

FIGURE 6.18. Space group *Imma*.

*Crystal Data**
System: orthorhombic.
Unit-cell dimensions: $a = 8.10$, $b = 5.16$, $c = 8.77$ Å.
V_c: 366.6 Å^3.
D_m: 7.95 g cm^{-3}.
M: 440.3.
Z: 3.99, or 4 to the nearest integer.
Unit-cell contents: 4 K and 8 Hg atoms.
Absent spectra: hkl: $h + k + l = 2n + 1$
$hk0$: $h = 2n + 1$, $(k = 2n + 1)$.

* E. J. Duwell and N. C. Baenziger, *Acta Crystallographica* **8**, 705 (1955).

TABLE 6.5. Special Positions in *Imma*

4	(a)	$2/m$	$0,0,0;$ $\frac{1}{2},0,0;$ $\frac{1}{2},\frac{1}{2},\frac{1}{2};$ $0,\frac{1}{2},\frac{1}{2}$
4	(b)	$2/m$	$0,\frac{1}{2},0;$ $\frac{1}{2},\frac{1}{2},0;$ $\frac{1}{2},0,\frac{1}{2};$ $0,0,\frac{1}{2}$
4	(c)	$2/m$	$\frac{1}{4},\frac{1}{4},\frac{1}{4};$ $\frac{1}{4},\frac{3}{4},\frac{1}{4};$ $\frac{3}{4},\frac{3}{4},\frac{3}{4};$ $\frac{3}{4},\frac{1}{4},\frac{3}{4}$
4	(d)	$2/m$	$\frac{1}{4},\frac{3}{4},\frac{3}{4};$ $\frac{1}{4},\frac{1}{4},\frac{3}{4};$ $\frac{3}{4},\frac{1}{4},\frac{1}{4};$ $\frac{3}{4},\frac{3}{4},\frac{1}{4}$
4	(e)	$mm2$	$\frac{1}{4},0,z;$ $\frac{3}{4},0,\bar{z};$ $\frac{3}{4},\frac{1}{2},\frac{1}{2}+z;$ $\frac{1}{4},\frac{1}{2},\frac{1}{2}-z$
8	(f)	2	$\pm\{0,y,0;$ $\frac{1}{2},y,0;$ $\frac{1}{2},\frac{1}{2}+y,\frac{1}{2};$ $0,\frac{1}{2}+y,\frac{1}{2}\}$
8	(g)	2	$\pm\{x,\frac{1}{4},\frac{1}{4};$ $x,\frac{3}{4},\frac{1}{4};$ $\frac{1}{2}+x,\frac{3}{4},\frac{3}{4};$ $\frac{1}{2}+x,\frac{1}{4},\frac{3}{4}\}$
8	(h)	m	$\pm\{x,0,z;$ $\frac{1}{2}-x,0,z;$ $\frac{1}{2}+x,\frac{1}{2},\frac{1}{2}+z;$ $\bar{x},\frac{1}{2},\frac{1}{2}+z\}$
8	(i)	m	$\pm\{\frac{1}{4},y,z;$ $\frac{1}{4},\bar{y},z;$ $\frac{3}{4},\frac{1}{2}+y,\frac{1}{2}+z;$ $\frac{3}{4},\frac{1}{2}-y,\frac{1}{2}+z\}$

Possible space groups: *Im2a*, *I2ma*, or *Imma*.

In the absence of further information on the space group, we shall proceed with the analysis in *Imma* (Figures 6.18a and 6.18b). The reader may like to consider how easily these diagrams may be derived from *Pmma* (origin on $\bar{1}$) $+I$.

Symmetry and Packing Analyses. Since Z is 4 and there are 16 general equivalent positions in *Imma*, all atoms must lie in special positions. Table 6.5 lists these positions for this space group, with a center of symmetry (equivalent to $2/m$) as origin.*

This list presents a quite formidable number of alternatives for examination. The eight Hg atoms could lie in (f), (g), (h), or (i). However, further consideration of sets (f), (g), and (i) and sets (c) and (d) shows that they would all involve pairs of Hg atoms being separated by distances less than $b/2$ (2.58 Å). This value is much shorter than known Hg—Hg bond distances in other structures, and we shall reject these sets. (The positions in these sets may be plotted to scale in order to verify the spatial limitations.)

Of the remaining sets, (a) and (b) together would again place neighboring Hg atoms too close to one another. There are three likely models:

Model I: four Hg in (a) + four Hg in (e).
Model II: four Hg in (b) + four Hg in (e).
Model III: eight Hg in (h).

The Patterson function enables us to differentiate among these alternative models.

Vector Analysis of the Alternative Hg Positions. Model I would produce, among others, an Hg—Hg vector at $u=\frac{1}{2}$, $w=0$, from the atoms in

---

* In the original paper, the origin in *Imma* was chosen on a center of symmetry displaced by $\frac{1}{4}, \frac{1}{4}, \frac{1}{4}$ from this origin.

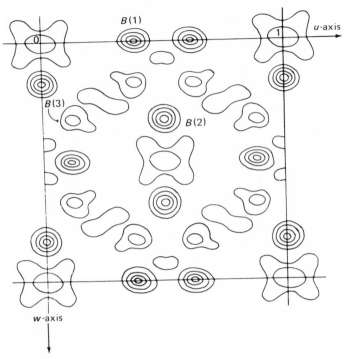

FIGURE 6.19. Patterson projection $P(u, w)$ for KHg_2; the origin peak has not been contoured.

set (a). The b axis Patterson projection (Figure 6.19) shows no peak at that position, and we eliminate model I. For a similar reason, with the atoms of set (b), model II is rejected. It is necessary to show next that model III is consistent with the Patterson function. The a-axis projection is shown in Figure 6.20.

Interpretation of $P(u, w)$. In this projection, no reference is made to the y coordinates, and we look for vectors of the type $\pm\{\frac{1}{2}+2x, 0\}$ and $\pm\{\frac{1}{2}, 2z\}$, and four vectors related by $2mm$ symmetry $\pm\{2x, 2z\}$ and $\pm\{2\bar{x}, 2z\}$. The double-weight peak labeled $B(1)$ is on the line $w = 0$, and $B(2)$ is on the line $u = 1/2$. Hence, $x_{Hg} = 0.064$ and $z_{Hg} = 0.161$. These values are corroborated by measurements from the single-weight peak $B(3)$.

Interpretation of $P(v, w)$. Vectors like A (Figure 6.20) are of the type $\pm\{0, 2z\}$. We deduce $z_{Hg} = 0.161$, in excellent agreement with the value obtained from the b-axis projection.

Superposition techniques applied to the a-axis projection indicate that

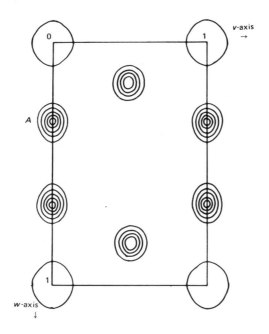

FIGURE 6.20. Patterson projection $P(v, w)$ for KHg_2; the
origin peak has not been contoured.

the K atoms are in special positions (b), but this result is not supported by the
b-axis projection. Evidently, the Patterson results can give only a partial
structure, and supplementary methods are needed to carry the analysis to
completion. In summary, we have determined the positions of the mercury
atoms to be in set (h),[*] Table 6.5, with $x = 0.064$, $z = 0.161$.

6.4.4 Absolute Scale of $|F_o|$ and Overall Temperature Factor

In any structure analysis it soon becomes necessary to calculate structure factors for a trial structure for comparison with the experimental data
To do this successfully we need a temperature factor for the atomic
scattering factors and a scale factor for $|F_o|$. They can be derived approxi-
mately by Wilson's method.[†] He showed that for a unit cell containing a

[*] In the work of Duwell and Baenziger (*loc. cit.*), the positions listed are 8 (i), with $x = 0.186$ and
$z = 0.089$, each being $-1/4$ *plus* the value given here (see footnote to page 255).
[†] A. J. C. Wilson, *Nature* **150**, 152 (1942).

random distribution of N atoms, the local average value of $|F(hkl)|^2$ is given by*

$$\overline{|F(hkl)|^2} \approx \sum_{j=1}^{N} g_j^2(hkl) \qquad (6.62)$$

This approximation has been found to hold satisfactorily for a wide range of structures, provided that the values of $|F|^2$ are averaged over small ranges (local) of $(\sin\theta)/\lambda$, so that f is not varying rapidly within any range.

We assume a single isotropic temperature factor B for all atoms and a scale factor K for the experimental $|F_o|$ data [see (4.44) and (4.57)]. Hence, from (6.62)

$$K^2\overline{|F_o(hkl)|^2} = \exp[-2B(\sin^2\theta_r)/\lambda^2] \sum_{j=1}^{N} f_{j,\theta_r}^2 \qquad (6.63)$$

where θ_r is a representative value of each range of $(\sin\theta)/\lambda$. Rearranging, and taking $\log_e$ we obtain,

$$\log_e q_r = 2\log_e K + 2B(\sin^2\theta_r)/\lambda^2 \qquad (6.64)$$

where q_r is given by

$$q_r = \left(\sum_{j=1}^{N} f_{j,\theta_r}^2 \right) \Big/ \overline{|F_o(hkl)|_{\theta_r}^2} \qquad (6.65)$$

If $\log_e q_r$ is plotted against $(\sin^2\theta)/\lambda^2$ and the best straight line drawn, the slope is equal to $2B$ and the intercept on the ordinate is equal to $2\log_e K$.

This graph is often called a Wilson plot, and may be carried out in the following way.

(a) Three-dimensional reciprocal space is divided into a number r spherical shells of approximately equal volume, with 80–100 reflections per shell; the shells decrease in thickness with increasing values of $(\sin\theta)/\lambda$.

(b) The average value of $|F_o(hkl)|^2$ is calculated over the available reciprocal space by expanding the data to include the symmetry-related reflections with their correct multiplicities. It is necessary to allocate values to "unobserved" reflections. Wilson has shown that $0.5|F_m|$ for centrosymmetric crystals and $0.67|F_m|$ for noncentrosymmetric crystals are the most probable values; $|F_m|$ is the minimum value of $|F_o|$ in the immediate locality of the shell under consideration. Systematically absent reflections

---

* See also the discussion of the ε-factor on pages 296 and 461.

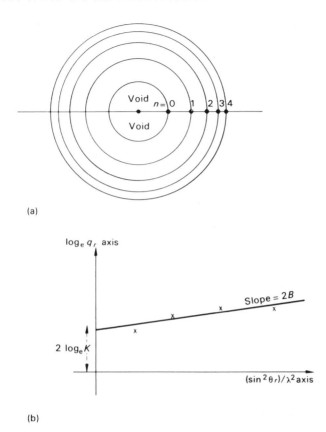

FIGURE 6.21. Scale and temperature factor: (a) division of reciprocal space into spherical shells, (b) Wilson plot.

and those in a region up to about the second-order reflection on each axis are ignored in taking the averages, as they are atypical of the general distribution of intensities, leaving r shells for (6.66).

(c) The mean values of $(\sin^2\theta_r)/\lambda^2$ are given, without sensible error, by

$$(\sin^2\theta_r)/\lambda^2 = (\sin^2\theta_n + \sin^2\theta_{n+1})/2\lambda^2 \qquad (6.66)$$

where $n+1$ is the number of the outer boundary defining the rth shell, starting at $n=0$, where $\sin\theta = \sin\theta_{min}$ (Figure 6.21a).

(d) $\sum f_{j,\theta_r}^2$ is calculated for each shell, using tabulated f data.

(e) The Wilson plot is drawn and B and K determined (Figure 6.21b).

6.4.5   Heavy-Atom Method and Partial Fourier Synthesis

The heavy-atom method was conceived originally as a method for determining the positions of light atoms in a structure containing a relatively small number of heavier atoms. However, the technique can be applied to most situations where a partial structure analysis has been effected, provided that certain conditions are met.

Imagine a situation where N_k of the N atoms in a unit cell have been located; N_k may be only one atom, if it is a heavy atom. There will be N_u atoms remaining to be located, and we may express the structure factor (4.47) in terms of known and unknown atoms:

$$\mathbf{F}(hkl) = \sum_{j=1}^{N_k} g_j \exp[i2\pi(hx_j + ky_j + lz_j)]$$

$$+ \sum_{u=1}^{N_u} g_u \exp[i2\pi(hx_u + ky_u + lz_u)] \qquad (6.67)$$

or

$$\mathbf{F}(hkl) = \mathbf{F}_c(hkl) + \mathbf{F}_u(hkl) \qquad (6.68)$$

In practice, $|F_o|$ data, appropriately scaled, replace $|F(hkl)|$, and $\mathbf{F}_c(hkl)$ refers to the known (N_k) atomic positions. As more of the structure becomes known, the values of $|F_c(hkl)|$ approach $|F_o(hkl)|$ and the phase angle ϕ_c approaches the unobservable but required value $\phi(hkl)$. Figure 6.22 illustrates this argument for any given reflection. The values of ϕ_c may provide

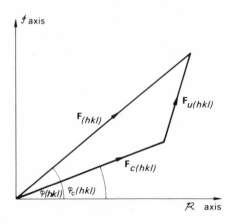

FIGURE 6.22. Partial-structure phasing; $\mathbf{F}(hkl)$ is the true structure factor of modulus $|F_o(hkl)|$ and phase $\phi(hkl)$.

sufficiently reasonable approximations to $\phi(hkl)$ for an electron density map to be calculated with some confidence. The nearer $\mathbf{F}_c$ is to $\mathbf{F}(hkl)$, the better the value of the phase angle, and this is clearly dependent upon the percentage of the scattering power which is known. As a guide to the effective phasing power of a partial structure, the quantity r is calculated:

$$r = \sum_{j=1}^{N_k} Z_j^2 \Big/ \sum_{u=1}^{N_u} Z_u^2 \qquad (6.69)$$

where Z refers to the atomic number of a species. A value of r near unity is considered to provide a useful basis for application of the heavy-atom method. However, values of r quite different from unity have produced successful results, because for a given reflection the important quantity is really r', given by

$$r' = \sum_{j=1}^{N_k} g_j^2 \Big/ \sum_{u=1}^{N_u} g_u^2 \qquad (6.70)$$

If r is large, however, the heavy-atom contributions tend to swamp those from the lighter atoms, which may then not be located very easily from electron density maps. On the other hand, if r is small, the calculated phase will deviate widely from the desired value, and the resulting electron density map may be very difficult to interpret. These extreme situations are found in bisdiphenylmethyldiselenide ($r = 2.4$) and papaverine hydrochloride ($r = 0.28$), based, in each case, on the heavy atoms alone in N_k.

The underlying philosophy of the heavy-atom method depends on the acceptance of calculated phases, even if they contain errors, for the computation of the electron density synthesis. Large phase errors give rise to high background features, which mask the image of the correct structure. The calculated phases ϕ_c contain errors arising from inadequacies in the model, but the $|F_o|$ data, although subject to experimental errors, hold information on the complete structure. Phase errors may be counteracted to some extent by weighting the Fourier coefficients according to the degree of confidence in a particular phase. For centrosymmetric structures, the weight $w(hkl)$ by which $|F_o(hkl)|$ is multiplied is given by*

$$w(hkl) = \tanh(\chi/2) \qquad (6.71)$$

* M. M. Woolfson, *Acta Crystallographica* **9**, 804 (1956).

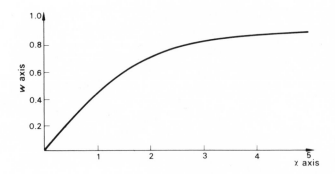

FIGURE 6.23. Weighting factor $w(hkl)$ as a function of χ in noncentrosymmetric crystals.

where χ is given by

$$\chi = 2|F_o|\,|F_c|/\sum g_u^2 \qquad (6.72)$$

The subscripts c and u refer, respectively, to the known and unknown parts of the structure. In noncentrosymmetric structures, $w(hkl)$ can be obtained from the graph in Figure 6.23.* Weighting factors should be applied to $|F_o|$ values that have been placed on an absolute, or approximately absolute, scale.

Bearing these points in mind, it follows that the best electron density map one can calculate with phases determined from a partial structure is given by

$$\rho(x, y, z) = \frac{2}{V_c} \sum_{h=0} \sum_{k-\infty}^{\infty} \sum_l w(hkl)|F_o(hkl)| \cos[2\pi(hx + ky + lz) - \phi_c(hkl)]$$

$$(6.73)$$

where

$$\phi_c(hkl) = \tan^{-1}[B_c'(hkl)/A_c'(hkl)] \qquad (6.74)$$

$A_c'(hkl)$ and $B_c'(hkl)$ are the real and imaginary components, respectively, of the calculated structure factor, which is included in the right-hand side of (6.68).

* G. A. Sim, *Acta Crystallographica* **13**, 511 (1960).

Electron density maps calculated from partial-structure phasing contain features which characterize both the true structure and the partial, or trial, structure. This fact may be illustrated by writing the x-, y-, and z-independent terms from the electron density synthesis (6.37) as $(w|F_o|/|F_c|)|F_c|\exp[i\phi_c(hkl)]$. The terms $|F_c|\exp[i\phi_c(hkl)]$ alone would synthesize the partial structure. This synthesis is modified, in practice, by the term $w|F_o|/|F_c|$; evidently, the electron density synthesis is biased toward the trial structure.

If the model includes atoms in reasonably accurate positions, we can expect two important features in the electron density map: (a) Atoms of the trial structure should appear, possibly in corrected positions, and (b) additional atoms should be revealed by the presence of peaks in stereochemically sensible positions.

If neither of these features is observed in the electron density synthesis, it may be concluded that the trial structure contains very serious errors. Correspondingly, there would be poor agreement in the pattern of relationship between $|F_o|$ and $|F_c|$.

Pseudosymmetry in Electron Density Maps

The electron density map calculated with phases derived from the heavy-atom positions may not exhibit the true space-group symmetry, but rather that of a space group of higher symmetry. As an example, consider a structure having space group $P2_1$ with one heavy atom per asymmetric unit. The origin is defined with respect to the x and z axes by the 2_1 axis along $[0, y, 0]$, but the y coordinate of the origin is determined with respect to an arbitrarily assigned y coordinate for *one* of the atoms. Consider the heavy atoms at x, y, z and, symmetry-related, at $\bar{x}, \frac{1}{2}+y, \bar{z}$. This arrangement has the symmetry of $P2_1/m$, with the m planes cutting the y axis at whatever y coordinate is chosen for the heavy atom, say y_H, and at $\frac{1}{2}+y_H$. If $y_H = \frac{1}{4}$, a center of symmetry is at the origin, and the calculated phases will be 0 or π. This situation is illustrated in Figure 6.24, which indicates that an unscrambling of the images must be carried out. If the heavy atom is given any general value for y_H, $B'(hkl)$ will not be zero and the phase angles will not be 0 or π, but the pseudosymmetry will still exist.

Successive Fourier Refinement

A single application of the method described above does not usually produce a complete set of atomic coordinates. It should lead to the inclusion

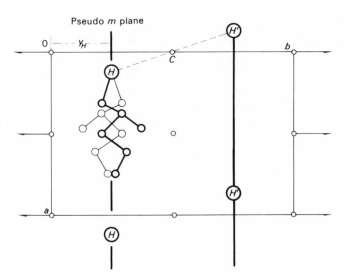

FIGURE 6.24. Introduction of pseudosymmetry into space group $P2_1$ by single heavy-atom phasing. H is the heavy atom and C is a center of symmetry introduced between H and its $P2_1$ equivalent H'. The space group (for the heavy atoms alone) thus appears as $P2_1/m$ with mirror (and $\bar{1}$) pseudosymmetry. The electron density map, phased on the H and H' species, will contain contours for two mirror-related images in the asymmetric unit, with a certain degree of confusion between them.

of more atoms into subsequent structure factor calculations and so to a better electron density map, and so on. This iterative process of Fourier refinement should, after several cycles, result in the identification of all nonhydrogen atoms in the structure to within about 0.1 Å of their true positions. Further improvement of the structure would normally be carried out by the method of least squares, which is described in Chapter 7.

6.4.6 Difference-Fourier Synthesis

Some errors present in the trial structure may not be revealed by Fourier synthesis. In particular, the following situations are important.

(i) Atoms in completely wrong positions tend to be returned by the Fourier process with the same fractional coordinates, but sometimes with a comparatively low electron density.

(ii) Correctly placed atoms may have been assigned either the wrong atomic number, for example, C for N, or an incorrectly estimated temperature factor.

(iii) Small corrections to the fractional coordinates may be difficult to assess from the Fourier map.

In these circumstances, a difference-Fourier synthesis is valuable. We shall symbolize the Fourier series with $|F_o|$ coefficients as $\rho_o(x, y, z)$ and the corresponding synthesis with $|F_c|$ instead as $\rho_c(x, y, z)$; the difference-Fourier synthesis $\Delta\rho(x, y, z)$ may be obtained in a single-stage calculation from the equation

$$\Delta\rho(x, y, z) = \frac{2}{V_c} \sum_h \sum_k \sum_l (|F_o| - |F_c|) \cos[2\pi(hx + ky + lz) - \phi_c] \quad (6.75)$$

Since the phases are substantially correct at this stage, it is, in effect, a subtraction, point by point, of the "calculated," or trial, Fourier synthesis from that of the "observed," or experimentally based, synthesis. The difference synthesis has the following useful properties.

(a) Incorrectly placed atoms correspond to regions of high electron density in $\rho_c(x, y, z)$ and low density in $\rho_o(x, y, z)$; $\Delta\rho(x, y, z)$ is therefore negative in these regions.

(b) A correctly placed atom with either too small an atomic number or too high a temperature factor shows up as a small positive area in $\Delta\rho$. The converse situations produce negative peaks in $\Delta\rho$.

(c) An atom requiring a small positional correction tends to lie in a negative area at the side of a small positive peak. The correction is applied by moving the atom into the positive area.

(d) Very light atoms, such as hydrogen, may be revealed by a $\Delta\rho$ synthesis when the phases are essentially correct, after least-squares refinement has been carried out. There may be some advantage in using only reflections for which $(\sin\theta)/\lambda$ is less than about 0.35. Hydrogen scatters negligibly above this value.

(e) As one final test of the validity of a refined structure, the $\Delta\rho$ synthesis should be effectively featureless within 2–3 times the standard deviation of the electron density (page 356).

6.4.7 Limitations of the Heavy-Atom Method

The Patterson and heavy-atom techniques are effective for structures containing up to about 100 atoms in the asymmetric unit. It is sometimes necessary to introduce heavy atoms artificially into structures. This process may not be desirable because a possible structural interference may arise and there will be a loss in the accuracy of the light-atom positions. An introduction to "direct methods," capable of solving the phase problem for such structures, is given in the next chapter. Very large structures, such as proteins, containing many more than 100 atoms in the asymmetric unit, may be investigated by a variation of the heavy-atom method.

6.4.8 Isomorphous Replacement

A common feature of biologically important substances is their high molecular weight. Proteins and enzymes, for example, are polymers built up from various amino acid residues and forming very large assemblies with molecular weights greater than 5000. The study of the conformations of these giant molecules is necessary for the understanding of their biological functions, and the principal method of obtaining structural detail is by X-ray analysis.

Because of their high molecular weight, protein structures do not yield to analysis by the heavy-atom method. The value of r, from (6.69), is typically 0.03 for a protein molecule of molecular weight 5000 containing one mercury atom. This value of r is too small to be useful. Another difficulty is that most proteins and enzymes contain neither very heavy atoms nor easily replaceable groups to facilitate the introduction of heavy atoms. In spite of these difficulties, if a heavy-atom derivative of a large molecule can be prepared, it may be possible to induce it to crystallize in a similar size of unit cell and with the same space group as the native compound. Such pairs of compounds are said to be isomorphous.

The structure factor of the heavy-atom derivative may be expressed vectorially as $\mathbf{F}_{PH}$, where

$$\mathbf{F}_{PH} = \mathbf{F}_P + \mathbf{F}_H \qquad (6.76)$$

$\mathbf{F}_P$ and $\mathbf{F}_H$ are the structure factors for the parent protein and the heavy atoms alone, respectively, for the same reflection. This relationship is shown in Figure 6.25.

Assuming that the positions of the N_H heavy atoms in the unit cell can be determined, their contribution can be calculated:

$$F_H = \sum_{j=1}^{N_H} g'_j \exp[i2\pi(hx_j + ky_j + lz_j)] \qquad (6.77)$$

where $g'_j = f'_j \exp(-B_j \sin^2\theta/\lambda^2)$, and $f'_j = K_j f_j$; K_j is a site occupation factor,

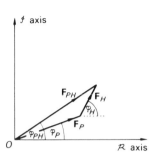

FIGURE 6.25. Graphical interpretation of the isomorphous replacement equation. In practice, the phases ϕ_P and ϕ_{PH} are unknown initially. F_H may be known with a fair degree of accuracy if the heavy-atom positions in the isomorphous derivative are known. This enables a solution, as illustrated in Figure 6.26, to be obtained.

less than or equal to unity, depending on the degree of substitution at the heavy-atom site j (some heavy-atom binding sites of the protein molecules in the crystal may not be substituted).

To obtain an idea of the effect of a heavy atom on the intensities of X-ray reflections from a protein, we shall carry out a simple calculation for a crystal containing one protein molecule per unit cell in space group $P1$. Assuming that it has a molecular weight of about 13,000, about 1000 nonhydrogen atoms would comprise the molecule; we shall assume that they are all carbon ($Z_c = 6$). Accepting Wilson's approximation (6.62),

$$\overline{|F_P|^2} \approx \sum_{j=1}^{1000} f_C^2 \tag{6.78}$$

At $\sin \theta = 0$, $\overline{|F_P|^2}$ is 36,000. If the derivative contains one mercury atom ($Z_{\text{Hg}} = 80$),

$$\overline{|F_{PH}|^2} \approx \sum_{j=1}^{1000} f_C^2 + \sum_{j=1}^{N_H} Z_j'^2 \tag{6.79}$$

which has the value 42,400 at $\sin \theta = 0$. Hence, the maximum change in intensity is about 18%, which is a surprisingly high value.

Experimentally, two sets of data $|F_P(hkl)|$ and $|F_{PH}(hkl)|$ are obtained, and, because of the comparative nature of the phase-determining procedure with isomorphous compounds, they must be placed on the same relative scale, which can be achieved by Wilson's method.

Rewriting (6.76), we have

$$|F_P| \exp(i\phi_P) = |F_{PH}| \exp(i\phi_{PH}) - \mathbf{F}_H \tag{6.80}$$

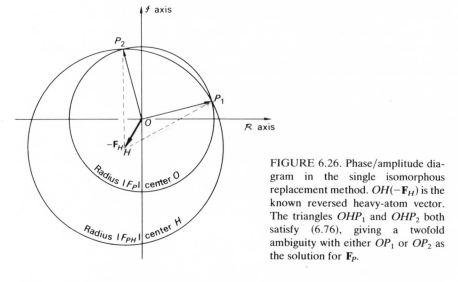

FIGURE 6.26. Phase/amplitude diagram in the single isomorphous replacement method. $OH(-\mathbf{F}_H)$ is the known reversed heavy-atom vector. The triangles OHP_1 and OHP_2 both satisfy (6.76), giving a twofold ambiguity with either OP_1 or OP_2 as the solution for $\mathbf{F}_P$.

This equation involves two unknown quantities, ϕ_P and ϕ_{PH}, and cannot yield a unique solution. However, Figure 6.26 shows that only two solutions for ϕ_P are real, corresponding to the vectors OP_1 and OP_2, one of which is the true $\mathbf{F}_P$ vector. Another isomorphous derivative, PH', with a *different set* of heavy-atom positions, will also have two solutions for $\mathbf{F}_P$, say, OP_3 and

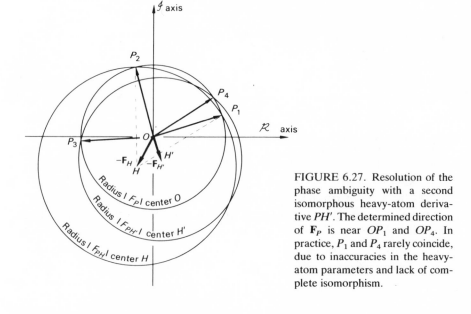

FIGURE 6.27. Resolution of the phase ambiguity with a second isomorphous heavy-atom derivative PH'. The determined direction of $\mathbf{F}_P$ is near OP_1 and OP_4. In practice, P_1 and P_4 rarely coincide, due to inaccuracies in the heavy-atom parameters and lack of complete isomorphism.

OP_4, one of which should agree with either OP_1 or OP_2 within experimental error, thus resolving the ambiguity (Figure 6.27). With a more extensive series of isomorphous derivatives, it is possible to obtain phases capable of yielding interpretable electron density maps. Many protein structures have been investigated successfully by this technique.

Centrosymmetric Projections

Proteins always crystallize in noncentrosymmetric space groups because the amino acid residues in the polypeptide have "left-handed" configurations about the α-carbon atoms. Amino acid residues with "right-handed" configurations are very rare in nature. Although noncentrosymmetric structures usually present more difficulties than centrosymmetric structures, there is a compensation in the relative ease of determination of the space group; ambiguities such as $P2_1$ and $P2_1/m$ do not exist for the protein crystallographer. Most noncentrosymmetric space groups have at least one centrosymmetric zone. In such a case, (6.76) becomes

$$s_{PH}|F_{PH}| = s_P|F_P| + s_H|F_H| \tag{6.81}$$

where s refers to the sign of the structure factor, and is ± 1 (see page 181).

Unless both $|F_{PH}|$ and $|F_P|$ are very small compared with $|F_H|$, it is unlikely that s_{PH} will differ from s_P. Generally $\mathbf{F}_P$ and $\mathbf{F}_{PH}$ are pointing in the same direction. Accepting this statement, we may substitute s_P for s_{PH} in (6.81):

$$s_P(|F_{PH}| - |F_P|) = s_H|F_H| \tag{6.82}$$

or

$$s_P = s_H|F_H|/\Delta F \tag{6.83}$$

where $\Delta F = |F_{PH}| - |F_P|$. Since we are interested only in the signs, (6.83) may be rewritten as

$$s_P = s_H s_\Delta \tag{6.84}$$

where s_Δ is $+1$ if $|F_{PH}| > |F_P|$ and -1 if $|F_{PH}| < |F_P|$. In this way, signs can often be determined for centric reflections in a protein crystal with only a single isomorphous derivative, and we shall illustrate the method by the following example.

TABLE 6.6. $h0l$ Data for Ribonuclease

| hkl | Observed data | | | | Calculated data | | Deduced sign |
	$\lvert F_P \rvert$	$\lvert F_{PH} \rvert$	$\lvert \Delta F \rvert$	s_Δ	$\lvert F_H \rvert$	s_H	$s_P = s_H s_\Delta$
003	437	326	111	−1	50	+1	−1
006	59	48	11	−1	27	−1	+1
007	182	109	73	−1	90	−1	+1
$10\overline{17}$	144	196	52	+1	31	−1	−1
$10\overline{13}$	146	82	64	−1	52	+1	−1
$10\overline{9}$	97	165	68	+1	55	−1	−1
106	183	242	59	+1	45	+1	+1
$30\overline{4}$	746	861	115	+1	72	+1	+1
405	103	57	46	−1	56	+1	−1

Sign Determination for Centric Reflections in Protein Structures

We shall consider data for both the enzyme ribonuclease and a heavy-atom derivative prepared by soaking pregrown crystals of the enzyme in $K_2[PtCl_6]$ solution.*

Crystal Data for Ribonuclease.

System: monoclinic.

Unit-cell dimensions: $a = 30.31$, $b = 38.26$, $c = 52.91$ Å, $\beta = 105.9°$.

M: 13,500 (ribonuclease).

Z_P: two molecules of ribonuclease plus an unknown number of water molecules.

Z_{PH}: as for $Z_P + N_H$ $[PtCl_6]^{2-}$ groups per unit cell.

Absent spectra: $0k0: k = 2n + 1$.

Space group: $P2_1$. The $h0l$ zone is centrosymmetric.

Table 6.6 shows how the signs for some $h0l$ reflections have been determined. Notice that experimental errors in $\lvert F_P \rvert$ and $\lvert F_{PH} \rvert$, together with errors in the calculated $\lvert F_H \rvert$ arising from inaccuracies in the heavy-atom model, are reflected in the inequality of ΔF and $\lvert F_H \rvert$. The validity of (6.84) is upheld by these data.

Location of Heavy-Atom Positions in Proteins

In a centrosymmetric zone, it follows from (6.83) that

$$\lvert F_H \rvert = \lvert \Delta F \rvert \tag{6.85}$$

* C. H. Carlisle *et al.*, *Journal of Molecular Biology* **85**, 1 (1974).

where $|\Delta F| = ||F_{PH}| - |F_P||$. A Patterson function calculated with $|F_H|^2$ as coefficients would give the vector set of the substituted heavy atoms in the protein molecule. Since $|F_H|$ cannot be observed, the next best procedure is to calculate a Patterson map with $(\Delta F)^2$ as coefficients. If the experimental errors in $|F_P|$ and $|F_{PH}|$ are not significant, and not too many sign "cross-overs" with s_P and s_{PH} occur, then the $(\Delta F)^2$ Patterson projection would be expected to reveal the heavy-atom vectors (in projection). In the case of general noncentrosymmetric reflections, we note from (6.80) that

$$|F_H|^2 = ||F_{PH}| \exp(i\phi_{PH}) - |F_P| \exp(i\phi_P)|^2 \qquad (6.86)$$

Since $\phi_{PH} \neq \phi_P$, the relation (6.85) does not hold, but instead

$$|F_H| \approx |\Delta F|/\cos \delta \qquad (6.87)$$

where δ is an unknown angle ($\phi_H - \phi_P$ in Figure 6.25). Hence

$$|F_H|^2 \approx (\Delta F)^2/\cos^2\delta \qquad (6.88)$$

In practice, the angle δ is undeterminable, and the best one can do is to calculate a Patterson function with $(\Delta F)^2$ coefficients as for centro-symmetric reflections, but, as an added precaution to ensure that $(\phi_{PH} - \phi_P)$ is small, use only those terms for which both $|F_P|$ and $|F_{PH}|$ are large. Although the noncentrosymmetric $(\Delta F)^2$ is not a true Patterson function, it has been used successfully to determine the heavy-atom distribution in proteins.

The most useful derivatives contain a small number of highly substi-tuted sites. Unlike the structure analysis of smaller molecules, it is not known initially how many heavy-atom sites have been incorporated into the molecule.

As an example, we shall consider the $(\Delta F)^2$ Patterson map for the Pt derivative of ribonuclease (space group $P2_1$). The vectors between symmetry-related atoms occur on the Harker section $(u, \frac{1}{2}, w)$. Eight peaks occur on the Harker section and four at $v = 0$ (Figures 6.28a and 6.28b). This result suggests that there is more than one heavy-atom site per protein molecule. The most obvious choice is two, since four heavy atoms per unit cell would give rise to 12 nonorigin peaks. If the two sites are labeled 1 and 2, their Harker peaks will be of the form $\pm\{2x_1, \frac{1}{2}, 2z_1\}$ and $\pm\{2x_2, \frac{1}{2}, 2z_2\}$.

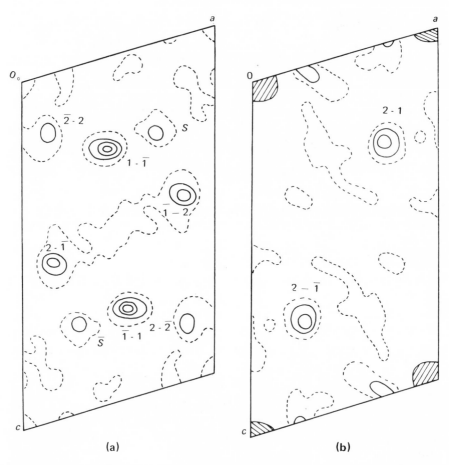

FIGURE 6.28. $(\Delta F)^2$ Patterson sections for the Pt derivative of ribonuclease: (a) $P(u, \frac{1}{2}, w)$, (b) $P(u, 0, w)$.

Interpretation of the Patterson function is best undertaken in terms of the Harker section, assuming that the peaks represent nonoverlapping vectors and ignoring the possibility that some peaks could be non-Harker peaks. Since the true Harker peaks are of the form $2x, 2z$, values of x and z can be obtained by dividing by 2 the fractional coordinates on the Harker section.

This analysis may be carried out graphically. The peak positions from the Harker section are replotted, on tracing paper, on a unit-cell projection in which the a and c dimensions are each reduced by a factor of $\frac{1}{2}$. This procedure results in one quadrant of Figure 6.29a. The diagram is com-

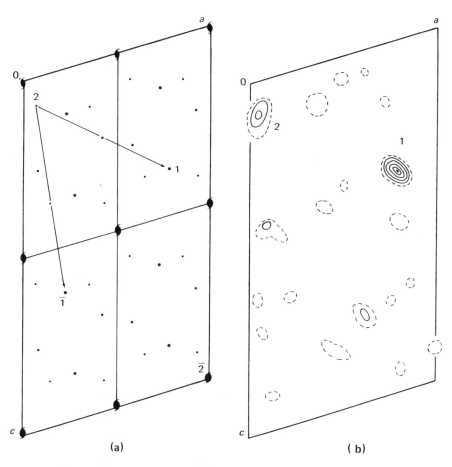

FIGURE 6.29. Interpretation of the $(\Delta F)^2$ Paterson sections: (a) implication diagram, (b) electron density map showing the Pt atom sites.

pleted by operating on the first quadrant with the translation of $a/2$, and then on both quadrants by $c/2$, thus completing an area the size of the true unit-cell projection.

All points marked on this map locate potential (x, z) coordinates for the heavy atoms. In fact, it contains four equivalent solutions with respect to the four unique 2_1 axes in the unit cell, marked ◗. Cross-vector peaks are found by moving this implication diagram* to other sections of the Patterson function, using pairs of potential sites to generate potential vectors. To see

* See Bibliography.

FIGURE 6.30. Stereoviews of the polypeptide chain in ribonuclease; the main site in the Pt
derivative is shown as a simulated octahedrally coordinated group.

how this mechanism operates, place the site marked 2 on the tracing paper over the origin of the section $v = 0$ and note the coincidence of site 1 with the peak 2-1. Similarly, the peak 2-$\bar{1}$ and others on the section $v = \frac{1}{2}$ can be generated from the sites 1, 2, $\bar{1}$, and $\bar{2}$ on the implication diagram. Peaks S and $\bar{S}$ are not explained in this way; they may be assumed to be spurious [remember $(\Delta F)^2$ is not a true representation of $|F_H|^2$].

Figure 6.29b shows a composite electron density map of the Pt atom sites which were prepared by an independent method, and confirm the Patterson analysis. The y coordinates of the two heavy-atom sites are almost equal, which accounts for the presence of the non-Harker peaks $\pm(2\text{-}\bar{1})$ on the Harker section. Figure 6.30 is a stereo pair showing the course of the polypeptide chain in ribonuclease and the position of the main site in the Pt derivative.

6.4.9 Further Details of the Isomorphous Replacement Phasing Procedure

In single isomorphous replacement (SIR), the ambiguity in ϕ_P is best resolved (see Figure 6.31) by taking $\phi_P = \phi_H$ with $\mathbf{F}_P$ along the median OM between P_1 and P_2; $|F_P|$ should be weighted* by $m = \cos \psi$ where ψ is the semi-angle between P_1 and P_2. By the cosine rule

$$m = \cos \psi = (|F_{PH}|^2 - |F_P|^2 - |F_H|^2)/2|F_P||F_H| \qquad (6.89)$$

* M. G. Rossman and D. M. Blow, *Acta Crystallographica* **14**, 641–646 (1961).

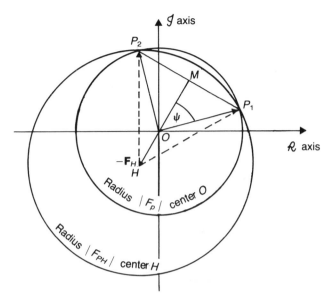

FIGURE 6.31. Phase-amplitude diagram in the single isomorphous replacement method; **OH** $(-\mathbf{F}_H)$ is the known reversed heavy-atom vector. The triangles OHP_1 and OHP_2 both satisfy (6.76), giving twofold ambiguity with either $\mathbf{OP}_1$ or $\mathbf{OP}_2$ as the solution for $\mathbf{F}_P$; OM defines the weighted SIR solution for $\mathbf{F}_P$.

we see that m would have a maximum value of unity in the special case for which $|F_H| = \|F_{PH}| - |F_P\|$, where the two circles are tangential. The coefficients in the SIR electron density map would be composed of $m|F_P|$ and ϕ_H. The electron density of such a map would be subject to the pseudosymmetry effects discussed in Section 6.4.5. Taking $\phi_P = \phi_H$ is thus the SIR equivalent of the initial stage of the heavy-atom method, in which we take $\phi = \phi_H$ (p. 261).

Analytical Calculation of Phases in SIR and MIR

The geometrical determination of phases by the isomorphous replacement method using Harker's construction is impractical for several reasons:

(i) In MIR (multiple isomorphous replacement), phase-circle intersections, due to accumulated errors, do not usually give the absolutely clear indications of ϕ_P idealized in Figure 6.32. Actual phase determination in MIR, exemplified by Figure 6.33, contains a complexity of multiple-derivative phase indications, the phase circles intersecting in rather ill-defined regions (see also problem 6.8).

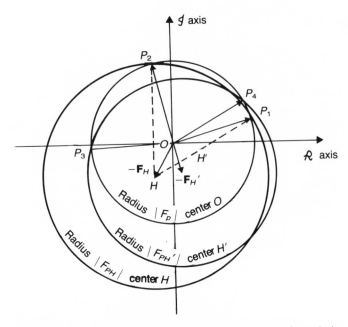

FIGURE 6.32. Multiple isomorphous replacement (MIR) resolution of the phase ambiguity with a second isomorphous derivative PH'. The determined direction is near $\mathbf{OP_1}$ and $\mathbf{OP_4}$. In practice P_1 and P_4 rarely coincide, owing to inaccuracies in the heavy-atom parameters and a lack of true isomorphism.

(ii) The size of the task of estimating thousands of $\phi_P(hkl)$ values in a typical protein analysis necessitated the development of an analytical formula suitable for computer programming, as outlined below.

A computer-generated alternative to the Harker construction for SIR is shown in Figure 6.34, in which the inner circle represents $|F_P|$ and the spokes represent ϕ_T, a series of trial values of ϕ_P for $\phi_T = 0$ to $360°$ in steps of $30°$. The vector $\mathbf{F}_H$ (which would be *calculated* from (6.77), using the known heavy-atom parameters) is plotted at the end of each spoke. In order to simplify the drawing, the third side of the isomophous replacement triangle, representing $\mathbf{F}_{PH}$, has not been joined up. The SIR solutions (corresponding to P_1 and P_2 in Figure 6.31) would occur when $\mathbf{F}_H$ just touches the F_{PH} circle, which is plotted concentric with the F_P circle and is the outer circle in Figure 6.34. These two positions are indicated in the diagram, which should be compared with Figure 6.31. Because of the method of selecting ϕ_T, neither is generated exactly in this mehod. Now consider Figure 6.35, which shows a more detailed representation of the case where

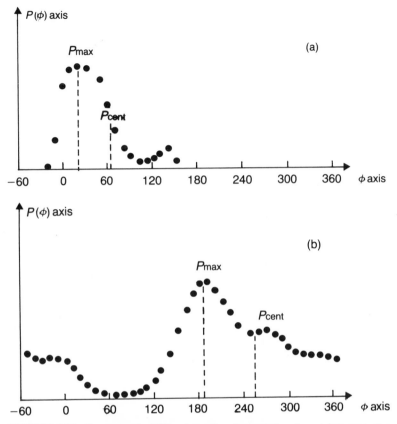

FIGURE 6.33. Two cases in MIR where the phase circles do not intersect at a point. The most probable value of ϕ and the centroid of $P(\phi)$ are indicated: (a) $P(\phi)$ calculated for the two-derivative case in Figure 6.32 and (b) an example of three-derivative phasing (see problem 6.8).

$\phi_T = 30°$ (Figure 6.34). This is one of the general cases (not P_1 or P_2) where the $\mathbf{F}_{PH}$ trial vector would not close the third side of the phase triangle properly. In the case shown $\mathbf{F}_{PH}$ is too short; other situations evident in Figure 6.34 would correspond to $\mathbf{F}_{PH}$ being too long. In these situations there is a *lack-of-closure error* denoted by ε_{ϕ_T} (Fig. 6.35). For the SIR solutions, $\varepsilon_{\phi_T} = 0$. In general, ε_{ϕ_T} may be calculated as follows:

$$D_{\phi_T}^2 = |F_P|^2 + |F_H|^2 + 2|F_P||F_H|\cos(\phi_T - \phi_H) \qquad (6.90)$$

and

$$\varepsilon_{\phi_T}^2 = (|F_{PH}| - D_{\phi_T})^2 \qquad (6.91)$$

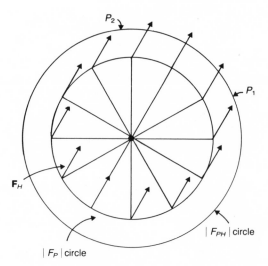

FIGURE 6.34. The concept of lack of closure; the inner circle represents $|F_P|$ and the outer circle $|F_{PH}|$. Trial values of ϕ_P are plotted at 30° intervals, each carrying the known $\mathbf{F}_H$ vector. At P_1 and P_2, $\mathbf{F}_H$ ends exactly on the F_{PH} circle; otherwise it fails to close, being too long for the smaller region spanning $P_1 - P_2$ and too short for the rest (see Figure 6.26 for Harker's construction of this SIR case).

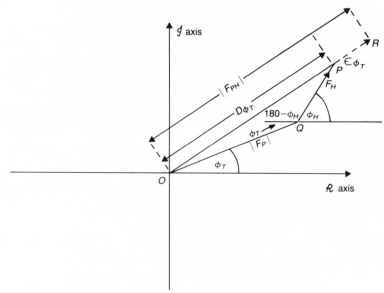

FIGURE 6.35. Calculation of the lack-of-closure error ε_{ϕ_T}.

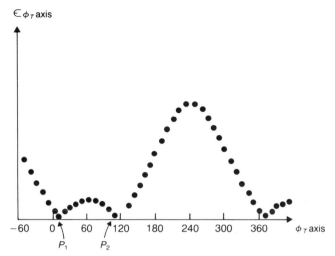

FIGURE 6.36. Values of ε_{ϕ_T} plotted against ϕ_T; P_1 and P_2 are the two positions for which $\varepsilon_{\phi_T} = 0$, corresponding to the SIR solutions (see Figures 6.31 and 6.34).

The SIR solutions could be determined to a satisfactory degree of precision by plotting ε_{ϕ_T} against ϕ_T and locating the two ϕ_T values for which $\varepsilon_{\phi_T} = 0$. This is shown, for example, in Figures 6.31 and 6.34, and the resulting graph is shown in Figure 6.36. Both solutions P_1 and P_2 are of course equally probable in the SIR method. In the theory of phase analysis by the MIR method, errors may be assumed to reside in $|F_{PH}|$,[*] which simplifies the calculations. For a given trial value of ϕ_T, the probability that ϕ_T is the correct value is

$$P(\phi_T) = \exp(-\varepsilon_{\phi_T}/2E^2) \qquad (6.92)$$

where E is the root-mean-square error in $|F_{PH}|$ arising from data errors.

In MIR there would be one value of ε_{ϕ_T} per derivative. Let $\varepsilon_i(\phi_T)$ be the value for derivative i, where $i = 1, 2, \ldots$ to the total number of derivatives. Then the probability for the ith derivative is

$$P_i(\phi_T) = \exp[-\varepsilon_i^2(\phi_T)/2E^2] \qquad (6.93)$$

and the combined or joint probability over all derivatives is

$$P(\phi_T) = P_1(\phi_T) \cdot P_2(\phi_T) \ldots$$

[*] D. M. Blow and F. H. C. Crick, *Acta Crystallographica* **12**, 794–799 (1959).

or

$$P(\phi_T) = \exp[-\sum \varepsilon_i^2(\phi_T)/2E_i^2] \tag{6.94}$$

Typical examples of probability distributions met in practice are given in Figure 6.33. Generally the distributions are bimodal, indicating a stronger preference for one maximum over the other. The most probable electron density map uses coefficients $\{|F_P|, \phi_M\}$, where ϕ_M is the phase angle corresponding to the maximum probability in the range 0–360°. However, the electron density map with the least overall root-mean-square error uses coefficients $\{|F_P|, \phi_B\}$, where ϕ_B is the 'best' phase angle, corresponding to the centroid (centre of gravity) of the probability distribution, and m is a weighting function (or figure of merit), given by

$$m \cos \phi_B = \sum_{\phi_T} P(\phi_T) \cos \phi_T / \sum P(\phi_T) \tag{6.95}$$

$$m \sin \phi_B = \sum_{\phi_T} P(\phi_T) \sin \phi_T / \sum P(\phi_T) \tag{6.96}$$

It is convenient in practice to evaluate these expressions by stepping from 0 to 360° in regular intervals of 5° or 10°. The probability distributions and corresponding phases may be readily evaluated by suitable programming.[*]

For each derivative the root-mean-square estimate of error may be taken initially as

$$E_j^2 = \langle (|\Delta F_i| - |F_{H_i}|)^2 \rangle_{hkl} \tag{6.97}$$

where

$$\Delta F_i = |F_{PH_i}| - |F_P| \tag{6.98}$$

evaluated for centric reflections only.

The error in a phase angle may be defined as $\Delta\phi = \phi_B - \phi_M$ and $m = \cos \Delta\phi$. A value of $m = 1$ corresponds to $\Delta\phi \cong 42°$. The average value of m is a measure of the average of $\cos \Delta\phi$. In a typical protein analysis at

* R. E. Dickerson, J. E. Weintzierl, and R. A. Palmer, *Acta Crystallographica* **B24**, 997–1003 (1968).

resolution $2\mathring{A}$* ($\sin \theta_{max} = \lambda/4$), an average m of 0.6–0.7 would be acceptable.

Electron Density Maps Used in Large-Molecule Analysis

(i) The correlation of heavy-atom sites between derivatives requires one to establish the coordinates of heavy atoms in derivative i with respect to those of another derivative or combination of derivatives for which phases $\phi_{P\neq i}$ have been determined. A difference electron density map may be calculated as:

$$\rho\Delta_i(xyz) = \frac{1}{V}\sum_h \sum_k \sum_l (|F_{PH_i}| - |F_P|) \cos[2\pi(hx + ky + lz) - \phi_{P\neq i}]$$

(6.99)

This should reveal the heavy atoms in derivative i with respect to the same origin as in the other heavy-atom derivatives. Derivative i can then be added into the MIR procedure.

(ii) For a trial structure in which phases have been calculated, as in small-molecule analysis, a difference electron density map may be used in order to effect corrections to the structure:

$$\Delta\rho(xyz) = \frac{1}{V}\sum_h \sum_k \sum_l (|F_P| - |F_c|) \cos[2\pi(hx + ky + lz) - \phi_c]$$

(6.100)

(iii) Alternatively a double difference map

$$\rho'(xyz) = \frac{1}{V}\sum_h \sum_k \sum_l (2|F_P| - |F_C|) \cos[2\pi(hx + ky + lz) - \phi_c]$$

(6.101)

where $\rho'(xyz) = \rho_c(xyz) + \Delta\rho(xyz)$, may be used since new features may be more easily recognized in $\Delta\rho(xyz)$ against the background of the known $\rho_c(xyz)$ structure. This map is very useful in computer graphics analysis.

* The resolution of a protein X-ray analysis is loosely defined as d_{min}, where $d_{min} = \lambda/2 \sin \theta_{max}$, θ_{max} being the maximum Bragg angle associated with the analysis: θ_{max} may be restricted in order to limit the labor required, at the expense of the quality of the electron density image.

(iv) In MIR the most error-free electron density is calculated as:

$$\rho_P(xyz) = \frac{1}{V}\sum_h \sum_k \sum_l m|F_P| \cos[2\pi(hx+ky+lz)-\phi_B] \qquad (6.102)$$

where ϕ_B is the MIR phase corresponding to the centroid of the phase probability distribution (the best phase) and m is the figure of merit—see equations (6.95) and (6.96).

6.4.10 Anomalous Scattering

Friedel's law (page 169) is not an exact relationship, and becomes less so as the atomic numbers of the constituent atoms in a crystal increase. The law breaks down severely if X-rays are used that have a wavelength just less than that of an absorption edge (page 424) of an atom in the crystal. However, this criterion is not essential for anomalous scattering to be used in two important aspects of crystal structure analysis, namely, the determination of absolute stereochemical configurations and the phasing of reflections.

Anomalous scattering introduces a phase change into a given atomic scattering factor, which becomes complex:

$$\mathbf{f} = \mathbf{f}_o + \Delta\mathbf{f}' + i\,\Delta\mathbf{f}'' \qquad (6.103)$$

$\Delta\mathbf{f}'$ is a real correction, usually negative, and $\Delta\mathbf{f}''$ is an imaginary component which is rotated anticlockwise through 90° in the complex plane with respect to $\mathbf{f}_o$ and $\Delta\mathbf{f}'$.

A possible situation is illustrated in Figure 6.37. In Figure 6.37a, atom A is assumed to be scattering in accordance with Friedel's law, and it is clear that $|F(\mathbf{h})| = |F(\bar{\mathbf{h}})|$, where $\mathbf{h}$ stands for hkl. In Figure 6.37b, atom A is represented as an anomalous scatterer, with its three components, according to (6.89). In this situation, $|F(\mathbf{h})| \neq |F(\bar{\mathbf{h}})|$, and intensity measurements of Friedel pairs of reflections produce different values.

A given set of atomic coordinates can be used to calculate $|F_c|$ for a number of Friedel pairs, because the coordinates of the enantiomorph may be obtained, for example, by inversion through the origin. One of the enantiomorphs will show the better fit, and thus the correct absolute configuration can be deduced. Some typical results are listed in Table 6.7, from which it may be deduced that the structure giving $|F_c|_{\bar{x},\bar{y},\bar{z}}$ corresponds

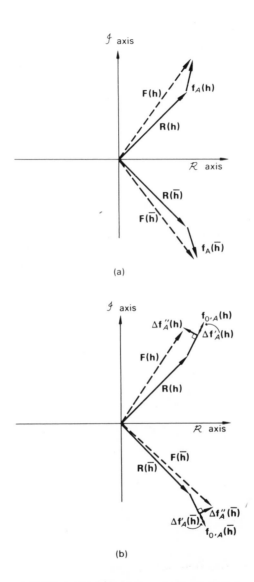

FIGURE 6.37. Anomalous scattering of atom A with respect to the rest of the structure R: (a) normal case—$|F(\mathbf{h})| = |F(\bar{\mathbf{h}})|$; (b) anomalous case—$|F(\mathbf{h})| \neq |F(\bar{\mathbf{h}})|$.

TABLE 6.7. Example of Some Friedel Pairs and the
Corresponding $|F|$ Values

hkl	$|F_o|$	$|F_c|_{x,y,z}$	$|F_c|_{\bar{x},\bar{y},\bar{z}}$
121	17.0	19.1	18.3
122	21.2	22.9	21.9
123	41.4	44.4	42.8
341	36.7	38.7	35.5
342	7.8	9.5	8.2
413	14.2	15.3	13.5

to the absolute configuration. An equivalent procedure would be to measure
the values of both $|F(hkl)|$ and $|F(\bar{h}\bar{k}\bar{l})|$ and compare them with $|F_c|_{x,y,z}$.
The technique can be used only with crystals which are noncentrosymmetric,
because $F(\mathbf{h}) = F(\bar{\mathbf{h}})$ in centrosymmetric crystals, but this limitation is not
important because molecules which crystallize with a single enantiomorph
cannot do so in a space group containing any form of inversion symmetry.

Hamilton's R Ratio

The reliability of the determination of the absolute configuration by
the technique just described should be checked by means of Hamilton's R
ratio test.* The generalized weighted R factor is defined by

$$R_g = \left(\frac{\sum_{hkl} w(hkl)[|F_o(hkl)| - k^{-1}|F_c(hkl)|]^2}{\sum_{hkl} w(hkl)|F_o(hkl)|^2} \right)^{1/2} \qquad (6.104)$$

where the summations are taken over the unique set of reflections. Let the
generalized weighted R factors for a structure and its enantiomorph be R_1
and R_2, respectively. The R factor ratio is compared with $R'_{p,n-p,\alpha}$ where

$$R = R_2/R_1; \qquad R' = \left| \frac{p}{n-p} F_{p,n-p,\alpha} + 1 \right|^{1/2} \qquad (6.105)$$

p is the number of parameters and n is the number of reflections, so that
$n - p$ is the number of degrees of freedom. The null hypothesis for parameter

* W. C. Hamilton, *Acta Crystallographica* **18**, 502 (1965).

sets r_1 and r_2, corresponding to R_1 and R_2,

$$H_0 = r_2 - r_1 \qquad (6.106)$$

can be tested at the $100\alpha\%$ significance level, where α defines a chosen level of significance. If $R > R'$, then we can reject the hypothesis at the $100\alpha\%$ level. When the number of degrees of freedom $(n-p)$ is large,

$$F_{p,n-p} = \chi_p^2/p \qquad (6.107)$$

and R may be tested against

$$R' = R'_{p,n-p,\alpha} = \left(\frac{\chi_p^2}{n-p} + 1\right)^{1/2} \qquad (6.108)$$

The absolute configuration of stercuronium iodide[*] was confirmed by applying Hamilton's ratio test at the 0.01 level. The data are as follows:

$$n = 2420 \qquad p = 304$$
$$n - p = 2116 \qquad \alpha = 0.01$$

Following Hamilton, we obtain

$$R' = R'_{304,2116,0.01} = \left(\frac{367.5}{2116} + 1\right)^{1/2} = 1.083$$

$$R = R_2/R_1 = 0.1230/0.1037 = 1.186$$

Since $R > R'$, the structure with $R_g = R_2$ was rejected at the 1% level (99% certain).

Phasing

Anomalous scattering can be used in phasing reflections. We saw in the previous section that the isomorphous replacement technique in noncentrosymmetric crystals leads to an ambiguity in phase determination (Figure 6.26). The ambiguity cannot be resolved unless the replaceable site is

---

[*] J. Husain, R. A. Palmer, and I. J. Tickle, *Journal of Crystal and Molecular Structure* **11**, 87 (1981).

changed. Merely using a third derivative with the same replaceable site would lead to a situation comparable with that in Figure 6.26. The heavy atom vector would still be directed along OH, and its different length would be just balanced by the change in $|F_o|$, so that *three* circles would intersect at P_1 and P_2.

The vector change in the heavy atom contribution (Figure 6.27) can be brought about through anomalous scattering in a given derivative, instead of invoking a different replaceable site. Following Figure 6.31b, we see that two different $|F_0|$ values can arise for $\mathbf{h}$ and $\bar{\mathbf{h}}$. Consequently, $\phi(\mathbf{h}) \neq \phi(\bar{\mathbf{h}})$, and the ambiguity can be resolved by the experimental data. This technique is particularly important with synchrotron radiation (see Appendix 5), where the wavelength can be tuned to the absorption edge (q.v.) of a relatively heavy atom in the structure so as to obtain the maximum difference between $|F_o(\mathbf{h})|$ and $|F_o(\bar{\mathbf{h}})|$.

Bibliography

General Structure Analysis

BUERGER, M. J., *Vector Space*, New York, Wiley (1959).

LIPSON, H. S., *Crystals and X-Rays*, London, Wykeham Publications (1970).

STOUT, G. H., and JENSEN, L. H., *X-Ray Structure Determination—A Practical Guide*, New York, Macmillan (1968).

WOOLFSON, M. M., *An Introduction to X-Ray Crystallography*, Cambridge, Cambridge University Press (1970).

Protein Crystallography

BLUNDELL, T. L., and JOHNSON, L. N., *Protein Crystallography*, New York/London/San Francisco, Academic Press (1977).

EISENBERG, D., in *The Enzymes*, Vol. I, X-ray crystallography and enzyme structure, New York, Academic Press (1970).

PHILLIPS, D. C., *Advances in Structure Research by Diffraction Methods 2*, 75 (1966).

Chemical Data

SUTTON, L. E. (Editor), *Tables of Interatomic Distances and Configuration in Molecules and Ions*, London, The Chemical Society (1958; supplement, 1965).

Problems

6.1. Write down the symmetry-equivalent amplitudes of $|F(hkl)|$, $|F(0kl)|$, and $|F(h0l)|$ in (a) the triclinic, (b) the monoclinic, and (c) the orthorhombic crystal systems. Friedel's law may be assumed.

6.2. (a) Determine the orientations of the Harker lines and sections in Pa, $P2/a$, and $P222_1$.

(b) A monoclinic, noncentrosymmetric crystal with a primitive space group shows concentrations of peaks on $(u, 0, w)$ and $[0, v, 0]$. How might this situation arise?

6.3. Diphenyl sulfoxide, $(C_6H_5)_2SO$, is monoclinic, with $a = 8.90$, $b = 14.08$, $c = 8.32$ Å, $\beta = 101.12°$, and $Z = 4$. The conditions limiting possible X-ray reflections are as follows.

$$hkl: \quad \text{none;} \qquad h0l: \quad h+l=2n; \qquad 0k0: \quad k=2n$$

(a) What is the space group?

(b) Figures P6.1a–c are the Patterson sections at $v = \frac{1}{2}$, $v = 0.092$, and $v = 0.408$, respectively, and contain S—S vector peaks. Write the coordinates of the nonorigin S—S vectors (in terms of x, y, and z), and from the sections provided determine the best values for the S atoms in the unit cell. Plot these atomic positions as seen along the b axis, with an indication of the heights of the atoms with respect to the plane of the diagram.

6.4. Figure P6.2 shows an idealized vector set for a hypothetical structure C_6H_5S in space group $P2$ with $Z = 2$, projected down the b axis. Only the S—S and S—C vector interactions are considered.

(a) Determine the x and z coordinates for the S atoms, and plot them to the scale of this projection.

(b) Use the Patterson superposition method to locate the carbon atom positions on a map of the same projection.

6.5. Hafnium disilicide, $HfSi_2$, is orthorhombic, with $a = 3.677$, $b = 14.55$, $c = 3.649$ Å, and $Z = 4$. The space group is $Cmcm$, and the Hf and Si atoms occupy three sets of special positions of the type

$$\pm\{0, y, \tfrac{1}{4}; \quad \tfrac{1}{2}, \tfrac{1}{2}+y, \tfrac{1}{4}\}$$

The contributions from the Hf atoms dominate the structure factors.

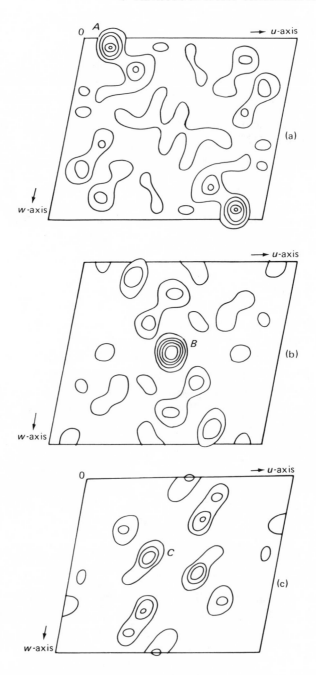

FIGURE P6.1 Patterson sections at (a) $v = \frac{1}{2}$, (b) $v = 0.092$, (c) $v = 0.408$.

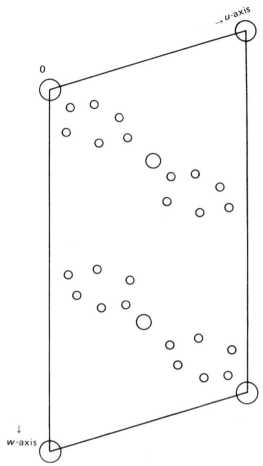

FIGURE P6.2. Idealized vector map for "C$_6$H$_5$S."

By combining the terms $\cos 2\pi ky$ and $\cos 2\pi(ky + k/2)$, show that
the geometric structure factor $A(0k0)$ is approximately proportional
to $\cos 2\pi y_{Hf}$. The $|F(0k0)|$ data are listed below, from which the
values of $|F(0k0)|^2$, divided by 10 and rounded to the nearest integer,
have been derived.

$0k0$	020	040	060	080	010,0	012,0	014,0	016,0
$|F(0k0)|$	7	14	18	13	12	<1	20	<1
$|F(0k0)|^2$	5	20	32	17	14	0	40	0

(a) Calculate the one-dimensional Patterson function $P(v)$, using the

equation
$$P(v) \propto \sum_k |F(0k0)|^2 \cos 2\pi kv$$

The multiplying factor $2/b$ and the $F(000)$ term have been omitted to simplify the calculation; they can never change the form of the synthesized function, although the neglect of the term involving $F(000)$ gives rise to negative values in the calculated $P(v)$.

One-dimensional summations may be carried out conveniently by means of Fourier summation tables,* a technique similar to that first put forward in 1936 by Lipson and Beevers.† Each line contains the values of (amplitude) $\cos 2\pi$(index)(interval), evaluated to the nearest integer. The intervals are 60ths of the repeat unit, and the lines run from 0 to 30/60. If there is reflection symmetry at $\frac{1}{4}$ along the repeat (i.e., the index is even only), as well as at $\frac{1}{2}$, then only the values 0–15 need be summed. The columns are added vertically to form the sum over the index. (See Table P6.1.)

Plot the function, extend it to one repeat unit, interpret the four highest nonorigin peaks, and determine y_{Hf}.

(b) Using the value of y_{Hf} and the form of the geometric structure factor $A(0k0)$, determine the signs for the $0k0$ reflections. Hence, compute the electron density:

$$\rho(y) \propto \sum_k \pm |F(0k0)| \cos 2\pi ky$$

Again the $2/b$ factor and $F(000)$ have been omitted.‡ Use the summation table given in Table P6.2. if a negative amplitude is indicated for any reflection, the signs of all the numbers, except the index, in the corresponding line are changed. Plot the function and determine y_{Hf}. What can be deduced about the positions of the Si atoms? In the light of your results, study $P(v)$ again.

*6.6. The alums, $MAl(NO_4)_2 \cdot 12H_2O$, where $M = NH_4$, K, Rb, Tl, and $N = S$, Se, are isomorphous. They crystallize in the cubic centrosym-

---

* A set of tables in the form of a booklet can be obtained from one of the authors (M.F.C.L.). They enable summations to be calculated with relative ease for one-dimensional, centrosymmetric structures. The index range is 0–15 and the amplitude range is 0–20. The amplitude range can be extended through multiplication by positive or negative integers.

† H. Lipson and C. A. Beevers, *Proceedings of the Physical Society* **48**, 772 (1936).

‡ These omissions give rise to the proportionality signs, rather than equality signs, in the equations on pages 290 and 292.

TABLE P6.1

Amplitude	Index	$\frac{0}{60}$	$\frac{1}{60}$	$\frac{2}{60}$	$\frac{3}{60}$	$\frac{4}{60}$	$\frac{5}{60}$	$\frac{6}{60}$	$\frac{7}{60}$	$\frac{8}{60}$	$\frac{9}{60}$	$\frac{10}{60}$	$\frac{11}{60}$	$\frac{12}{60}$	$\frac{13}{60}$	$\frac{14}{60}$	$\frac{15}{60}$
5	2	5	5	5	4	3	2	2	1	$\bar{1}$	$\bar{2}$	$\bar{2}$	$\bar{3}$	$\bar{4}$	$\bar{5}$	$\bar{5}$	$\bar{5}$
20	4	20	18	13	6	$\bar{2}$	$\overline{10}$	$\overline{16}$	$\overline{20}$	$\overline{20}$	$\overline{16}$	$\overline{10}$	$\bar{2}$	6	13	18	20
32	6	32	26	10	$\overline{10}$	$\overline{26}$	$\overline{32}$	$\overline{26}$	$\overline{10}$	10	26	32	26	10	$\overline{10}$	$\overline{26}$	$\overline{32}$
17	8	17	11	$\bar{2}$	$\overline{14}$	$\overline{17}$	8	5	16	16	5	$\bar{8}$	$\overline{17}$	$\overline{14}$	$\bar{2}$	11	17
14	10	14	7	$\bar{7}$	$\overline{14}$	$\bar{7}$	7	14	7	$\bar{7}$	$\overline{14}$	$\bar{7}$	7	14	7	$\bar{7}$	$\overline{14}$
40	14	40	4	$\overline{39}$	$\overline{12}$	37	20	$\overline{32}$	$\overline{27}$	27	32	$\overline{20}$	$\overline{37}$	12	39	$\bar{4}$	$\overline{40}$

Add columns for $\sum_k$

128 71 $\overline{20}$ ...

TABLE P6.2

7	2	7	7	6	6	5	3	2	1	$\bar{1}$	$\bar{2}$	$\bar{3}$	$\bar{5}$	$\bar{6}$	$\bar{6}$	$\bar{7}$	$\bar{7}$
14	4	14	13	9	4	$\bar{1}$	$\bar{7}$	$\overline{11}$	$\overline{14}$	$\overline{14}$	$\overline{11}$	$\bar{7}$	$\bar{1}$	4	9	13	14
18	6	18	15	6	$\bar{6}$	$\overline{15}$	$\overline{18}$	$\overline{15}$	$\bar{6}$	6	15	18	15	6	$\bar{6}$	$\overline{15}$	$\overline{18}$
13	8	13	9	$\bar{1}$	$\overline{11}$	$\overline{13}$	$\bar{6}$	4	12	12	4	$\bar{6}$	$\overline{13}$	$\overline{11}$	$\bar{1}$	9	13
12	10	12	6	$\bar{6}$	$\overline{12}$	$\bar{6}$	6	12	6	$\bar{6}$	$\overline{12}$	$\bar{6}$	6	12	6	$\bar{6}$	$\overline{12}$
20	14	20	2	$\overline{20}$	$\bar{6}$	18	10	$\overline{16}$	$\overline{13}$	13	16	$\overline{10}$	$\overline{18}$	6	20	$\bar{2}$	$\overline{20}$

metric space group $Pa3$, with the unit-cell side a in the range 12.2–12.4 Å and $Z = 4$. A symmetry analysis leads to the following atomic positions:

4 M: $0, 0, 0$; $0, \frac{1}{2}, \frac{1}{2}$; $\frac{1}{2}, 0, \frac{1}{2}$; $\frac{1}{2}, \frac{1}{2}, 0$

4 Al: $\frac{1}{2}, \frac{1}{2}, \frac{1}{2}$; $\frac{1}{2}, 0, 0$; $0, \frac{1}{2}, 0$; $0, 0, \frac{1}{2}$

8 N: $\pm\{x, x, x;\quad \frac{1}{2}+x, \frac{1}{2}-x, x;\quad \bar{x}, \frac{1}{2}+x, \frac{1}{2}-x;\quad \frac{1}{2}-x, \bar{x}, \frac{1}{2}+x\}$

The N atoms lie on cube diagonals, and x_N may be obtained by a one-dimensional Fourier synthesis along the line [111], using $F(hhh)$ data. Table P6.3 lists these data for four alums ($N = S$). Tl may be assumed to be sufficiently heavy to make all F values positive in this derivative. The same sites in each crystal are occupied by the replaceable atoms.

(a) Use the isomorphous replacement technique to determine the signs of the reflections in Table P6.3.

TABLE P6.3. $|F(hhh)|$ for Isomorphous Alums

hkl	NH_4^+ (10 electrons)	K^+ (18 electrons)	Rb^+ (36 electrons)	Tl^+ (80 electrons)
111	86	38	19	113
222	0	19	79	195
333	111	125	158	236
444	25	6	55	125
555	24	49	64	131
666	86	86	122	164
777	53	34	0	18
888	0	16	22	56

(b) Compute $\rho[111]$ for K alum, using the following equation:

$$\rho(D) \propto \sum_h \pm |F(hhh)| \cos 2\pi hD$$

where D is the sampling interval along [111], again in 60ths. Table P6.4 is a summation table for the range 0–30/60; the signs of the terms must be adjusted in accordance with your findings in the table of $|F(hhh)|$ data. Plot the function and determine a probable value for x_S.

(c) The corresponding hhh data for the isomorphous K/Se alum are listed below. The signs have been allocated by a similar isomorphous replacement procedure. Calculate and plot $\rho(D)$ for these data. Compare the two electron density plots and comment upon the results. Table P6.5 is the appropriate summation table.

hkl	111	222	333	444	555	666	777	888
$\pm|F|$	−48	−52	64	0	116	100	−16	0

6.7. A crystal contains five atoms per unit cell. Four of them contribute together $100e^{i\phi}$ to F(010). The fifth atom has fractional coordinates 0.00, 0.10, 0.00, and its atomic scattering factor components f_o, $\Delta f'$, and $\Delta f''$ are 52.2, −2.7, and 8.0, respectively. If $\phi = 60°$, determine, graphically or otherwise, $|F(010)|$, $|F(0\bar{1}0)|$, $\phi(010)$, and $\phi(0\bar{1}0)$.

6.8. A protein crystal structure is to be solved using MIR. Three isomorphous derivatives are prepared using platinum, uranium, and iodine compounds. For the reflection 060 the following measurements were

TABLE P6.4

F	h																
38	1	38	37	36	35	33	31	28	25	28	31	33	35	36	37	38	38
19	2	19	17	15	13	9	6	2	$\overline{6}$	9	13	15	17	19	19		
125	3	125	101	74	39	0	$\overline{39}$	74	101	119	119	125					
49	5	49	24	0	$\overline{24}$	42	49	42	24	0	$\overline{24}$	42	49				
86	6	86	27	$\overline{27}$	70	86	70	27	$\overline{27}$	70	86						
34	7	34	25	4	$\overline{20}$	33	29	11	14	29	33	20	4	25	34		
16	8	16	11	$\overline{2}$	13	16	8	5	15	15	5	8	16	13	2	11	16

TABLE P6.5

F	h														
48	1	48	47	46	44	42	39	36	32	28	24	20	15	10	5
52	2	52	51	48	42	35	26	16	5	$\overline{5}$	16	26	35	42	48
64	3	64	61	52	38	20	0	20	38	52	61	64			
116	5	116	101	58	0	58	101	116	101	58	0	58	101	116	
100	6	100	81	31	31	81	100	81	31	31	81	100			
16	7	16	12	2	9	16	14	5	3	8	15	15	8	3	5

recorded:

For protein $|F_P| = 858$

For Pt derivative $|F_{PH_1}| = 756$, $|F_{H_1}| = 141$, $\phi_{H_1} = 78°$
  U derivative $|F_{PH_2}| = 856$, $|F_{H_2}| = 154$, $\phi_{H_2} = 63°$
  I derivative $|F_{PH_3}| = 940$, $|F_{H_3}| = 100$, $\phi_{H_3} = 146°$

Use Harker's construction to obtain an estimate for ϕ_P for this reflection from the native protein crystal.

7

Direct Methods and Refinement

7.1 Introduction

In this chapter we shall consider direct (phase probability) methods of solving the phase problem, and certain other techniques which are usually involved in the overall investigation of crystal and molecular structure.

7.2 Direct Methods of Phase Determination

Direct methods of solving the phase problem are now an important technique, particularly in their ability to yield good phase information for structures containing no heavy atoms.

One feature common to the structure-determining methods that we have encountered so far is that values for the phases of X-ray reflections are derived initially by structure factor calculations, albeit on only part of the structure. Since the data from which the best phases are ultimately derived are the $|F_o|$ values, we may imagine that the phases are encoded in these quantities, even though their actual values are not recorded experimentally. This philosophy led to the search for analytical methods of phase determination, which are independent of trial structures, and initiated the development of direct methods, or phase probability techniques.

7.2.1 Normalized Structure Factors

A simplification in direct phase-determining formulae results by replacing $|F(hkl)|$ by the corresponding normalized structure factor $|E(hkl)|$, which is given by the equation*

$$|E_o(hkl)|^2 = \frac{K^2 |F_o(hkl)|^2}{\varepsilon \sum\limits_{j=1}^{N} g_j^2} \qquad (7.1)$$

The E values have properties similar to those of the sharpened F values derived for a point-atom model (page 237); they are largely compensated for the fall-off of f with $\sin \theta$. High-order reflections with comparatively small $|F|$ values can have quite large $|E|$ values, an important fact in the application of direct methods. We may note in passing that $|E|^2$ values can be used as coefficients in a sharpened Patterson function, and since $\overline{|E|^2} = 1$ [see Table 7.2 and equation (6.62)], the coefficients $(|E|^2 - 1)$ produce a sharpened Patterson function with the origin peak removed. This technique is useful because, in addition to the general sharpening effect, vectors of small magnitude which are swamped by the origin peak may be revealed.

ε-Factor

Because of the importance of individual reflections in direct phasing methods, care must be taken to obtain the best possible $|E|$ values. The factor ε in the denominator of (7.1) takes account of the fact that reflections in certain reciprocal lattice zones or rows may have an average intensity which is greater than that for the general reflections. The ε-factor depends upon the crystal class, and its values for some crystal classes are listed in Table 7.1. Some further considerations of ε and of the $|E|$ statistics in the next section will be found in Appendix A9.

$|E|$ Statistics

The distribution of $|E|$ values holds useful information about the space group of a crystal. Theoretical quantities derived for equal-atom structures in space groups $P1$ and $P\bar{1}$ are listed in Table 7.2, together with the experimental results for two crystals.

Crystal 1 is pyridoxal phosphate oxime dihydrate, $C_8H_{11}N_2O_6P \cdot 2H_2O$, which is triclinic. The values in Table 7.2 favor the centric distribution C, and the structure analysis† confirmed the assignment of space

---

* We shall generally omit the subscript o in the $|E|$ symbol, since such terms almost always refer to observed data.

† A. N. Barrett and R. A. Palmer, *Acta Crystallographica* **B25**, 688 (1969).

TABLE 7.1. ε-Factors for Some Crystal Classes

		hkl	0kl	h0l	hk0	h00	0k0	00l	hhl	hh0
Triclinic	1	1	1	1	1	1	1	1		
	$\bar{1}$	1	1	1	1	1	1	1		
Monoclinic	2	1	1	1	1	1	2	1		
	m	1	1	2	1	2	1	2		
	2/m	1	1	2	1	2	2	2		
Orthorhombic	222	1	1	1	1	2	2	2		
	mm2	1	2	2	1	2	2	4		
	mmm	1	2	2	2	4	4	4		
Tetragonal	4	1	1	1	1	1	1	4		
	$\bar{4}$	1	1	1	1	1	1	2		
	4/m	1	1	1	2	2	2	4		
	$\bar{4}2m$	1	1	1	1	2	2	4	2	2
	4mm	1	2	2	1	2	2	8	2	2
	422	1	1	1	2	2	2	4	2	1
	4mm m	1	2	2	2	4	4	8	4	2

TABLE 7.2. Some Theoretical and Experimental Values Related to $|E|$ Statistics

	Theoretical values		Experimental values and conclusions			
Mean values	$P\bar{1}$ (C)	P1 (A)	Crystal 1		Crystal 2	
$|E|^2$	1.00	1.00	0.99		0.98	
$\overline{|E|}$	0.80	0.89	0.85	A/C	0.84	A/C
$\overline{||E|^2-1|}$	0.97	0.74	0.91	C	0.82	A
Distribution	%	%	%		%	
$|E|>3.0$	0.30	0.01	0.20	C	0.05	A
$|E|>2.5$	1.24	0.19	0.90	C	0.98	C
$|E|>2.0$	4.60	1.80	2.70	A/C	2.84	A/C
$|E|>1.75$	8.00	4.71	7.14	C	6.21	A/C
$|E|>1.5$	13.4	10.5	12.9	C	10.5	A
$|E|>1.0$	32.0	36.8	33.7	C	37.1	A

group $P\bar{1}$. Crystal 2 is a pento-uloside sugar; the results correspond, on the whole, to an acentric distribution A, as expected for a crystal of space group $P2_12_12_1$.[*]

It should be noted that the experimentally derived quantities do not always have a completely one-to-one correspondence with the theoretical values, and care should be exercised in using these statistics to select a space group.

7.2.2 Structure Invariants and Origin-Fixing Reflections

The formulae used in direct phasing require, initially, the use of a few reflections with phases known, either uniquely or symbolically. In centrosymmetric crystals, the origin[†] is taken on one of the eight centers of symmetry in the unit cell, and we speak of the sign $s(hkl)$ of the reflection; $s(hkl)$ is $F(hkl)/|F(hkl)|$ and is either $+$ or $-$. We shall show next that, in any primitive, centrosymmetric space group in the triclinic, monoclinic, or orthorhombic systems, arbitrary signs can be allocated to three reflections in order to specify the origin at one of the centers of symmetry. These signs form a basic set, or "fountainhead," from which more and more signed reflections emerge as the analysis proceeds.

From (4.69) it follows that

$$F(hkl)_{0,0,0} = \sum_{j=1}^{N} g_j \cos 2\pi(hx_j + ky_j + lz_j) \qquad (7.2)$$

where $F(hkl)_{0,0,0}$ indicates an origin of coordinates at the point $0, 0, 0$. If this origin is moved to a center of symmetry at $\frac{1}{2}, \frac{1}{2}, 0$, the point that was originally x_j, y_j, z_j becomes $x_j - \frac{1}{2}, y_j - \frac{1}{2}, z_j$ (Figure 7.1), with $p = q = \frac{1}{2}$. The structure factor equation is now

$$F(hkl)_{1/2,1/2,0} = \sum_{j=1}^{N} g_j \cos 2\pi[(hx_j + ky_j + lz_j) - (h+k)/2] \qquad (7.3)$$

Expanding the cosine term, and remembering that $\sin[2\pi(h+k)/2]$ is zero, we obtain

$$F(hkl)_{1/2,1/2,0} = (-1)^{h+k} F(hkl)_{0,0,0} \qquad (7.4)$$

Equation (7.4) demonstrates that $|F(hkl)|$ is invariant under change of origin, as would be expected, but that a change of sign may occur, depending on the parity of the indices hkl. The complete results are listed in Table 7.3.

[*] H. T. Palmer, Personal communication (1973).
[†] The origin of the x, y, z coordinates of the structure.

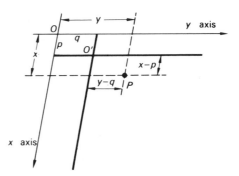

FIGURE 7.1. Transformation of the point $P(x, y)$, with respect to two-dimensional axes, by moving the origin from O to $O'(p, q)$; the transformed coordinates of P are $(x - p, y - q)$.

The use of this table will be illustrated by the following examples. Reflection 312 belongs to parity group 7 (ooe, in short). If $s(312)$ is given a plus sign, the origin could be regarded as being restricted to one from the following list:

$$0, 0, 0; \quad 0, 0, \tfrac{1}{2}; \quad \tfrac{1}{2}, \tfrac{1}{2}, 0; \quad \tfrac{1}{2}, \tfrac{1}{2}, \tfrac{1}{2}$$

Similarly, if $s(322)$, parity group 2 (oee), is also given a plus sign, the possible origins are

$$0, 0, 0; \quad 0, \tfrac{1}{2}, 0; \quad 0, 0, \tfrac{1}{2}; \quad 0, \tfrac{1}{2}, \tfrac{1}{2}$$

Combining these two sign allocations, the common origins are

$$0, 0, 0; \quad 0, 0, \tfrac{1}{2}$$

TABLE 7.3. Effect of a Change of Origin of Coordinates, among Centers of Symmetry, on the Sign of a Structure Factor

Parity group	1	2	3	4	5	6	7	8
Centers of symmetry	h even k even l even	h odd k even l even	h even k odd l even	h even k even l odd	h even k odd l odd	h odd k even l odd	h odd k odd l even	h odd k odd l odd
$0, 0, 0$	+	+	+	+	+	+	+	+
$\tfrac{1}{2}, 0, 0$	+	−	+	+	+	−	−	−
$0, \tfrac{1}{2}, 0$	+	+	−	+	−	+	−	−
$0, 0, \tfrac{1}{2}$	+	+	+	−	−	−	+	−
$0, \tfrac{1}{2}, \tfrac{1}{2}$	+	+	−	−	+	−	−	+
$\tfrac{1}{2}, 0, \tfrac{1}{2}$	+	−	+	−	−	+	−	+
$\tfrac{1}{2}, \tfrac{1}{2}, 0$	+	−	−	+	−	−	+	+
$\tfrac{1}{2}, \tfrac{1}{2}, \tfrac{1}{2}$	+	−	−	−	+	+	+	−

In order to fix the origin uniquely at, say, 0, 0, 0, we select another reflection with a plus sign with respect to 0, 0, 0. Reference to Table 7.3 shows that parity groups 4, 5, 6, and 8 each meet this requirement.

Parity groups 1 and 3 are excluded from the choice as the third origin-specifying reflection. Group 1 is a special case discussed below. Group 3 (eoe) is related to groups 2 and 7 through an addition of indices:

$$312 + 322 \rightarrow 634 \tag{7.5}$$

or, more generally,

$$ooe + oee \rightarrow eoe \tag{7.6}$$

since $o + o$ or $e + e \rightarrow e$, and $e + o \rightarrow o$.

Parity groups 2, 3, and 7 are said to be linearly related, and cannot be used together in defining the choice or origin.

Structure factors belonging to parity group 1 do not change sign on change of origin, as is evident from both the development of (7.4) and Table 7.3. Reflections in this group are called structure invariants; their signs depend on the actual structure and cannot be chosen at will.

7.2.3   Sign Determination—Centrosymmetric Crystals

Over the past twenty years, many equations have been proposed which are capable of providing sign information for centrosymmetric crystals. Two of these expressions have proved to be outstandingly useful, and it is to them that we first turn our attention.

Triple-Product Sign Relationship

In 1952, Sayre* derived a general formula for structures containing identical resolved atoms. For centrosymmetric crystals, it may be given in the form

$$s(hkl)s(h'k'l')s(h - h', k - k', l - l') \approx +1 \tag{7.7}$$

where the sign $\approx$ means "is probably equal to." The vectors associated with these reflections, $d^*(hkl)$, $d^*(h'k'l')$, and $d^*(h - h', k - k', l - l')$ form a

---

* D. Sayre, *Acta Crystallographica* **5**, 60 (1952).

closed triangle, or vector triplet, in reciprocal space. In practice, it may be possible to form several such vector triplets for a given hkl; Figure 7.2a shows two triplets for the vector 300. If two of the signs in (7.7) are known, the third can be deduced, and we can extend the sign information beyond that given in the starting set.

A physical meaning can be given to equation (7.7) by drawing the traces, in real space, of the three planes that form a vector triplet in reciprocal space (Figures 7.2a and 7.2b). For a centrosymmetric crystal, (6.41) becomes

$$\rho(x, y, z) = \frac{2}{V_c} \sum_h \sum_k \sum_l \pm |F(hkl)| \cos 2\pi(hx + ky + lz) \qquad (7.8)$$

The $|F(hkl)|$ terms in this equation take a *positive* sign if the traces of the corresponding planes pass through the origin, like the full lines in Figure 7.2b, and a *negative* sign if they lie midway between these positions, like the dashed lines in Figure 7.2b. The combined contributions from the three planes in question will thus have maxima at the points of their mutual intersections, which are therefore potential atomic sites, and correspond to regions of high electron density.

This argument is particularly strong if the three planes have high $|E|$ values. It may be seen from the diagram that triple intersections occur only at points where either three full lines ($+ + +$) meet, or two dashed lines and one full line meet (some combination of $+ - -$). This result is in direct agreement with (7.7). It is interesting to note that the structure of hexamethylbenzene was solved in 1929 by Lonsdale* through drawing the traces of three high-order, high-intensity reflection planes, $7\bar{3}0$, 340, and $4\bar{7}0$, and placing atoms at their intersections. These planes form a vector triplet, and this structure determination contained, therefore, the first but apparently inadvertent use of direct methods.

Σ_2 Formula

Hauptman and Karle† have given a more general form of (7.7):

$$s[E(hkl)] \approx s\left[\sum_{h'k'l'} E(h'k'l')E(h - h', k - k', l - l') \right] \qquad (7.9)$$

---

* K. Lonsdale, *Proceedings of the Royal Society A* **123**, 494 (1929).
† H. Hauptman and J. Karle, *Solution of the Phase Problem, I. The Centrosymmetric Crystal*, American Crystallographic Association Monograph No. 3 (1953).

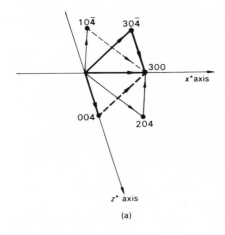

(a)

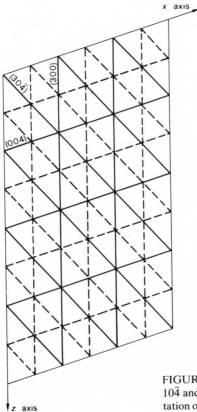

(b)

FIGURE 7.2. (a) Vector triplets 300, 204, 10$\bar{4}$ and 300, 30$\bar{4}$, 004; (b) physical interpretation of (7.7); the points of triple intersection are possible atomic sites.

the summation being over all vector pairs with known signs which form a triplet with hkl. The probability associated with (7.9) is given by

$$P_+(hkl) = \tfrac{1}{2} + \tfrac{1}{2}\tanh[(\sigma_3/\sigma_2^{3/2})\alpha'] \qquad (7.10)$$

where α' is given by

$$\alpha' = |E(hkl)| \sum_{h'k'l'} E(h'k'l')E(h-h', k-k', l-l') \qquad (7.11)$$

and σ_n by

$$\sigma_n = \sum_j Z_j^n \qquad (7.12)$$

where Z_j is the atomic number of the jth atom.

For a structure containing N identical atoms, $\sigma_3/\sigma_2^{3/2}$ is equal to $N^{-1/2}$. From (7.11), we see that the probability is strongly dependent upon the magnitudes of the $|E|$ values. Furthermore, unless glide-plane or screw-axis symmetry is present [see, for example, (7.14) and (7.15)], or there exists some other means of generating negative signs, (7.9) will produce only positive signs for all $E(hkl)$. Such a situation would correspond to a structure with a very heavy atom at the origin, and would, in general, lead to an incorrect solution.

If the combination of signs under the summation in (7.11) produces a large and negative value for α', the corresponding value of $P_+(hkl)$ may tend to zero. This result indicates that $s(hkl)$ is negative, with a probability that tends to unity.

Probability curves for different numbers N of atoms in the unit cell as a function of α' are shown in Figure 7.3. Since the most reliable signs from (7.9) are, from (7.11), associated with large $|E|$ values, we can now add to the origin-specifying criteria (page 298) the requirements of both large $|E|$ values and a large number of Σ_2 interactions for each reflection in the starting set. In this way, strong and reliable sign propagation is encouraged.

To illustrate the operation of the Σ_2 relationship, we shall consider the two vector triplets in Figure 7.2. The sign to be determined is $s(300)$, the others are assumed to be known. It may be noted that sometimes we speak of a sign as + or − and at other times as +1 or −1. The latter formulation is clearly more appropriate to computational methods. The data are tabulated as follows.

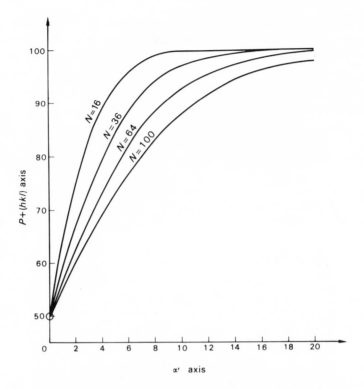

FIGURE 7.3. Percentage probability of a single triple-product (Σ_2) sign relationship as a function of α' for different numbers N of atoms in a unit cell.

hkl	$|E(hkl)|$					
300	2.40					
$h'k'l'$	$E(h'k'l')$	$h-h', k-k', l-l'$	$E(h-h', k-k', l-l')$	α'	$s(hkl)$	$P_+(hkl)$, %
$10\bar{4}$	+2.03	204	−2.22	19.3	−1	0.4
004	−1.95	$30\bar{4}$	+1.81		−1	

Assuming that N is 64, the indication given is that $s(300)$ is negative with a probability of 99.2%.

7.2.4 Amplitude Symmetry and Phase Symmetry

In space group $P\bar{1}$, the only symmetry-related structure factors are $F(hkl)$ and $F(\bar{h}\bar{k}\bar{l})$. According to Friedel's law (4.58), the intensities and, hence, the amplitudes* of these stucture factors are equal, and in centrosymmetric space groups $s(hkl) = s(\bar{h}\bar{k}\bar{l})$. Thus, the amplitude symmetry and the phase symmetry follow the same law, but this will not necessarily be true in other space groups.

From the geometric structure factor for space group $P2_1/c$, (4.116), and (4.117),

$$|F(hkl)| = |F(\bar{h}\bar{k}\bar{l})| = |F(h\bar{k}l)| = |F(\bar{h}k\bar{l})| \qquad (7.13)$$

while for the signs there are two possibilities:

$$k + l = 2n: \qquad s(hkl) = s(\bar{h}\bar{k}\bar{l}) = \ \ s(h\bar{k}l) = \ \ s(\bar{h}k\bar{l}) \qquad (7.14)$$

$$k + l = 2n + 1: \qquad s(hkl) = s(\bar{h}\bar{k}\bar{l}) = - \ s(h\bar{k}l) = -s(\bar{h}k\bar{l}) \qquad (7.15)$$

These relationships provide enhanced opportunities for Σ_2 relationships to be developed, and in this way space-group symmetry can improve the chances of successful phase determination. The amplitude symmetry and phase symmetry for all space groups are contained in the *International Tables for X-Ray Crystallography*, Vol. I.†

7.2.5 Σ_2 Listing

Because of both the increased probability in relationships developed for reflections with high $|E|$ values and the existence of many vector triplets in a complete set of data, the initial application of direct methods is limited to reflections with large $|E|$ values, say, greater than 1.5. As a more stringent condition, however, it must be remembered that an electron density map needs about eight symmetry-independent reflections per atom in the asymmetric unit in order to provide reasonable resolution of the electron density image.

A Σ_2 listing is prepared by considering each value of $|E(hkl)|$ greater than the preset limit, in order of decreasing magnitude, as a basic hkl vector, and searching the data for all interactions with $h'k'l'$ and $h - h', k - k', l - l'$. Some reflections will enter into many such interactions, while others will produce only a small number.

---

* **E** and **F** have the same symmetry relationships in any space group.
† See Bibliography, Chapter 1.

7.2.6 Symbolic-Addition Procedure

Karle and Karle* have described a technique for the systematic application of the Σ_2 formula for building up a self-consistent sign set. The various steps involved are outlined below, using results obtained with pyridoxal phosphate oxime dihydrate.†

Crystal Data

> Formula: $C_8H_{11}N_2O_6P \cdot 2H_2O$.
> System: triclinic.
> Unit-cell dimensions: $a = 10.94$, $b = 8.06$, $c = 9.44$ Å, $\alpha = 57.18$, $\beta = 107.68$, $\gamma = 116.53°$.
> V_c: 627 Å^3.
> D_m: 1.57 g cm^{-3}.
> M: 261.
> Z: 2.01, or 2 to the nearest integer.
> Absent spectra: none.
> Possible space groups: $P1$ or $P\bar{1}$. $P\bar{1}$ was chosen on the basis of intensity statistics (Table 7.2).
> All atoms are in general positions.

Sign Determination

(a) A total of 163 reflections for which $|E| \geq 1.5$ were arranged in descending order of magnitude, and a Σ_2 listing was obtained using a computer program.

(b) From a study of the Σ_2 listing, three reflections were allocated + signs (Table 7.4); they are the origin-fixing reflections, selected according to the procedures already discussed.

(c) Equation (7.9) was used by searching, initially, between members of the origin-fixing set and other reflections. To maintain a high probability, only the highest $|E|$ values were used. For example, $9\bar{5}\bar{5}$ ($|E| = 2.31$) is generated by the combination of $8\bar{1}\bar{5}$ and $\bar{1}40$:

$$s(9\bar{5}\bar{5}) \approx s(8\bar{1}\bar{5})s(\bar{1}40) = (+1)(+1) = +1 \qquad (7.16)$$

From (7.10), α' is 16.5, and Figure 7.3 tells us ($N = 38$, excluding hydrogen)

* J. Karle and I. L. Karle, *Acta Crystallographica* **21**, 849 (1966).
† Barrett and Palmer (see footnote to page 296).

TABLE 7.4. Starting Set for the Symbolic-
Addition Procedure

hkl	$\lvert E\rvert$	Sign
$9\bar{1}\bar{4}$	2.97	$+$ ⎱
$8\bar{1}5$	3.00	$+$ ⎰ a
$\bar{1}40$	2.38	$+$ ⎭
020	4.50	A ⎱
253	2.24	B ⎪
822	2.71	C ⎬ b
303	2.69	D ⎪
023	2.28	E ⎭

^{a} Origin-fixing reflections.
^{b} Letter symbols, each representing $+$ or $-$.

that the probability of this indication is about 99.7%. The new sign was accepted and used to generate more signs. This process was continued until no new signs could be developed with high probability.

(d) At this stage, it is usually found that the number of signs developed with confidence is small. This situation arose with pyridoxal phosphate oxime dihydrate, and the Σ_2 formula was then applied to reflections with symbolic signs. In this technique, a reflection was selected, again by virtue of its high $\lvert E\rvert$ value and long Σ_2 listing, and allocated a letter symbol (Table 7.4). Generally, less than five symbolic phases are sufficient, and there are no necessary restrictions on the parities of these reflections. However, it is desirable that there are no redundancies in the complete starting set, that is, no three reflections in the set should themselves be related by a triple product relationship.

As a symbol became involved in a sign of a reflection, it was written into the Σ_2 listing. The example in Table 7.5 shows a Σ_2 entry for $98\bar{6}$. Reading across the table, sign combinations are seen to be generated by multiplying $s(h'k'l')$ by $s(h-h', k-k', l-l')$, which are then written as $s(98\bar{6})$ in the penultimate column. Recurring combinations, such as ABD, gave rise to consistent indications. If the probability that $s(98\bar{6}) = s(ABD)$ is sufficiently large, this sign value is entered for $s(98\bar{6})$ wherever these indices occur. In the final column of the table, the probability of each sign indication is listed. Although they are small individually, the combined probability that $s(98\bar{6})$ was ABD is 100% [see (7.10)].

(e) When this process, too, had been exhausted with all letter signs, the results were examined for agreement among sign relationships. For

TABLE 7.5. Σ_2 Listing for the Reflection $9\bar{8}\bar{6}$ of Pyridoxal Oxime Phosphate with Appropriate Phase Symbols Added[a]

$h'k'l'$	$s(h'k'l')$	$\dfrac{|E_2|}{|E(h'k'l')|}$	$h-h', k-k', l-l'$	$s(h-h', k-k', l-l')$	$E(h-h', k-k', l-l'),$ $\dfrac{|E_3|}{}$	$|E_1||E_2||E_3|$	$s(9\bar{8}\bar{6})$	$P_+(9\bar{8}\bar{6})$, %
$1\,\bar{5}\,0$	BD	2.16	$8\,\bar{3}\,\bar{6}$	A	1.63	6.40	ABD	90
$10,\bar{2}\,\bar{2}$	AB	2.04	$1\,6\,4$	D	1.88	6.97	ABD	91
$10,\bar{7}\,\bar{1}$	D	1.87	$1\,1\,5$	AB	1.63	5.54	ABD	87
$4\,\bar{8}\,\bar{3}$	D	1.83	$5\,0\,\bar{3}$	ECD	1.58	5.25	EC	85
$3\,\bar{9}\,\bar{4}$		1.76	$6\,1\,2$		1.58	5.03		
$3\,\bar{5}\,\bar{6}$		1.70	$6\,\bar{3}\,0$		1.51	4.66		
$6\,\bar{7}\,\bar{2}$		1.68	$3\,\bar{1}\,\bar{4}$		1.63	4.98		
$10,\bar{4}\,\bar{2}$	B	1.62	$1\,4\,4$	AD	1.67	4.93	ABD	84
$0\,\bar{2}\,0$	$-A$	4.50	$9\,\bar{6}\,\bar{6}$		1.73	14.08		
$0\,\bar{8}\,0$	$+$	2.48	$9\,0\,\bar{6}$		1.85	8.30		

[a] $|E(9\bar{8}\bar{6})| = |E_1| = 1.89.$

example, in Table 7.5 there is a weak indication that $ABD = EC$. The most significant relationships found overall were $AC = E$, $C = EB$, $B = ED$, $AD = E$, and $AB = CD$. Multiplying the first by the second, and the first by the fourth, and remembering that products such as A^2 equal $+1$, reduces this list to $A = B$, $C = D$, and $E = AC$.

The five symbols were reduced, effectively, to two, A and C. The sign determination was rewritten in terms of signs and the symbols A and C; reflections with either uncertain or undetermined signs were rejected from the first electron density calculation.

7.2.7 Calculation of E Maps

The result of the above analysis meant that four possible sign sets could be generated by the substitutions $A = \pm 1$, $C = \pm 1$. The set with $A = C = +1$ was rejected because this phase assignment implies a very heavy atom at the origin of the unit cell. The three other sign combinations were used to calculate E maps. These maps are obtained by Fourier syntheses, using (7.8), but with $|E|$ replacing $|F|$ as the coefficients. The "sharp" nature of E [see (7.1)] is advantageous when using a limited number of data to resolve atomic peaks in the electron density map.

The sign combination for pyridoxal phosphate oxime dihydrate that led to an interpretable E map was $A = C = -1$. The atomic coordinates from this map (Figure 7.4a) were used in a successful refinement of the structure, and Figure 7.4b shows the conformation of this molecule.

If there are n symbolic signs in the final centrosymmetric phase solution, thee will be 2^n combinations, each of which can give rise to an E map, and it is desirable to set up criteria that will seek the most probable set. We shall consider such criteria during our discussion of the noncentrosymmetric case, where they are of even greater importance.

7.2.8 Phase Determination—Noncentrosymmetric Crystals

The noncentrosymmetric case is more difficult, both in theory and in practice. Much of this difficulty stems from the fact that the phase angle can take on any value between 0 and 2π, with a consequent imprecision in its determination. Nevertheless, direct methods are now being used regularly to solve such structures, with up to 100 atoms in the asymmetric unit. The methods for noncentrosymmetric crystals are in a state of active development, and the reader is referred to the Bibliography for accounts of some recent advances.

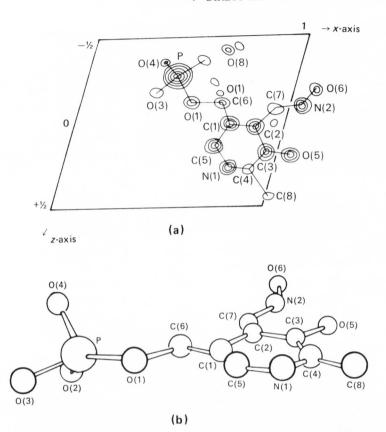

FIGURE 7.4. Pyridoxal phosphate oxime dihydrate: (a) composite three-dimensional E map on the ac plane; (b) molecular conformation viewed at about 10° to the plane of the six-membered ring.

Equations for the noncentrosymmetric crystal are, not surprisingly, more general forms of some of those for centrosymmetric crystals, such as (7.9). Using the fact that the electron density distribution is a nonnegative function, Karle and Hauptman* derived a set of inequality relationships, the first three of which can be written as

$$F_{000} \geq 0 \tag{7.17}$$

$$|F_{\mathbf{h}}| \leq F_{000} \tag{7.18}$$

$$\mathbf{F_h} - \delta_{\mathbf{h,k}} \leq \mathbf{r} \tag{7.19}$$

* J. Karle and H. Hauptman, *Acta Crystallographica* **3**, 181 (1950).

where

$$\delta_{h,k} = \mathbf{F}_{h-k}\mathbf{F}_k / F_{000} \tag{7.20}$$

and

$$\mathbf{r} = \frac{\begin{vmatrix} F_{000} & \mathbf{F}^*_{h-k} \\ \mathbf{F}_{h-k} & F_{000} \end{vmatrix} \begin{vmatrix} F_{000} & \mathbf{F}^*_k \\ \mathbf{F}_k & F_{000} \end{vmatrix}}{F_{000}} \tag{7.21}$$

We shall use here the convenient notation $\mathbf{h}$ for hkl, $\mathbf{k}$ for $h'k'l'$, $\mathbf{h-k}$ for a third reflection that forms a vector triplet with $\mathbf{h}$ and $\mathbf{k}$, and, in order to avoid excessive parentheses, $\mathbf{F_h}$ for $\mathbf{F(h)}$. Equations (7.17) and (7.18) are immediately acceptable in terms of earlier discussions in this book.

The structure factor $\mathbf{F_h}$, given only $|F_h|$, must lie on a circle of that radius on an Argand diagram (Figure 7.5). Equation (7.19) then indicates that $\mathbf{F_h}$ lies within a circle, center $\delta_{h,k}$ and radius $|r|$, between the points P and Q. Expansions of the determinants in (7.21), remembering that

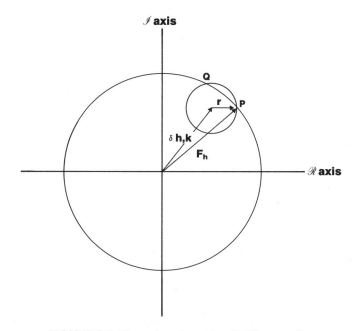

FIGURE 7.5. Illustration of equation (7.19) at equality.

$\mathbf{F} \cdot \mathbf{F}^* = |F|^2$, shows that the larger the values of $|F_{\mathbf{k}}|^2$ and $|F_{\mathbf{h-k}}|^2$, the closer $\mathbf{F_h}$ approaches $\delta_{\mathbf{h,k}}$. For a given $\mathbf{h}$, as $\mathbf{k}$ is varied, $\mathbf{F_h}$ is proportional to the average over $\mathbf{k}$

$$\mathbf{F_h} \propto \langle \mathbf{F_k F_{h-k}} \rangle_{\mathbf{k}} \tag{7.22}$$

the proportionality constant being F_{000}. We can see how equation (7.22) can give rise to (7.9), or to (7.7) for a single interaction.

Using the general relation

$$\mathbf{F_h} = |F_h| \exp(i\phi_{\mathbf{h}}) \tag{7.23}$$

we obtain the phase addition, or triple product relationship (tpr), formula

$$\phi_{\mathbf{h}} \approx \phi_{\mathbf{k}} + \phi_{\mathbf{h-k}} \tag{7.24}$$

The sign $\approx$ indicates an approximation which is better the larger the values of the corresponding structure factors. Where several triplets are involved with a given $\mathbf{h}$, (7.24) becomes

$$\phi_{\mathbf{h}} \approx \langle \phi_{\mathbf{k}} + \phi_{\mathbf{h-k}} \rangle_{\mathbf{k}} \tag{7.25}$$

where $\langle \ \rangle_{\mathbf{k}}$ implies an average, taken over a number of tpr's common to $\mathbf{h}$.

The $|F_o|$ data derived experimentally are converted to $|E|$ values, as already described. Again, we commence phase determination with $|E|$ values greater than about 1.5, in order to maintain acceptable probability limits. Equation (7.25) is illustrated by an Argand diagram in Figure 7.6 for four values of $\mathbf{k}$; $\phi_{\mathbf{h}}$ is the estimated phase angle associated with the resultant vector $\mathbf{R_h}$. Each vector labeled κ depends on a product $|E_{\mathbf{k}}| |E_{\mathbf{h-k}}|$ and may be resolved into components A and B along the real and imaginary axes, respectively, such that

$$A = |E_{\mathbf{k}}| |E_{\mathbf{h-k}}| \cos(\phi_{\mathbf{k}} + \phi_{\mathbf{h-k}}) \tag{7.26}$$

and

$$B = |E_{\mathbf{k}}| |E_{\mathbf{h-k}}| \sin(\phi_{\mathbf{k}} + \phi_{\mathbf{h-k}}) \tag{7.27}$$

It follows from (7.25)–(7.27) that

$$\tan \phi_{\mathbf{h}} \approx \frac{\sum_{\mathbf{k}} w_{\mathbf{h}} |E_{\mathbf{k}}| |E_{\mathbf{h-k}}| \sin(\phi_{\mathbf{k}} + \phi_{\mathbf{h-k}})}{\sum_{\mathbf{k}} w_{\mathbf{h}} |E_{\mathbf{k}}| |E_{\mathbf{h-k}}| \cos(\phi_{\mathbf{k}} + \phi_{\mathbf{h-k}})} \tag{7.28}$$

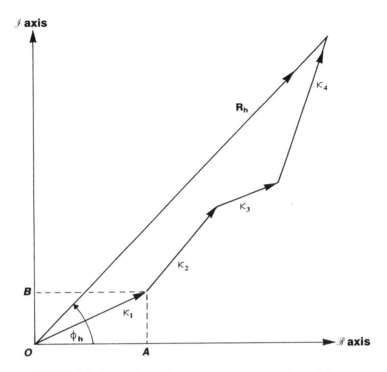

FIGURE 7.6. Summation of four vectors κ_1–κ_4 on an Argand diagram.

Equation (7.28) is a weighted tangent formula, where weights w_h may be either unity or given values as explained in Section 7.2.11. Current phase-determining procedures are based largely on (7.28). The reliability of (7.28) can be measured by the variance $V(\phi_h)$. The graph of Figure 7.7 may be obtained for $V(\phi_h)$ as a function of α_h, where

$$\alpha_h{}^2 = \left[\sum_k \kappa_{hk} \cos(\phi_k + \phi_{h-k})\right]^2 + \left[\sum_k \kappa_{hk} \sin(\phi_k + \phi_{h-k})\right]^2 \quad (7.29)$$

and

$$\kappa_{hk} = 2\sigma_3\sigma_2{}^{-3/2} \, |E_h| \, |E_k| \, |E_{h-k}| \quad (7.30)$$

with

$$\sigma_n = \sum_{j=1}^{N} Z_j{}^n \quad (7.31)$$

Z_j being the atomic number of the jth atom in a unit cell containing a total of N atoms. The parameter α_h gives a measure of the reliability with which

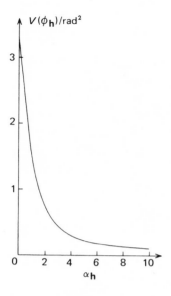

FIGURE 7.7. Variance $V(\phi_\mathbf{h})$ as a function of $\alpha_\mathbf{h}$.

$\phi_\mathbf{h}$ is determined by the tangent formula. When (7.29) contains only one term, as it may in the initial stages of phase determination, then $\alpha_\mathbf{h} = \kappa_{\mathbf{hk}}$ and is strongly dependent on the product $|E_\mathbf{h}|\,|E_\mathbf{k}|\,|E_{\mathbf{h}-\mathbf{k}}|$. Figure 7.7 shows clearly that $V(\phi_\mathbf{h})$ has acceptably small values when $\alpha_\mathbf{h}$ is greater than about 4 ($<30°$), but increases rapidly for $\alpha_\mathbf{h}$ decreasing below about 3 ($>40°$): $\alpha_\mathbf{h}$ depends also on $\sigma_3\sigma_2^{-3/2}$, which value depends on the number and types of atoms in the unit cell. This dependence may be illustrated by a hypothetical structure containing different numbers N of identical atoms. $\alpha_\mathbf{h}(=\kappa_{\mathbf{hk}})$ is then given by

$$\alpha_\mathbf{h} = \frac{2}{N^{1/2}}\,|\,E_\mathbf{h}\,|\,|\,E_\mathbf{k}\,|\,|\,E_{\mathbf{h}-\mathbf{k}}\,| \qquad (7.32)$$

Table 7.6 lists the values of $|\,E_{\min}\,|$ needed to obtain $\alpha_\mathbf{h} = 3$ for selected values of N from 25 to 100. The table illustrates clearly an important limitation of direct methods: the required $|\,E_{\min}\,|$ increases dramatically as a function of N whereas, as indicated earlier, the distribution of $|E|$ values is largely independent of structural complexity. Therefore it becomes more and more difficult to form a good starting set as N becomes larger and larger.

Calculation of $\alpha_\mathbf{h}$ from (7.29) is possible only when phases are available. In the initial stages of phase determination this is not practicable, and the following formula for the expectation value $(\alpha_E^{\,2})$ of $\alpha_\mathbf{h}^{\,2}$, which uses only

TABLE 7.6. Values of $|E_{min}|$ for $\alpha_h = 3.0$ in Structures Containing N Identical Atoms per Unit Cell

N	$|E_{min}|$
25	1.96
36	2.08
49	2.19
64	2.29
81	2.38
100	2.47

the values of κ_{hk}, has been developed:

$$\alpha_E{}^2 = \sum_k \kappa_{hk}^2 + \sum_k \sum_{k'} \kappa_{hk}\kappa_{hk'} \frac{I_1(\kappa_{hk})}{I_0(\kappa_{hk})} \frac{I_1(\kappa_{hk'})}{I_0(\kappa_{hk'})} \qquad (7.33)$$
$$k \neq k'$$

where I_0 and I_1 are modified Bessel functions of the zero and first orders, respectively. $I_1(\kappa)/I_0(\kappa)$ has the form shown in Figure 7.8 and may be expressed as the polynomial

$$I_1(\kappa)/I_0(\kappa) \approx 0.5658\kappa - 0.1304\kappa^2 + 0.0106\kappa^3$$

in the range $0 \leqslant \kappa \leqslant 6$; for $\kappa > 6$ the value of the function is essentially unity. These principles, used in conjunction with those discussed earlier for selecting the origin-determining reflections, may help a direct methods analysis to be established on a sound basis right from the beginning, and so lead to a number of sufficiently accurate phases to give an interpretable

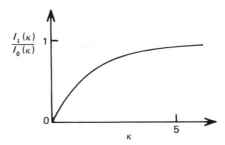

FIGURE 7.8. Variation of $I_1(\kappa)/I_0(\kappa)$ with κ.

TABLE 7.7. Crystal Data for Tubercidin

Formula	$C_{11}H_{14}N_4O_4$
M_r	266.3
Space group	$P2_1$
a	9.724(9) Å
b	9.346(11)
c	6.762(10)
β	94.64(10)°
V_c	610.4 Å³
D_m	1.449 g cm^{-3}
D_x	1.443
Z	2
$F(000)$	280

E map. Experience shows, however, that even with great care, the development of phases may not always be successful. In such an event the remedy is often to try again with a different starting set of reflections.

7.2.9 Phase Determination in Space Group $P2_1$

The structure of tubercidin was determined by Stroud*: Table 7.7 lists the crystal data for this compound.

In space group $P2_1$, $|E(hkl)|$ has the following symmetry equivalence:

$$| E(hkl) | = | E(\bar{h}k\bar{l}) | = | E(h\bar{k}l) | = | E(\bar{h}\bar{k}\bar{l}) | \qquad (7.34)$$

The phases of the symmetry-related reflections in this space group are also linked, but in a different way, according to the parity of k:

$$k = 2n: \qquad \phi(hkl) = \phi(\bar{h}k\bar{l}) = -\phi(h\bar{k}l) = -\phi(\bar{h}\bar{k}\bar{l}) \qquad (7.35)$$

$$k = 2n + 1: \qquad \phi(hkl) = \pi + \phi(\bar{h}k\bar{l}) = \pi - \phi(h\bar{k}l) = -\phi(\bar{h}\bar{k}\bar{l}) \qquad (7.36)$$

Although ϕ_h can, in general, have a value anywhere in the range 0–2π, the $h0l$ reflections are restricted to the values 0 or π in this space group; in other words the $h0l$ zone is centric.

* R. M. Stroud, *Acta Crystallographica* **B29**, 60 (1973).

TABLE 7.8. Origin-Specifying Phases for Tubercidin

hkl	$|E_\mathbf{h}|$	$\phi_\mathbf{h}$
$10\bar{6}$	1.95	0
$40\bar{1}$	2.09	0
$71\bar{4}$	2.45	0

The origin was specified by assigning phases to three reflections, according to the known rules, as shown by Table 7.8. Next, new phases were determined according to (7.24) or (7.25). In order to maintain an expected variance $V(\phi_\mathbf{h})$ (Figure 7.7) of no more than 0.5 rad^2, the product $|E_\mathbf{h}|\,|E_\mathbf{k}|\,|E_{\mathbf{h}-\mathbf{k}}|$ must be greater than 8.5 for this structure. Two new phases $\phi(80\bar{2})$ and $\phi(612)$ were thus determined from the origin set (Table 7.10), further phases being determined in terms of symbols (Table 7.9). Eleven phases were generated in terms of the origin phases and symbol a, 20 after adding letter b, and 47 after adding the third symbolic phase c. Table 7.10 illustrates the initial stages of this process. The criteria for accepting a phase were as follows:

(i)   that $V(\phi_\mathbf{h})$, irrespective of the actual choice for c and a (b is a structure invariant with phase 0 or π), should be less than 0.5 rad^2, no matter how many contributors there were to the sum in (7.25);

(ii) that where there were two or more different indications for a phase, the phase would be accepted only when indications of one type predominated strongly.

TABLE 7.9. Course of the Phase Determination Procedure for Tubercidin

hkl	$\phi_\mathbf{h}$	$|E_\mathbf{h}|$	Number of numerical or symbolic phases
Origin set	See Tables 7.8 and 7.10		5
$13\bar{8}$	a	2.99	11
206	b	2.20	20
790	c	2.76	47

TABLE 7.10. Initial Development of Phases for Tubercidin

$\mathbf{h}$ [a]	$|E_{\mathbf{h}}|$	$\phi_{\mathbf{h}}$	$|E_{\mathbf{h}}||E_{\mathbf{k}}||E_{\mathbf{h-k}}|$
Origin set			
* $40\bar{1}$	2.09	0	
* $10\bar{6}$	1.95	0	
* $71\bar{4}$	2.45	0	
New phases			
$40\bar{1}$		0	
$40\bar{1}$		0	
* $80\bar{2}$	2.33		10.2
$\bar{1}06$ [b]		0 [b]	
$71\bar{4}$		0	
* 612	1.83	0	8.7
First letter			
$13\bar{8}$	2.99	a	
612		0	
* $74\bar{6}$	2.20	a	12.0
$7\bar{1}\bar{4}$		π [c]	
$\bar{1}38$		$\pi + a$ [c]	
* 624	2.19	a	16.0

[a] An asterisk denotes a new phase in the list.
[b] Symmetry-related phase used from (7.35).
[c] Symmetry-related phase used from (7.36).

During the phase analysis it soon became clear that in order to avoid a large number of inconsistencies, the structure invariant b (ϕ_{206}) should have a value of zero. This result is shown by Table 7.11, which illustrates also an example of phase determination for the case of $|E(63\bar{3})|$. The indications

$$\phi(63\bar{3}) = c - 2a - b \tag{7.37}$$

and

$$b = 0 \tag{7.38}$$

TABLE 7.11. Phase Indications for $|E(6\bar{3}\bar{3})|$ for Tubercidin: $\mathbf{h} = 6\bar{3}\bar{3}$, $|E_{\mathbf{h}}| = 2.37$ [a]

k	$|E_{\mathbf{k}}|$	$\phi_{\mathbf{k}}$	h − k	$|E_{\mathbf{h-k}}|$	$\phi_{\mathbf{h-k}}$	$|E_{\mathbf{h}}||E_{\mathbf{k}}||E_{\mathbf{h-k}}|$	$\phi_{\mathbf{h}} \approx \phi_{\mathbf{k}} + \phi_{\mathbf{h-k}}$
790	2.76	c	$\bar{1}\,\bar{6}\,\bar{3}$	2.08	$-2a - b$	13.6	$c - 2a - b$
$\bar{2}\bar{0}\bar{6}$	2.20	b	$8\,3\,3$	2.09	$c - 2a - b$	10.9	$c - 2a$
624	2.19	a	$0\,1\,7$	1.59			*
$\bar{2}\bar{2}\bar{5}$	2.13	a	$8\,1\,2$	1.49			*
840	1.67	$c - 2a$	$\bar{2}\,1\,\bar{3}$	1.69	0	6.7	$c - 2a$
422	1.60	$a + b$	$10\,1\,\bar{5}$	1.82			*
$\bar{7}\bar{1}\bar{4}$	2.45	π	$\bar{1}\,4\,1$	1.78	$\pi + c - 2a - b$	10.3	$c - 2a - b$
$\bar{7}\bar{4}\bar{6}$	2.20	$-a$	$\bar{1}\,7\,3$	1.96	$c - a - b$	10.2	$c - 2a - b$
$0\bar{2}\bar{1}$	2.12	$-a - b$	$6\,5\,\bar{2}$	2.06	$c - a$	10.4	$c - 2a - b$
$\bar{6}\bar{1}\bar{2}$	1.83	π	$0\,4\,\bar{5}$	1.61	$\pi + c - 2a - b$	7.0	$c - 2a - b$
$\bar{1}\bar{4}0$	1.64	$a - b$	$5\,7\,\bar{3}$	2.49	$c - a$	9.7	$c - 2a - b$
$\bar{4}\bar{2}\bar{2}$	1.60	$a - b$	$2\,5\,\bar{1}$	1.66	$c - a - b$	6.3	$c - 2a$
$\bar{5}\bar{1}\bar{3}$	1.59		$1\,4\,0$	1.64	$a + b$		*
$\bar{5}\bar{1}\bar{1}$	1.53		$1\,4\,\bar{2}$	1.66	$c - a$		*

[a] There are six indications that $\phi(6\bar{3}\bar{3}) = c - 2a - b$ and three indications that $\phi(6\bar{3}\bar{3}) = c - 2a$. If we accept $\phi(6\bar{3}\bar{3}) = c - 2a - b$, there are three indications that $b = 0$. Five interactions marked with an asterisk do not contribute.

are strong, because they both come from multiple indications (6 and 3, respectively). By reiteration of the phase addition procedure described above, the results in Table 7.12 indicate relationships between a and c.

Bearing in mind that an objective is a self-consistent set of phases, it is well to consider how this might now be achieved. Refinement of phases could in principle be achieved by application of (7.28). However, this would be possible only if numerical values for a and c (taking b as zero) were available. Alternatively, if a working formula relating a and c could be found, (7.28) could be implemented by substitution of values for one symbol only. Table 7.12 shows that there were 41 indications that

$$c = \pi + pa \qquad\qquad (7.39)$$

where the best numerical value for p was found to be 3.29. Hence,

$$c = \pi + 3.29a \qquad\qquad (7.40)$$

The symbol a was then limited to the range

$$0 < a < \pi \qquad\qquad (7.41)$$

in order to fix the enantiomorph (Appendix 10). Values for a were chosen such that

$$a = n\pi/8 \qquad (n = 1, 2, \ldots, 8) \qquad\qquad (7.42)$$

TABLE 7.12. Relationships between Letter Symbols

Form of relationship	Number of indications
$c = \pi + 2a$	7
$c = \pi + 3a$	15
$c = \pi + 4a$	19
$c = 3a$	5
$c = 4a$	2
$c = -3a$	4 or 5
$a = 0$	2
$a = \pi$	2
$b = 0$	Many
$b = \pi$	None

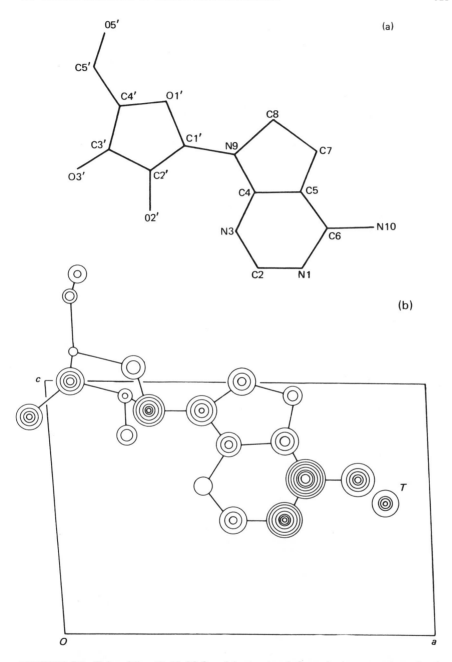

FIGURE 7.9. Tubercidin, $C_{11}H_{14}N_4O_4$: (a) structural formula in approximately the same orientation as in the E map (b) composite E map; contours (idealized) are drawn at arbitrary equal intervals. Some peaks are heavier than others because of the limited data set used; peak T was the only significant spurious peak.

TABLE 7.13. Summary of Phase Indications[a] for Tubercidin from Phase
Addition, Assuming $b = 0$

$\mathbf{h}$ [b]	$\lvert E_{\mathbf{h}} \rvert$	$\phi_{\mathbf{h}}$	Number of indications with $\lvert E_{\mathbf{h}} \rvert \lvert E_{\mathbf{k}} \rvert \lvert E_{\mathbf{h-k}} \rvert \geq 8.5$	Number in disagreement
** 1 3 $\bar{8}$	2.99	a	7	0
** 7 9 0	2.76	c	10	
7 7 1	2.54	$c - a$	7	
5 7 $\bar{3}$	2.49	$c - a$	5	
* 7 1 $\bar{4}$	2.45	0	13	
6 3 $\bar{3}$	2.37	$c - 2a$	7	
8 0 $\bar{2}$	2.33	0	8	
7 7 0	2.22	$\pi + c - a$	7	1
** 2 0 6	2.20	0	6	
7 4 $\bar{6}$	2.20	a	4	
6 2 4	2.19	a	8	
8 0 $\bar{3}$	2.18	π	4	
8 6 1	2.18	$c - a$	3	
7 0 0	2.17			
7 5 5	2.16	$\pi + 2a$	2	
2 2 5	2.13	a	6	
0 2 1	2.12	a	4	
8 2 $\bar{3}$	2.12	a	6	
3 3 4	2.12	a	1	
9 2 $\bar{1}$	2.11	0	2	
* 4 0 $\bar{1}$	2.09	0	3	
8 3 3	2.09	$c - 2a$	4	1
1 6 3	2.08	$2a$	5	
6 5 $\bar{2}$	2.06	$c - a$	2	
6 2 $\bar{5}$	2.04		0	6
5 3 $\bar{4}$	2.03	a	1	
10 2 0	2.02	$\pi + a$	1	
5 0 0	2.01	0	1	2
8 6 0	2.01	$2a$		1
1 3 $\bar{4}$	1.98	$c - 2a$	2	1
7 0 3	1.97	π	1	
6 2 $\bar{7}$	1.97	a	2	
0 7 3	1.97	$\pi + c - a$	2	
8 6 $\bar{4}$	1.96	$\pi + 2a$	1	
1 7 $\bar{3}$	1.96	$\pi + c - a$	5	1
* 1 0 6	1.95	0	1	
8 5 $\bar{5}$	1.95	$\pi + c - 2a$	1	

TABLE 7.13—*cont.*

h ^{b}	$\|E_\mathbf{h}\|$	$\phi_\mathbf{h}$	Number of indications with $\|E_\mathbf{h}\|\|E_\mathbf{k}\|\|E_{\mathbf{h}-\mathbf{k}}\| \geq 8.5$	Number in disagreement
0 2 6	1.94			
6 0 $\bar{4}$	1.88			
7 0 $\bar{3}$	1.87	0	0	2
6 4 $\bar{4}$	1.86			
8 5 $\bar{1}$	1.85	$\pi + 2a$	0	2
6 4 $\bar{5}$	1.84	a	2	1
9 5 $\bar{2}$	1.84	$\pi + 2a$	0	
9 6 $\bar{3}$	1.84			
6 2 $\bar{6}$	1.83	$\pi + a$	0	1
6 1 2	1.83	0	2	1
5 7 3	1.83			
2 0 $\bar{7}$	1.82	0	1	
10 1 $\bar{5}$	1.82			
5 3 $\bar{5}$	1.81			
12 0 $\bar{2}$	1.81	π	1	
1 5 $\bar{4}$	1.81	$\pi + c - 2a$	3	1
1 5 5	1.81	$c - a$	2	1
8 3 $\bar{2}$	1.80	$\pi + a$	0	2
6 0 7	1.79			
1 6 6	1.79	$\pi + c - a$	0	1
1 4 $\bar{1}$	1.78	$\pi + c - 2a$	3	1
6 6 0	1.78	$\pi + 2a$	0	1
2 4 2	1.77			
9 4 $\bar{5}$	1.77	$\pi + c - a$	0	1
7 0 4	1.76	0	1	0
7 2 2	1.75			
4 2 $\bar{7}$	1.75			
1 8 0	1.75			
7 2 5	1.74			
7 2 $\bar{4}$	1.71			
3 3 $\bar{6}$	1.71	$\pi + a$	0	
1 4 7	1.71	$\pi + c - 2a$	1	
5 5 4	1.71	$c - a$	0	1
11 0 $\bar{1}$	1.70			
1 2 $\bar{6}$	1.70	π or $\pi + a$	0	
2 1 3	1.69	0	1	1
0 4 7	1.69			

continued overleaf

TABLE 7.13—*cont.*

$\mathbf{h}$[b]	$\lvert E_{\mathbf{h}} \rvert$	$\phi_{\mathbf{h}}$	Number of indications with $\lvert E_{\mathbf{h}} \rvert \lvert E_{\mathbf{k}} \rvert \lvert E_{\mathbf{h-k}} \rvert \geq 8.5$	Number in disagreement
5 3 $\bar{2}$	1.68			
6 0 5	1.67	0	0	
9 0 $\bar{1}$	1.67	π	0	
8 4 0	1.67			
8 3 1	1.66			
1 4 $\bar{2}$	1.66	$c - 2a$	0	2
2 4 $\bar{5}$	1.66			
2 5 $\bar{1}$	1.66	$c - a$	0	
7 3 2	1.65	$\pi + c - 2a$	0	1
1 5 $\bar{3}$	1.65	$c - \dfrac{3a}{\bar{2}}$	1	
1 3 1	1.64	$\pi + a$	3	
1 4 0	1.64	a	1	
9 3 $\bar{5}$	1.63			
2 4 $\bar{7}$	1.63			
7 8 0	1.63	$c - a$	0	1
12 0 $\bar{3}$	1.62	0	1	
3 2 $\bar{8}$	1.62			
0 6 4	1.62	$\pi + c - a$	2	
3 8 0	1.62			
0 4 5	1.61	$\pi + c - 2a$	2	
4 2 $\bar{2}$	1.60	a	0	2
4 2 6	1.60			
6 3 3	1.60			
5 7 $\bar{1}$	1.60			
0 1 7	1.59			
5 1 $\bar{3}$	1.59			
9 4 4	1.59	$c - 2a$	0	
6 5 4	1.59			
4 6 $\bar{3}$	1.59			

[a] Seventy-two of the 103 phases were assigned symbolic or numerical values.
[b] One asterisk indicates a symbolic phase assignment; two asterisks indicate the origin-determining phase.

and converted into phases by (7.40); each set was expanded and refined by (7.28) (taking $w_h = 1$) for up to 419 reflections with $|E_{min}| \geq 1.0$. Some phases were rejected because of inconsistencies in their phase indications. An interpretable E map was obtained using the refined phase set with $a = 6\pi/8$; a composite diagram is given in Figure 7.9.

In conclusion, we give in Table 7.13 a list of phase indications obtained for 103 reflections with $|E_h| > 1.59$, using (7.24) and (7.25), and $b = 0$. The list contains about two or three times the number of terms normally recommended for an initial phase analysis; 72 of the 103 phases have symbolic or numerical phase assignments. The reader is invited to construct a Σ_2 listing and work through this analysis. It is an excellent test of one's expertise in handling the basic formulas used in direct methods, including the proper use of symmetry relationships for the space group in question.

7.2.10 Advantages and Disadvantages of Symbolic Addition

Symbolic addition has several advantages and disadvantages, summarized as follows:

(a) *Advantages*

   (i) The user is in control throughout the analysis. He has the responsibility of making sure that all formulas, including symmetry relationships, are applied correctly.

  (ii) The user can make decisions regarding criteria of acceptance of phase indications, the number of $|E|$ values to include, the number of symbolic phases, the choice of starting set, and so on.

(b) *Disadvantages*

   (i) The analysis can be carried out only by a specialist in crystallography.

  (ii) The procedure is slow, requiring many hours of preparation before meaningful results emerge.

 (iii) If a large number of symbols is required, many phase sets will be produced, each of which requires refinement by the tangent formula.

Not surprisingly, alternative rapid and more automatic methods of applying direct methods formulas were sought in the late 1960s, leading to development of the multisolution methods.*

---

*M. M. Woolfson and G. Germain, *Acta Crystallographica* **B24**, 91 (1968); G. Germain, P. Main and M. M. Woolfson, *Acta Crystallographica* **B26**, 274 (1970).

7.2.11 Multisolution Philosophy and Brief Description of the Program MULTAN

In symbolic addition, we saw that a new phase may be indicated several times by the same combination of symbols. Then, the individual indications reinforce one another to produce an improved joint probability, since α_h (7.29) is then given by

$$\alpha_h = \sum_{k_r} \kappa_{h,k} \tag{7.43}$$

taking $\kappa_{h,k}$ from (7.30).

The combination of indications involving entirely different symbols presents a problem. For instance, suppose two separate indications for ϕ_h are $a + b$ and $c + d$, where $a = \pi/4$, $b = 3\pi/4$, $c = -\pi/4$, and $d = -3\pi/4$, say. The individual indications $a + b$ and $c + d$ both predict $\phi_h = \pi$. Symbolic combination would yield an indication $\frac{1}{2}(a + b + c + d)$, which results in the false value of $\phi_h = 0$. Even if values of $c = -\pi/4 + 2\pi = 7\pi/4$, $d = -3\pi/4 + 2\pi = 5\pi/4$ are used the combination indication would again be false, since $\frac{1}{2}(a + b + c + d) \equiv 0$ modulo 2π. In such a situation, the usual practice with symbolic addition is to accept the strongest indication and put other indications aside for possible future use.

The multisolution philosophy gets around this problem by introducing numerical phases rather than symbols at an early stage. It is then always possible to combine individual phase indications by means of the tangent formula. This strategy was implemented in the program MULTAN.* This phase-determining procedure is based on a starting set of phases, formed in a similar way to that used in symbolic addition, that may be categorized mainly as follows:

(a) origin and enantiomorph definition, requiring up to four phase assignments according to the rules already discussed, and
(b) further phases required to initiate a continuous phase-determining process by tangent formula expansion; these phases are variables.

It is in category (b) where MULTAN differs fundamentally from the symbolic addition technique. Instead of introducing letter phases for these reflections, they are given numerical values. Specifically, according to space group symmetry, values such as 0, π, or $\pm\pi/2$ may be assigned. General

---

* P. Main *et al.*, Department of Physics, University of York, York, England (1974). The latest version is MULTAN 82.

noncentrosymmetric phases are assigned the values $\pm\pi/4$ and $\pm 3\pi/4$, or for enantiomorph specification merely $\pm\pi/4$. In this way, a total of p variable phases, including enantiomorph specification, would therefore yield $2 \times 4^{p-1}$ possible phase sets for a noncentrosymmetric crystal. The method is justified numerically in that it gives a maximum error of $45°$ for any of the initial variable phases with a mean error of only $22.5°$. The tangent formula is used to determine probable values of new phases, and the number of possible phase sets rises rapidly with p, as Table 7.14 shows.

The MULTAN program employs a modified tangent formula given by

$$\tan\phi_h = \frac{\sum_k Q_{h,k} \sin(\phi_k + \phi_{h-k})}{\sum_k Q_{h,k} \cos(\phi_k + \phi_{h-k})} = \frac{T_h}{B_h} \tag{7.44}$$

where

$$Q_{h,k} = w_k w_{h-k} \, | \, E_k \, | \, | \, E_{h-k} \, | / (1 - | \, U_h \, |^2) \tag{7.45}$$

with

$$w_h = \tan[\sigma_3 \sigma_2^{-3/2} \, | \, E_h \, | \, (T_h^2 + B_h^2)^{1/2}] \tag{7.46}$$

and

$$| \, U_h \, | = | \, F_h \, | \Big/ \sum_{j=1}^{N} f_j \qquad \text{(the unitary structure factor)} \tag{7.47}$$

Thus, each phase assignment carries a weight designed such that poorly determined phases have little effect in the generation of new phases, while the fact that all phases are included leads to efficient propagation of phase information throughout the data set.

TABLE 7.14. Number of Phase Sets Generated by the Multisolution Method

Number of variable phases p	Number of phase sets generated	
	for a noncentrosymmetric crystal $2 \times 4^{p-1}$	for a centrosymmetric crystal 2^p
1	2	2
2	8	4
3	32	8
4	128	32
5	512	64

The choice of starting-set reflections for MULTAN is made automatically by a subroutine called CONVERGE. In the case of an unsatisfactory choice by CONVERGE, facilities are available for the user to make his own origin selection. CONVERGE forms an ordered map of reflections such that, from the starting set, each reflection may be determined in terms of all those preceding it. In the initial stages of phase determination, the first 60 phases from the convergence map are used and only those phase relationships determined with α_h (7.29) greater than 5 are accepted. Several passes are made through the tangent formula, lowering this limit each time. As phases are developed the α_h's increase, self-consistency being defined as a change of less than 2% in $\sum_h \alpha_h$ from one cycle to the next. Up to this stage the phases of the starting reflections have been kept constant. They are now allowed to vary and refine to produce a trial phase set.

7.2.12 Figures of Merit

It is helpful to be able to choose the most probable set of phases prior to the calculation of an E map. In the centrosymmetric case, the following quantities may be determined:

$$M_1 = \sum s_h s_k s_{h-k} \tag{7.48}$$

$$M_2 = \sum |E_h E_k E_{h-k}| s_h s_k s_{h-k} \tag{7.49}$$

$$M_3 = \sum P_{hk} s_h s_k s_{h-k} \tag{7.50}$$

where P_{hk} is given by (7.10), written for a single tpsr, i.e., without the summation in (7.11). It is easy to see that M_1, M_2, and M_3 should all be large for a set of strong, consistent sign relationships. If the E map produced by the most clearly indicated set of signs does not contain a correct structure, the next set in the order of merit would be tried.

In MULTAN, phase sets are assessed by three figures of merit, ABS FOM, PSI ZERO, and RESID, computed for each phase set. In addition, a fourth combined figure of merit, COMBINED FOM, is now produced in order to provide an overall picture of the three individual quantities. ABS FOM is a measure of the internal consistency among the Σ_2 relationships and is given by Z where

$$Z = \sum_h (\alpha_h - \alpha_{R_h}) \bigg/ \sum_h (\alpha_{E_h} - \alpha_{R_h}) \tag{7.51}$$

$\alpha_{R_\mathbf{h}}$ is the value expected from random phases given by

$$\alpha_R = \left\{ \sum_\mathbf{k} \kappa^2 \right\}^{1/2} \tag{7.52}$$

with κ given by (7.30); $\alpha_{E_\mathbf{h}}$ is the estimated value of $\alpha_\mathbf{h}$ calculated from (7.33) during the convergence procedure. Thus, ABS FOM is zero for random phases and unity if $\alpha_\mathbf{h}$ is equal to its expectation value. For crystal structures containing translational symmetry elements, the correct set of phases should correspond to one of the higher values of ABS FOM, because the tangent formula tends to maximize phase relation consistencies. In practice, a correct set of phases has been found usually to correspond to values of ABS FOM in the range of 1.0–1.4, but phase sets with values as low as 0.7 have yielded interpretable E maps; there are also many instances of values of ABS FOM much larger than 1.5 leading to correct structures.

PSI ZERO is defined as

$$\psi_0 = \sum_\mathbf{h} \sum_\mathbf{k} | E_\mathbf{k} | | E_{\mathbf{h}-\mathbf{k}} | \tag{7.53}$$

where the $| E |$ values in this summation are either very small or zero. For small $| E_\mathbf{h} |$, ψ_0 should have a small value for the correct phase set. It is independent of the tangent formula and therefore may be useful as a discriminator when ABS FOM yields similar values for different phase sets.*
RESID corresponds to the Karle R_K parameter, and is calculated in a way similar to the familiar crystallographic R factor:

$$\text{RESID} = R_\mathrm{K} = \frac{\sum_\mathbf{h} | | E_\mathbf{h} | - | E_\mathbf{h} |_{\text{calc}} |}{\sum_\mathbf{h} | E_\mathbf{h} |} \tag{7.54}$$

where

$$| E_\mathbf{h} |_{\text{calc}} = K \langle | E_\mathbf{k} | | E_{\mathbf{h}-\mathbf{k}} | \rangle_\mathbf{h} \tag{7.55}$$

and K is a scale factor given by

$$K = \sum_\mathbf{h} | E_\mathbf{h} |^2 \Big/ \sum_\mathbf{h} \langle | E_\mathbf{k} | | E_{\mathbf{h}-\mathbf{k}} \rangle^2 \tag{7.56}$$

The correct set of phases should correspond to that with the lowest RESID.
Experience has shown that discrimination between phase sets is often possible in terms of either Z or R_K. However, sometimes both may fail to

---

*Often with space groups containing no translational symmetry elements.

enable the correct phase set to be selected easily. A further useful indicator is the combined figure of merit C, given by

$$C = W_1 \frac{(Z - Z_{\min})}{(Z_{\max} - Z_{\min})} + W_2 \frac{[(\psi_0)_{\max} - \psi_0]}{[(\psi_0)_{\max} - (\psi_0)_{\min}]}$$

$$+ W_3 \frac{[(R_K)_{\max} - R_K]}{[(R_K)_{\max} - (R_K)_{\min}]} \tag{7.57}$$

where W_1, W_2, and W_3 are weights, often unity, which may be changed to give more emphasis to ψ_0 and less to Z for space groups without translational symmetry elements.

A more complete description of MULTAN is not possible here. Recent features include the automatic production of E maps and their interpretation in terms of molecular geometry, the use of known atomic positions, and subtraction of contributions from heavy atoms in special positions. The success of the program in numerous structure determinations speaks for itself. Those structures that have failed to yield provide the incentive for further developments, perhaps advancing to the solution of small protein structures in the foreseeable future.

7.2.13 Example of the Use of MULTAN

Noncentrosymmetric Case: Methyl Warifteine (MEW)

Crystal data for methyl warifteine (Figure 7.10) are listed in Table 7.15. The starting-set reflections, generated automatically by CONVERGE, are given in Table 7.16. Using 228 data with $|E| > 1.70$, the number of Σ_2 interactions was limited (a facility available to the user) to 2000, and 16 phase sets were generated. The phase set having the second highest value

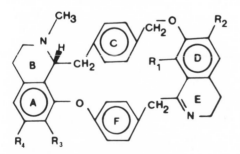

FIGURE 7.10. Structural formula of methylwarifteine, $R_1 = OH$; $R_2 = R_3 = R_4 = OCH_3$.

TABLE 7.15. Crystal Data for MEW

Formula	$C_{37}C_{38}N_2O_6$	V_c	3086 Å^3
M_r	606.36	D_x	1.31 g cm^{-3}
Space group	$P2_12_12_1$	Z	4
a	17.539(4) Å	$F(000)$	1288
b	12.224(3)	$\mu(Cu\,K\alpha)$	5.6 cm^{-1}
c	14.393(3)	Crystal size	0.2, 0.3, 0.3 mm

TABLE 7.16. Starting-Set Reflections for MEW

hkl	$\|E\|$	ϕ/deg
0 1 3	2.10	90 ⎤
1 0 3	2.30	90 ⎬ origin
14 3 0	4.73	360 ⎦
0 11 1	4.40	90, 270
1 1 7	2.36	45, 135, 225, 315
9 7 2	3.07	45, 315 (enantiomorph)

of Z (0.96) and lowest R_K (0.23) produced an E map from which 36 of the 45 nonhydrogen atoms were identified in geometrically acceptable positions. These 36 sites were drawn from the highest 60 peaks in the map. The remaining 9 atoms of MEW were located by an electron density synthesis using all reflections, the initial R factor being 0.35. After refinement the final R factor was 0.057 for 2508 observed reflections.* A stereoview of the molecule is shown in Figure 7.11.

7.2.14 Some Experiences

Direct methods have not yet been developed to the extent where all structures can be solved in a straightforward manner. The experience of many workers in this field does, however, permit the drawing up of a

---

* N. Borkakoti and R. A. Palmer, *Acta Crystallographica* **B34**, 482 (1978).

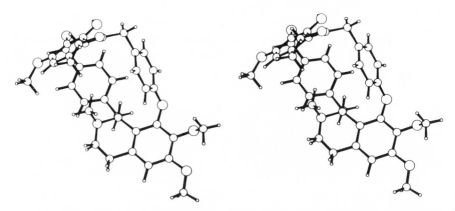

FIGURE 7.11. Stereoscopic view of MEW looking along c; the circles, in increasing order of size, represent H, C, N, and O.

number of strategies which have often proved significant in structure determinations.

Some Prerequisites for Success in Using Direct Methods

Some of the rules that have emerged can be summarized as follows:

(i)   As complete and accurate a set of $|F_o|$ data as possible must be used. Sometimes a small crystal size results in many weak-intensity measurements and consequent difficulty in fulfilling this condition.

(ii) If automatic solution of the phase problem fails, we must be prepared to intervene (a) by selecting origin and starting phases, (b) by varying the value of $|E_{min}|$, (c) by varying the number of interactions present, and (d) by employing special techniques, possibly involving tangent-formula recycling and so on.

(iii) Remember that a computer program may contain mistakes, even after considerable length of use. Always check the validity of results as far as possible, especially assignments of basic origin- and enantiomorph-fixing reflections and values of special phases.

(iv) Do not give in!

Figures of Merit: A Practical Guide

We have mentioned the four criteria calculated in MULTAN for assessment and ordering of phase sets. This facility is necessary in order to overcome the basic disadvantage inherent in the multisolution method, namely, that many possible phase sets may have to be considered. In principle, the phase sets are explored in order until the structure is found; normally, one does not investigate them further.

Experience suggests that two of the figures of merit currently calculated in MULTAN, Z and R_K, are often sufficient to enable the most likely E maps to be explored. C appears also to be very useful, and ψ_0 can be used in difficult cases. As a guide, and not as an absolute measure, we may use the following values in selecting the order of calculation of E maps: $Z > 1.0$, $R_K < 0.20$, and $C > 2.5$. Other programs may calculate different figures of merit, and one should check with the author for the corresponding recommended values.

Signs of Trouble, and Past Remedies When the Structure Failed to Solve

This section summarizes some of our own experiences, and records factors involved when a structure solution failed to emerge from direct phase-determining methods. Although these comments apply primarily to the use of MULTAN, similar considerations would apply to other like program systems.

(a) *All values of Z are too low* ($\ll 1.0$) *or all values of R_K are too high* ($\gg 0.20$)—*both conditions can arise.* The criterion R_K applies strictly speaking only when all possible Σ_2 interactions have been used. The usual maximum number of Σ_2 interactions allowed in MULTAN is 2000, but a cutoff can be applied by the user; some versions allow up to 4000 Σ_2 interactions. If a cutoff is applied, the correct phase set could have a much higher value of R_K (up to 0.3 has been noted).

(b) *All phase sets and their figures of merit are similar.* This situation may occur if too few Σ_2 interactions are being used. Try more.

(c) *The origin-defining set is poor or incorrect.* Try another one, choosing it with the aid of the rules given.

(d) *The E map contains one very large peak.* The phases are probably very inaccurate; the heavy peak may be located in the center of a closed ring. Start again; do not waste time trying to interpret the E map.

(e) *The E map is not interpretable or chemically sensible.* The phases are incorrect. Try again.

(f) *If heavy atoms are present in the structure, they alone may show up.* Proceed to Fourier methods using interpretable heavy-atom sites. A check against the Patterson function might prove useful here.

(g) *Only a small molecular fragment is discernible from the E map.* Try recycling, basing phases on the fragment found, or try to obtain more phases by increasing the initial data set.

(h) *The program selects an incorrect or poor starting set* (*too few Σ_2 interactions, for example*). Select your own starting set. If you suspect that the program may contain a fault, inform the author; do not attempt to correct it.

(i) *The solution still fails to emerge.* Review the calculation of the $|E|$ values; perhaps omit reflections that appear to have a bad influence on the phase-determining pathway.

(j) *All fails.* Go back to fundamentals. Check the space group, data collection, and processing and any other factor that might be at fault.

If you exhaust these possibilities without achieving success, try another method for determining the structure. Or give it a rest and try again later. Or study the *recent* Bibliography on direct methods.

7.3 Patterson Search Methods

In describing the heavy-atom method (Chapter 6) we showed that, for a structure containing a small number of relatively heavy atoms, the Patterson function can lead to successful determination of the atomic coordinates. Direct methods were discussed subsequently in order to be able to solve, particularly, light-atom structures having none of the predominant interatomic vectors necessary to the heavy-atom method. We saw in our introductory discussion on the Patterson function that the synthesis must contain the set of interatomic vectors for the given crystal structure. In other words, for a known light-atom structure, we could generate the set of interatomic vector coordinates which would match, within the limits of error in $|F_o|$, those of the function $P(u, v, w)$. The Patterson function is a map of the interatomic vectors in the unit cell all transferred to a common origin. Each atom thus forms an image of the structure in itself (see page 244), and the Patterson function for a unit cell containing N atoms comprises N such images,* together with their centrosymmetrically related counterparts. Since we can always calculate the vector set for a model structure, it would appear that even for light-atom structures, which include many important and interesting compounds, the Patterson function may be unscrambled in terms

---

* Buerger (Bibliography, Chapter 6).

of the atomic coordinates. These ideas have led to techniques in structure analysis that may be called Patterson search methods: the prerequisite for their application is a reasonably precise model of a part of the structure under investigation. Coordinates for assembling suitable molecular fragments may be derivable from the library of known crystal structures.[*] For example, a molecular fragment may be a cyclohexane ring; its coordinates could be obtained through tables of standard bond lengths and angles (see pages 359–360). The coordinates of the molecular fragment can be used to derive the coordinates of the corresponding vector set, which are then stored in a computer. Values of $P(u, v, w)$, calculated from the experimental $|F_o|^2$ data, are also stored in the computer. Two problems then have to be solved: one is to determine the *orientation* of the model fragment with respect to the crystallographic axes, and the other is to find the *position* of the correctly oriented model in the unit cell.

An approximately correctly oriented and translated molecular fragment can be used, either as the second stage of the heavy-atom method (page 260), or as an input to further identification of the vectors in $P(u, v, w)$. A computer program for carrying out the full implementation of the Patterson procedure has been described by Braun and his colleagues[†] and is known as Vector Verification. We shall describe this technique, and show its application to a crystal structure determination.

In order to reduce the effect of overlapping peaks in the Patterson function, the synthesis is sharpened (page 237); it is then stored at fine grid intervals of approximately 0.15 Å in the three axial directions. In order to accommodate the very large number of values thus produced, each $P(u, v, w)$ is given one of four values represented by two binary digits (bits), as set out below. The largest $P(u, v, w)$ values have the smallest coded value, for reasons given below. The assignment of the range limits t_1, t_2, t_3, and t_4 (see Table 7.17) is achieved by calculating the cumulative distribution curve, an example of which is shown in Figure 7.14. Fuller details may be found in the reference cited.

7.3.1 Orientation Search

In order to carry out the orientation search, the vector-set coordinates calculated from the orthonormal (see page 355) atomic coordinates of the

[*] O. Kennard, D. G. Watson, and F. Allen. Crystallographic Data Base, University of Cambridge, Cambridge, England.

[†] P. B. Braun, J. Hornstra, and J. I. Leenhouts, *Philips Research Reports* **42**, 85–118 (1969).

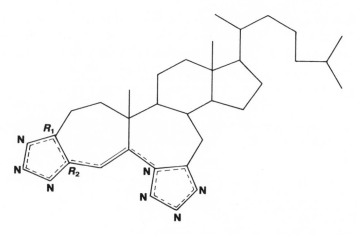

FIGURE 7.12. Chemical structural formula for HS649. HS650: $R_1 = $ N,
$R_2 = $ C. HS649: $R_1 = $ C, $R_2 = $ N.

model fragment are transformed by rotations α, β, and γ about the orthonormal axes x, y, and z. A sequence of values for these angles is used, generated
by approximately 9° stepwise variations of α, β, and γ. Patterson symmetry
and symmetry of the model may be used to limit the total number of
combinations generated. For example, with a monoclinic crystal we would
vary α from 0 to 360°, β from −90° to +90°, and γ from −90° to +90°,
making 16,000 orientations in all. For each α, β, γ triplet, the model
vector-set coordinates are transformed to the Patterson unit cell and both
origins superimposed.

　　A good fit between the vector-set coordinates of the model and the
Patterson function will correspond to the case where many vector-set
coordinates lie in low regions of *coded* Patterson density, that is, high values
of $P(u, v, w)$. The inversion of the numerical values by coding, as described
above, reduces the origin peak and other high Patterson values to zero;
they are thus easily recognized and classified. The program forms the sum
of the $P(u, v, w)$ values corresponding to the vectors of the model. This

TABLE 7.17. Vector Verification Codes for $P(u, v, w)$

Any $P(u, v, w)$	$P < t_1$	$t_1 \leq P < t_2$	$t_2 \leq P < t_3$	$t_3 \leq P$
Binary code	11	10	01	00
Patterson code	3	2	1	0

summation, called the measure of fit MOF, thus has smaller values for better fits. When all of the α, β, γ triplets have been tested, the program stores about 60 of the smallest MOFs and their corresponding angles.

7.3.2   Translation Search

The orientation search uses only those interatomic vectors that are confined to a restricted volume around the origin of the Patterson function. In all space groups other than $P1$ and $P\bar{1}$, it is necessary to take into account the vectors between symmetry-related molecules. For example, in space group $P2$ (page 88) the twofold related atomic positions are of the form x, y, z and $\bar{x}$, y, $\bar{z}$. If x, y, z is translated to $x + p_x$, $y + p_y$, $z + p_z$, the symmetry equivalent position is $-(x + p_x)$, $(y + p_y)$, $-(z + p_z)$. The vector between the jth atom and the symmetry-related (ith) atom is given by $u_{ij} = x_i - x_j + 2p_x$, $v_{ij} = y_i - y_j$, $w_{ij} = z_i - z_j + 2p_z$. In space group $P2$, p_x and p_z may be treated as parameters while p_y is initially indeterminate, as one would expect, and defines the actual location of the oriented model along y in the unit cell. The translation search thus involves variation of p_x, p_y, and p_z, as required by space group considerations, over a range of values and with a convenient interval, usually equal to three grid units, or 0.45 Å. For each translation, the coordinates of the calculated vector set are again verified by calculation of the MOF, and the procedure is repeated for each angle triplet passed on from the rotation search.

At the end of the stage, a list of 30 to 60 of the lower MOF values and the corresponding α, β, γ, p_x, p_y, p_z parameters is set up. These parameters are then optimized by a routine which subjects the six parameters to fine variations in order to obtain the best fit to the Patterson function. A significant reduction in an MOF value in this stage is usually taken to be an encouraging indication that a good fit has been obtained.

7.3.3   Example of a Structure Solution by Vector Verification

The heterocyclic synthetic bistetrazollo steroid structure HS649 (Figure 7.13) had been subjected to considerable investigation by direct methods, but without obtaining a solution of the structure. A set of coordinates for the search model was derived from the related structure HS650* (Table

---

* J. Husain, I. J. Tickle, and R. A. Palmer, *Acta Crystallographica* **C39**, 297–300 (1983).

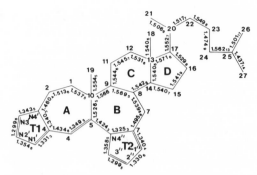

FIGURE 7.13. Bond lengths in HS650. The fragment used for vector verification comprises rings B,
C, D, and T_2.

7.18); Figure 7.13 identifies the atoms of the HS650 fragment that were
selected. The atom data are listed in Table 7.19: for the purposes of vector
verification, space group $C2$ may be treated as space group $P2$.

   The rotation search produced 14 angle triplets, listed in Table 7.20
with their MOF values; after translation, 60 coordinate sets were passed

TABLE 7.18. Crystal Data for HS649

Molecular formula	$C_{27}H_{42}N_8$
M_r	478.68
Crystal system	Monoclinic
Habit/form	Needle/needle axis [010]
a	35.537(4) Å
b	7.494(1)
c	10.319(1)
β	101.25°(1)
V_c	2695.2 Å^3
Z	4
D_m	1.143(6) g cm^{-3}
D_x	1.179 g cm^{-3}
$F(000)$	1040
$\lambda(\text{Cu }K\alpha)$	1.54178 Å
$\mu(\text{Cu }K\alpha)$	4.97 cm^{-1}
Crystal size	0.16, 0.20, 0.80 mm
Systematic absences	hkl: $h + k = 2n + 1$ $\begin{array}{l}(h0l: h = 2n + 1)\\(0k0:\ k = 2n + 1)\end{array}$
Space group	$C2$
Number of reflections measured	5866
Number of unique reflections	2700 [2318 with $I > 3\sigma(I)$]

TABLE 7.19. Fractional and Orthonormal Coordinates of the Model
Obtained from the Structure of HS650 and Used in Vector Verification

Atom	Fractional coordinates			Orthogonalized coordinates in Å		
	x	y	z	x	y	z
C(5)	0.3335	0.3741	0.0531	3.01	4.78	2.67
C(7)	0.5496	0.3131	0.0808	4.96	4.00	4.06
C(7a)	0.4776	0.2199	0.0909	4.31	2.81	4.57
C(8)	0.3269	0.2403	0.1063	2.95	3.07	5.34
C(9)	0.1961	0.2685	0.0893	1.77	3.43	4.49
C(10)	0.2116	0.3788	0.0746	1.91	4.84	3.75
C(11)	0.0510	0.2693	0.1067	0.46	3.44	5.36
C(12)	0.0277	0.1714	0.1212	0.25	2.19	6.09
C(13)	0.1540	0.1338	0.1395	1.39	1.71	7.01
C(14)	0.2903	0.1323	0.1204	2.62	1.69	6.05
C(15)	0.4144	0.0853	0.1389	3.74	1.09	6.98
C(16)	0.3402	0.0078	0.1574	3.07	0.10	7.91
C(17)	0.1607	0.0188	0.1508	1.45	0.24	7.58
N(4″)	0.4831	0.3820	0.0833	4.36	4.88	3.18
N(3″)	0.5817	0.4634	0.0571	5.25	5.92	2.87
N(2″)	0.6936	0.4461	0.0712	6.26	5.70	3.58
N(1″)	0.6837	0.3608	0.0846	6.17	4.61	4.25

TABLE 7.20. The 14 Most Favorable
Orientations (α, β, γ) of the Model Determined
by Rotation of the Calculated Vector Set about
the Patterson Origin

α	β	γ	MOF
216	−54	45	167
54	−45	66	197
90	−18	9	199
279	−36	20	200
189	27	−72	208
198	−72	30	209
342	18	−36	212
360	18	−36	216
63	−18	90	218
243	18	90	218
45	81	30	221
234	−36	80	222
279	45	−6	222
297	−45	−66	222

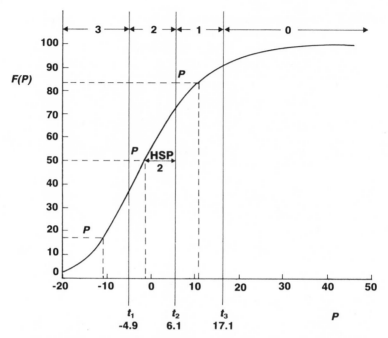

FIGURE 7.14. Cumulative distribution $F(P)$ of the sharpened Patterson function P for HS649. The 17, 50, and 83 percentiles, from which the test levels t_1, t_2, and t_3 are derived, are indicated; HSP is the expected height of a single-weight peak and is taken to be $P(000)/N$, where N is the number of nonhydrogen atoms in the unit cell.

on for optimization. Optimization was carried out by varying the angles by $\pm 4°$ and the translation parameters by $\pm 0.125\,\text{Å}$. The combined RTO (rotation/translation/optimization) MOFs are listed (Table 7.21). Of the 20 best R/T solutions, only sets 1 and 16 showed significant improvement.

The 17-atom fragment (Table 7.22), with coordinates given by the "best" set (16), was expanded by Fourier synthesis to give the complete 33-atom structure. It is recommended that the initial model fragment contain approximately $N^{1/2}$ atoms. The method has been used predominantly for polycyclic structures, and is not suitable generally for more flexible molecules such as peptides or nucleotides, although the structure of ethylenediamine tartrate was solved from a model based on the tartrate ion.[*] For HS649 there were 132 atoms in the unit cell, so that N is about

[*] M. F. C. Ladd and R. A. Palmer, *Journal of Crystal and Molecular Structure* **7**, 123–135 (1977).

TABLE 7.21. Twenty Possible Solutions of Which Only Two (Set 1 and Set 16) Have Optimized Significantly

Set No.	α	β	γ	x	y	z	ΣMOF^2 Before optimization	ΣMOF^2 After optimization
1	216	−54	45	11.11	0.0	1.33	152 105	103 732
2	45	81	30	17.32	0.0	9.73	192 781	189 397
3	216	−54	45	19.99	0.0	7.96	158 765	135 554
4	54	−45	66	19.55	0.0	8.40	181 136	179 437
5	90	−18	9	18.21	0.0	1.33	183 528	179 618
6	216	−54	45	18.66	0.0	1.33	162 170	142 468
7	216	−54	45	18.66	0.0	9.73	163 546	124 277
8	198	−72	30	18.66	0.0	9.73	185 434	129 706
9	297	−45	−66	15.99	0.0	7.08	199 805	188 240
10	63	−18	90	13.77	0.0	5.75	192 233	190 445
11	216	−54	45	20.88	0.0	9.29	165 625	147 321
12	189	27	−72	19.55	0.0	10.17	196 330	173 266
13	243	18	90	21.77	0.0	4.42	193 625	191 329
14	90	−18	9	16.44	0.0	0.88	189 765	185 608
15	54	−45	66	16.44	0.0	3.10	189 440	175 570
16	198	−72	30	11.11	0.0	1.33	188 929	102 965
17	360	18	−36	27.99	0.0	7.08	200 080	196 610
18	279	−36	20	17.77	0.0	9.29	192 209	181 717
19	90	−18	9	20.43	0.0	9.29	191 880	185 620
20	360	18	−36	23.99	0.0	0.44	201 492	194 324

12. The use of the 17-atom fragment in this example was slightly overplaying the method, although it need not prove a disadvantage, and is forgivable in view of its success (Table 7.22).

Other methods of automated Patterson interpretation have been reported from time to time, but it seems to us that Vector Verification has had the most general applicability.

7.4   Least-Squares Refinement

In Chapter 1, we used the equation of a line in two-dimensional space

$$y = mx + b \qquad (7.58)$$

If we have two pairs of values of x and y for measurements which are related by this equation, we can obtain a unique answer for the constants

TABLE 7.22. Fractional Coordinates of the Best
Solution for HS649 (Set 16) from Vector
Verification[a]

	x	y	z
C5	0.226	−0.188	0.132
	(0.2268)	(−0.1330)	(0.1346)
C7	0.195	−0.016	−0.070
	(0.1945)	(−0.0141)	(−0.0729)
C7a	0.173	0.103	−0.002
	(0.1707)	(0.1374)	(−0.0105)
C8	0.145	0.005	0.079
	(0.1435)	(0.0490)	(0.0706)
C9	0.164	−0.084	0.205
	(0.1640)	(−0.0432)	(0.2006)
C10	0.191	−0.250	0.183
	(0.1917)	(−0.2023)	(0.1824)
C11	(0.133	−0.148	0.283
	(0.1338)	(−0.1045)	(0.2804)
C12	0.107	−0.005	0.307
	(0.1067)	(−0.0432)	(0.3072)
C13	0.086	0.097	0.185
	(0.0864)	(0.1361)	(0.1823)
C14	0.119	0.159	0.116
	(0.1179)	(0.1988)	(0.1070)
C15	0.097	0.276	−0.002
	(0.0965)	(0.3199)	(−0.0011)
C16	0.064	0.366	0.047
	(0.0668)	(0.4157)	(0.0668)
C17	0.065	0.281	0.196
	(0.0677)	(0.3196)	(0.2000)
N4"	0.220	−0.146	−0.007
	(0.2203)	(−0.0971)	(−0.0041)
N3"	0.237	−0.237	−0.100
	(0.2382)	(−0.1884)	(−0.0887)
N2"	0.221	−0.171	−0.210
	(0.2230)	(−0.1279)	(−0.2064)
N1"	0.198	−0.044	−0.199
	(0.1960)	(−0.0028)	(−0.1989)

[a] The final refined coordinates for this part of the structure are given
in parentheses.

m and b. Sometimes, as in the Wilson plot (page 257), we have several pairs of values which contain random errors, and we need to obtain those values of m and b that best fit the complete set of observations.

In practical problems, we have often a situation in which the errors in the x values are negligible compared with those in y. Let the best estimates of m and b under these conditions be m_o and b_o. Then, the error of fit in the ith observation is

$$e_i = m_o x_i + b_o - y_i \qquad (7.59)$$

The principle of least squares states that the best-fit parameters are those that minimize the sum of the squares of the errors. Thus,

$$\sum_i e_i^2 = \sum_i (m_o x_i + b_o - y_i)^2 \qquad (i = 1, 2, \ldots, N) \qquad (7.60)$$

has to be minimized over the number N of observations. This condition corresponds to differentiating partially with respect to m_o and b_o, in turn, and equating the derivatives to zero. Hence,

$$m_o \sum_i x_i^2 + b_o \sum_i x_i = \sum_i x_i y_i \qquad (7.61)$$

$$m_o \sum_i x_i + b_o N = \sum_i y_i \qquad (7.62)$$

which constitute a pair of simultaneous equations (normal equations) easily solved for m_o and b_o.

In a crystal structure analysis, we are always manipulating more observations than there are unknown quantities; the system is said to be overdetermined. We shall consider some applications of the method of least squares.

7.4.1 Unit-Cell Dimensions

In Chapter 3, we considered methods for obtaining unit-cell dimensions with moderate accuracy. Generally, we need to enhance the precision of these measurements, which may be achieved by a least-squares analysis. Consider, for example, a monoclinic crystal for which the θ values of a number of reflections, preferably high-order, of known indices, have been measured to the nearest 0.01°. In the monoclinic system, $\sin \theta$ is given, through Table 2.4 and (3.32), by

$$4 \sin^2 \theta = h^2 a^{*2} + k^2 b^{*2} + l^2 c^{*2} + 2hla^* c^* \cos \beta^* \qquad (7.63)$$

In order to obtain the best values of a^*, b^*, c^*, and $\cos \beta^*$, we write, following (7.60),

$$\sum_i (h_i^2 a^{*2} + k_i^2 b^{*2} + l_i^2 c^{*2} + 2h_i l_i a^* c^* \cos \beta^* - 4 \sin^2 \theta_i)^2 \qquad (7.64)$$

and then minimize this expression, with respect to a^*, b^*, c^*, and $\cos \beta^*$, over the number of observations i. The procedure is a little more involved numerically; we obtain four simultaneous equations to be solved for the four variables, but the principles are the same as those involved with the straight line, (7.58)–(7.62).

7.4.2 Atomic Parameters

Correct trial structures are refined by the least-squares method. In essence, this process involves adjusting a scale factor and the positional and temperature parameters of the atoms in the unit cell so as to obtain the best agreement between the experimental $|F_o|$ values and the $|F_c|$ quantities derived from the structure model. In its most usual application, the technique minimizes the function

$$R' = \sum_{\mathbf{h}} w(|F_o| - G(|F_c|)^2 \qquad (7.65)$$

where the sum is taken over the set of crystallographically independent terms $\mathbf{h}$, w is a weight for each term, and G is the reciprocal of the scale factor K for $|F_o|$. Let p_j $(j = 1, 2, \ldots, n)$ be the variables in $|F_c|$ whose values are to be refined. Then

$$\frac{\partial R'}{\partial p_j} = 0 \qquad (7.66)$$

or

$$\sum_{\mathbf{h}} w \Delta \frac{\partial |F_c|}{\partial p_j} = 0 \qquad (7.67)$$

where Δ is $|F_o| - |F_c|$. For a trial set of parameters not too different from the correct values, Δ is expanded as a Taylor series to the first order:

$$\Delta(\mathbf{p}+\boldsymbol{\xi}) = \Delta(\mathbf{p}) - \sum_{i=1}^{n} \xi_i \frac{\partial |F_c|}{\partial p_j} \tag{7.68}$$

where ξ_i is the correction to be applied to parameter p_i; $\mathbf{p}$ and $\boldsymbol{\xi}$ represent the complete sets of variables and corrections. Substituting (7.68) in (7.67) leads to the normal equations:

$$\sum_{i=1}^{n} \left[\sum_{\mathbf{h}} w \frac{\partial |F_c|}{\partial p_i} \frac{\partial |F_c|}{\partial p_j} \right] \xi_i = \sum_{\mathbf{h}} w \Delta \frac{\partial |F_c|}{\partial p_j} \tag{7.69}$$

The n normal equations may be expressed neatly in matrix form:

$$\mathbf{A}\boldsymbol{\xi} = \mathbf{b} \qquad \text{or} \qquad \sum_i a_{ij}\xi_i = b_j \tag{7.70}$$

where

$$a_{ij} = \sum_{\mathbf{h}} w \frac{\partial |F_c|}{\partial p_i} \frac{\partial |F_c|}{\partial p_j} \tag{7.71}$$

and

$$b_j = \sum_{\mathbf{h}} w \Delta \frac{\partial |F_c|}{\partial p_j} \tag{7.72}$$

The solution of the normal equations is a well documented mathematical procedure that we shall not dwell upon. Instead, we draw attention to a few features of least-squares refinement. It is important to remember that least squares provides the best fit for the parameters that have been put into the model. Hence, it is essential to examine a final difference Fourier map (q.v.) at the completion of a least-squares refinement, after several cycles of calculations have led to negligible differences ξ_i. The techniques of least squares have been reported fully at Crystallographic Computing Conferences (see Bibliography).

Temperature Factors

An overall isotropic temperature factor T, as obtained from a Wilson plot, is the simplest approximation. A better procedure allots B parameters to atoms and refines them as least-squares parameters. We can write

$$T_\theta = \exp[-(B\lambda^{-2}\sin^2\theta)] \qquad (7.73)$$

and the equation

$$B = 8\pi^2 \overline{U^2} \qquad (7.74)$$

relates B to the mean square amplitude $\overline{U^2}$ of each atom: the surface of vibration is a sphere.

A more sophisticated treatment describes the vibrations of each atom by a symmetrical tensor **U** having six independent components in the general case. Now, we have

$$T_{hkl} = \exp[-2\pi^2(U_{11}h^2a^{*2} + U_{22}k^2b^{*2} + U_{33}l^2c^{*2}$$
$$+ 2U_{23}klb^*c^* + 2U_{31}lhc^*a^* + 2U_{12}hka^*b^*)] \qquad (7.75)$$

and the U_{ij} parameters are refined. The surface of vibration is now a biaxial (thermal) ellipsoid, and the mean square amplitude of vibration in the direction of a unit vector $\mathbf{l} = (l_1, l_2, l_3)$ is given by

$$\overline{U^2} = \sum_{i=1}^{3}\sum_{j=1}^{3} U_{ij}l_il_j \qquad (7.76)$$

Since **l** is defined with respect to the reciprocal lattice, the component of **U** with $\mathbf{l} = (1, 0, 0)$, parallel to a^*, is

$$\overline{U^2} = U_{11} \qquad (7.77)$$

In an orthorhombic crystal, for example, a direction 30° from a^* in the a^*b^* plane has $\mathbf{l} = (\sqrt{3}/2, 1/2, 0)$, and the component of **U** in that direction is

$$\overline{U^2} = U_{11}(\sqrt{3}/2)^2 + U_{22}(1/2)^2 + 2U_{12}(\sqrt{3}/2)(1/2) \qquad (7.78)$$

The following relationships among the values of B, $\overline{U^2}$, and the rms amplitude are often useful:

$B(\text{Å}^2)$	$\overline{U^2}(\text{Å}^2)$	rms amplitude (Å)
0.10	0.0013	0.036
0.50	0.0063	0.080
1.0	0.013	0.11
5.0	0.063	0.25
10	0.13	0.36

The smallest rms amplitudes encountered are circa 0.05 Å. Values of B between 3 and 10 Å^2 are found in organic structures at ambient temperatures. The corresponding rms amplitudes indicate that caution should be exercised in interpreting them in terms of bond lengths and their precision.

Scale Factor

In least-squares refinement, the $|F_o|$ data must not be adjusted, and so the parameter G in (7.65) is introduced. The inverse of the refined value of G may be applied to $|F_o|$ at the end of a refinement cycle. Several cycles of refinement may be needed before the parameters reach a sensibly constant value. Generally full-matrix least-squares refinement is to be preferred. However, where the number of parameters is very large or where computer availability is limited, an approximation may be used. One such method is the block-diagonal refinement, in which certain off-diagonal a_{ij} terms in (7.70) are neglected. Generally, more cycles are necessary in this procedure.

Weights

In the initial stages of refinement, weights may be set at unity or chosen so as to accelerate the process, such as downweighting reflections of small $|F_o|$ or of high order, or both. In the final stages, weights should be related to the precision of $|F_o|$, which can be achieved in two ways:

(a) $$w(hkl) = 1/\sigma^2(|F_o(hkl)|) \qquad (7.79)$$

where the estimated standard deviation, $\sigma|F_o(hkl)|$, is obtained from counting statistics in diffractometer data by the relationship

$$\sigma = \sqrt{N} \qquad (7.80)$$

N being related to the total counts, peak and background, for the given reflection. Sometimes $p|F_o|^2$ is added to the right-hand side of (7.79), where p is adjusted to achieve the condition given immediately below (7.81).

(b) $$w(hkl) = \Phi(|F_o|) \qquad (7.81)$$

where the function Φ is chosen so that $w\Delta^2$ is constant for small, significant ranges of $|F_o|$. One scheme is given by

$$w = (A + |F_o| + B|F_o|^2 + C|F_o|^3)^{-1} \qquad (7.82)$$

where the constants A, B, and C may be obtained by a least-squares fit of mean values of Δ, in ranges of $|F_o|$, to the inverse of the right-hand side of (7.82).

Precision

The choice of absolute weights (7.79) should yield parameters of lowest variances:

$$\sigma^2(p_j) = (a_{jj})^{-1} \qquad (7.83)$$

where $(a_{jj})^{-1}$ is an element of the matrix inverse to that of the a_{ij} elements (7.70). With weights related to $|F_o|$,

$$\sigma^2(p_j) = (a_{jj})^{-1}\frac{\sum_{\mathbf{h}} w \Delta^2}{m - n} \qquad (7.84)$$

where m is the number of reflections and n the number of parameters to be refined. Generally, $m \geqslant 5n$ is a satisfactory relationship. In a block-diagonal approximation, the standard deviations are usually underestimated by 15–20%.

Atoms in Special Positions

If any symmetry operation of the space group of a structure leaves an atom invariant, the atom is on a special position (q.v.). Consider one molecule of formula AB_2 in the unit cell of space group $P2$. The A atoms occupy, say, the special positions $0, y, 0$ and the B atoms occupy the general positions x, y, z and $\bar{x}$, y, $\bar{z}$. Several important points arise:

(a) The x and z coordinates of atoms A remain invariant at zero during refinement.

(b) In this space group, the origin is along the twofold axis y, and must be specified by fixing the y coordinate of an atom, the heavier the atom the better. It could be $y_B = 0$, in which case this parameter too remains invariant.

(c) With respect to the symmetry operations of the space group, atoms A must be given an atom multiplicity factor of $\frac{1}{2}$ so that a total of one A atom per unit cell obtains.

(d) In anisotropic refinement, one of the three twofold axes of the biaxial thermal ellipsoid must lie along the twofold axis y of the space group. In this case, the U_{12} and U_{23} elements of **U** remain invariant at zero value.

Calculations

The several aspects of least-squares refinement discussed above involve computations which, while not complex, are very lengthy. Fortunately, they have been implemented in well tested crystallographic program systems, such as XRAY and SHELX, that are available to workers in this field.

7.5 Molecular Geometry

7.5.1 Bond Lengths and Angles

When the formal structure analysis is complete, we need to express our results in terms of molecular geometry and packing. This part of the analysis includes the determination of bond lengths, bond angles, and intermolecular contact distances, with measures of their precision.

Consider three atoms with fractional coordinates $x_1, y_1, z_1, x_2, y_2, z_2$, and x_3, y_3, z_3 in a unit cell of sides a, b, and c (Figure 7.15). The vector $\mathbf{r}_j$ from the origin O to any atom j is given by

$$\mathbf{r}_j = x_j\mathbf{a} + y_j\mathbf{b} + z_j\mathbf{c} \tag{7.85}$$

The vector $\mathbf{r}_{12}$ is given by

$$\mathbf{r}_{12} = \mathbf{r}_2 - \mathbf{r}_1 \tag{7.86}$$

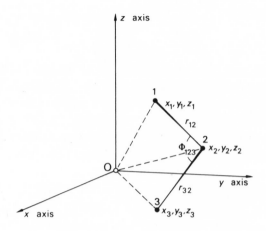

FIGURE 7.15. Geometry of the calculation of interatomic distances and angles.

or, using (7.85),

$$\mathbf{r}_{12} = (x_2 - x_1)\mathbf{a} + (y_2 - y_1)\mathbf{b} + (z_2 - z_1)\mathbf{c} \qquad (7.87)$$

Forming the dot product of each side with itself, remembering that

$$\mathbf{p} \cdot \mathbf{q} = pq \cos \widehat{pq} \qquad (7.88)$$

we obtain

$$r_{12}^2 = (x_2 - x_1)^2 a^2 + (y_2 - y_1)^2 b^2 + (z_2 - z_1)^2 c^2$$
$$+ 2(y_2 - y_1)(z_2 - z_1)bc \cos \alpha + 2(z_2 - z_1)(x_2 - x_1)ca \cos \beta$$
$$+ 2(x_2 - x_1)(y_2 - y_1)ab \cos \gamma \qquad (7.89)$$

This equation may be simplified for crystal systems other than triclinic. Thus,
if the atoms exist in a tetragonal unit cell,

$$r_{12}^2 = [(x_2 - x_1)^2 + (y_2 - y_1)^2]a^2 + (z_2 - z_1)^2 c^2 \qquad (7.90)$$

In a similar manner, we can evaluate r_{32} (Figure 7.12). Using (7.88), for the tetragonal system,

$$\cos \Phi_{123} = \frac{\{[(x_2-x_1)(x_2-x_3)+(y_2-y_1)(y_2-y_3)]a^2 + (z_2-z_1)(z_2-z_3)c^2\}}{(r_{12}r_{32})}$$

(7.91)

where r_{12} and r_{32} are evaluated following (7.89). Similar equations enable any distance or angle to be calculated, in any crystal system, in terms of the atomic coordinates and the unit-cell dimensions.

When the asymmetric unit of a crystal contains more than one copy of a given molecule, or when similar molecules occur in different crystals, the question arises of whether or not the several sets of molecular dimensions are significantly different. The statistical test applicable in this situation is the chi-square χ^2 test. It involves calculation of $\sum_{i=1}^{n}[\Delta_i/\sigma\Delta_i]^2$, which is distributed as χ^2 with n degrees of freedom: Δ_i is the difference between one measured property, e.g., a bond length, in a pair of molecules and $\sigma\Delta_i$ is the standard deviation in Δ_i, estimated typically as $[\sigma^2(d_{i1}) - \sigma^2(d_{i2})]^{1/2}$, assuming no correlation between d_{i1} and d_{i2}, and n is the number of pairs of measurements.

The significance of the result can be tested in the usual way* by making the null hypothesis that all of the differences can be accounted for by random errors in the experimental procedures, and then obtaining from statistical tables the significance level of the test, that is, the probability of incorrectly rejecting a good hypothesis. Normally, the test is not regarded as significant unless $\alpha \leq 0.05$.

7.5.2 Torsion Angles

A torsion angle is a useful conformational parameter of a molecule. In a freely rotating moiety, a torsion angle may be a function of the environment of the molecule. Consider an arrangement of four atoms 1, 2, 3, 4 (Figure 7.16). The torsion angle $\chi(1, 2, 3, 4)$ is defined by the angle between the planes 1, 2, 3 and 2, 3, 4, and lies in the range $-180° < \chi \leq 180°$; the sign is an important property of the parameter.

In the planar, eclipsed conformation shown in Figure 7.16, χ is zero. The torsion angle is the amount of rotation of 3, 4 about 2, 3 and, looking along the direction $2 \rightarrow 3$, a positive value of χ corresponds to a clockwise rotation. Let

$$\mathbf{p}_1 = \mathbf{r}_{23} \times -\mathbf{r}_{12}$$

(7.92)

* See, for example, the footnote to page 284.

and

$$\mathbf{p}_2 = \mathbf{r}_{23} \times \mathbf{r}_{24} \tag{7.93}$$

Then

$$\chi(1, 2, 3, 4) = \cos^{-1}\left(\frac{\mathbf{p}_1 \cdot \mathbf{p}_2}{p_1 p_2}\right) \tag{7.94}$$

and the sign of χ is that of $\mathbf{p}_1 \times \mathbf{p}_2 \cdot \mathbf{r}_{24}$.

7.5.3 Conformational Analysis

Confusion has arisen in the literature over the use of torsion angles in conformational analysis. It is often convenient to quote values of torsion angles as lying within certain ranges. For example, $\chi \approx 0°$ may be called *cis*, $\chi \approx 180°$ is *trans* (*t*), and $\chi \approx \pm 60°$ is $\pm$*gauche* ($\pm g$). However, because of changing conventions, it is best to quote the actual value of χ, and to state how it is defined (Figure 7.16). This procedure will minimize ambiguities in the future.

Ring Conformations

Two types of symmetry (or pseudosymmetry) must be considered in order to define ring conformations,* namely mirror planes perpendicular to the dominant ring plane and twofold axes lying in the ring plane. If there is an odd number (usually 5 or 7) of atoms in the ring, all symmetry elements pass through one of the ring atoms and bisect the opposite bond (Figure 7.17). In rings containing an even number of atoms (usually 6), symmetry elements may pass through two ring atoms located directly across the ring, or else bisect two opposite ring bonds.

Ten symmetry elements are possible in five-membered rings. The *planar* five-membered ring possesses all ten, five mirror planes and five twofold axes. The ideal *envelope* conformation has only a single *m* plane, and it passes through the out-of-plane atom. The ideal *half-chair* has one twofold axis bisecting the bond between the two out-of-plane atoms.

Six-membered rings possess twelve locations for symmetry elements. In determining the ring conformation, we can ignore the two-, three-, and

---

* W. L. Duax and D. A. Norton, *Atlas of Steroid Structure*, Vols. 1 and 2, New York, IFI/Plenum Data Company (1975, 1984).

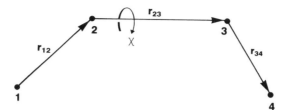

FIGURE 7.16. Torsion angle $\chi(1, 2, 3, 4)$: the torsion angle has a positive sign for a rotation of 3, 4 about 2, 3 as shown, that is, it is positive if, looking along r_{23}, a clockwise rotation is required to bring atom 1 into atom 4.

sixfold rotation axes perpendicular to the ring plane. Figure 7.18 illustrates the symmetry elements that define the ideal forms of commonly observed conformations. The planar ring, such as in benzene, has one m plane and one twofold axis at each of six locations ($6/mmm$). The *chair* form of cyclohexane has three m planes and three twofold axes ($\bar{3}m$). The *boat* and *twist-boat* have symmetry $mm2$ and 222 respectively, while the *sofa* has symmetry m and the *half-chair* symmetry 2.

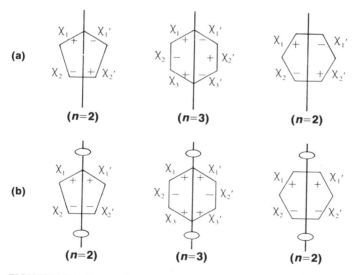

FIGURE 7.17. Ring conformations. (a) Torsion angles related by mirror planes (———) have opposite signs, and (b) angles related by twofold rotation axes ($\bar{\diamond}$) have the same sign.

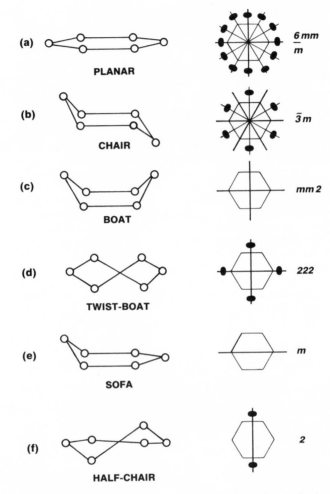

FIGURE 7.18. Commonly observed conformations of six-membered rings. The mirror and twofold rotational symmetries are indicated on the right.

Asymmetry Parameters

Once the atom coordinates are available, torsion angles can be calculated. Because of errors in the data and for stereochemical reasons, a particular cyclic structure will often depart from its ideal symmetry. The degree of this departure, its asymmetry, may be calculated in terms of asymmetry parameters. For this purpose, related or possibly related torsion angles are compared in a way that will result in a value of zero for a parameter if the corresponding symmetry is realized in the molecule. Mirror-related torsion angles have equal magnitude but opposite sign, and such torsion angles are compared by addition. The torsion angles related by twofold symmetry are equal in both magnitude and sign, and are compared by subtraction. The rms value of each discrepancy yields a measure of the deviation from ideal symmetry at the location in question. We calculate

$$\Delta C_s = \left(\sum_{i=1}^{n} (\chi_i + \chi_i')^2 / n \right)^{1/2} \tag{7.95}$$

in respect of m symmetry, and

$$\Delta C_2 = \left(\sum_{i=1}^{n} (\chi_i - \chi_i')^2 / n \right)^{1/2} \tag{7.96}$$

in respect of twofold symmetry; n is the number of individual comparisons, and χ_i and χ_i' are the related torsion angles in question.

7.5.4 Mean Planes

In discussing the geometry of a molecule, it is often desirable to test the planarity of a group of atoms. For a number n $(n > 3)$ of atoms, the best plane may be obtained by the method of least squares. Let the plane be given by

$$PX + QY + RZ + S = 0 \tag{7.97}$$

The constants P, Q, R, and S are obtained through a procedure similar to that expressed by (7.58)–(7.62). It is convenient to work with orthonormal coordinates X, Y, Z, given by the transformations

$$X = xa \sin \beta \sin \gamma^* \tag{7.98}$$

$$Y = -xa \sin \beta \cos \gamma^* + yb \sin \alpha \tag{7.99}$$

$$Z = xa \cos \beta + yb \cos \alpha + zc \tag{7.100}$$

which correspond to the orientation $Z \| c$ and Y coplanar with the plane bc.

7.6   Precision

Closely related to the calculations of bond lengths and angles is the expression of the precision of these quantities. The least-squares refinement procedure establishes values for an estimated standard deviation (esd) in each of the variables used in these calculations. Thus, a fractional coordinate of 0.3712 might have a standard deviation of 0.0003, written as 0.3712(3).

We need to know further how errors are propagated in a quantity which is a function of several variables, each of which contains some uncertainty arising from random errors. The answer is provided by the statistical principle of superposition of errors.

Let q be a function of several variables p_i $(i = 1, 2, 3, \ldots, N)$, with known standard deviations $\sigma(p_i)$. Then the esd in q is given by

$$\sigma^2(q) = \sum_{i=1}^{N} \left[\frac{\partial q}{\partial p_i} \sigma(p_i) \right]^2 \qquad (7.101)$$

A simple example may be given, through (7.87) and (7.101), for a bond between two atoms lying along the c edge of a tetragonal unit cell. Let c be 10.06(1) Å, z_1 be 0.3712(3), and z_2 be 0.5418(2). From (7.87),

$$r_{12} = (z_2 - z_1)c = (0.5418 - 0.3712)10.06 = 1.716 \text{ Å} \qquad (7.102)$$

and from (7.101),

$$\sigma^2(r_{12}) = (0.5418 - 0.3712)^2(0.01)^2$$
$$+ (10.06)^2(0.0002)^2 + (10.06)^2(0.0003)^2 \qquad (7.103)$$

Thus, $\sigma(r_{12})$ is 0.004 Å and we write $r_{12} = 1.716(4)$ Å. Similar calculations may be used for all distance and angle calculations in all crystal systems, but the general equations are quite involved numerically and best handled by computer methods.

The esd of a torsion angle can be calculated along the lines given in this section. Torsion angles are useful in comparing different, related molecules or, indeed, different conformations of one and the same molecule, but the significances of such differences are as important as the differences themselves.

The fitting of the best plane to a group of atoms is a simple matter. Again, in a discussion of the results it is essential to evaluate the perpendicular distances (deviations) of atoms from the plane, and their esd's. Let the jth atom have the orthonormal coordinates X_j, Y_j, Z_j. Then it is a simple exercise in coordinate geometry to show that the deviation of this atom from the best plane (7.97) is given by

$$\Delta_j = (PX_j + QY_j + RZ_j + S)/K \qquad (7.104)$$

where K is given by

$$K = (P^2 + Q^2 + R^2)^{1/2} \qquad (7.105)$$

To obtain the esd in Δ_j the esd's in X_j, Y_j, and Z_j are obtained from (7.98)–(7.100) by using (7.101). For example,

$$\sigma(X_j) = \{\sin^2 \beta \sin^2 \gamma^*[a^2\sigma^2(x_j) + x_j^2\sigma^2(a)]$$
$$+ x_j^2 a^2[\cos^2 \beta \sin^2 \gamma^*\sigma^2(\beta) + \sin^2 \beta \cos^2 \gamma^*\sigma^2(\gamma^*)]\}^{1/2} \qquad (7.106)$$

Then

$$\sigma(\Delta_j) = \{[P\sigma(X_j)]^2 + [Q\sigma(Y_j)]^2 + [R\sigma(Z_j)]^2\}^{1/2}K \qquad (7.107)$$

7.7 Correctness of a Structure Analysis

At this stage we may summarize four criteria of correctness of a good structure analysis. If we can satisfy these conditions in one and the same structure model, we shall have a high degree of confidence in it.

(a) There should be good agreement between $|F_o|$ and $|F_c|$, expressed through the R-factor (page 205). Ultimately, R depends upon the quality of the experimental data. At best, it will probably be about 1% greater than the average standard deviation in $|F_o|$.

(b) The electron density map should show neither positive nor negative density regions that are unaccountable, other than Fourier series termination errors (Figure 6.3).

(c) The difference-Fourier map should be relatively flat. This map eliminates series termination errors as they are present in both ρ_o and ρ_c. Random errors produce small fluctuations on a difference map, but they should lie within 2.5–3 times the standard deviation of the electron density

TABLE 7.23. Selected Ionic Radii (Å) Referred to Coordination
Number 6[a]

Ag^+	1.15	Hg^{2+}	1.02	Sr^{2+}	1.16
Al^{3+}	0.54	K^+	1.38	Th^{4+}	0.94
Ba^{2+}	1.35	La^{3+}	1.03	Ti^{2+}	0.86
Ca^{2+}	1.00	Li^+	0.76	Ti^{4+}	0.61
Cd^{2+}	0.95	Mg^{2+}	0.72	Tl^+	1.50
Ce^{3+}	1.01	Mn^{2+}	0.83	Zn^{2+}	0.74
Ce^{4+}	0.87	Na^+	1.02	NH_4^+	1.48
Co^{2+}	0.75	Ni^{2+}	0.69	Br^-	1.96
Co^{3+}	0.61	Pb^{2+}	1.19	Cl^-	1.81
Cr^{3+}	0.62	Pd^{2+}	0.86	F^-	1.33
Cs^+	1.67	Pt^{2+}	0.80	I^-	2.20
Cu^+	0.77	Pt^{4+}	0.63	O^{2-}	1.40
Cu^{2+}	0.73	Ra^{2+}	1.43	S^{2-}	1.84
Fe^{2+}	0.78	Rb^+	1.52	Se^{2-}	1.98
Fe^{3+}	0.65	Sn^{2+}	0.93	Te^{2-}	2.21

[a] The change in an ionic radius from coordination number 6 to coordination numbers 8, 4, 3, and 2 is approximately 1.5%, −1.5%, −3.0%, and −3.5%, respectively.

$\sigma(\rho_o)$:

$$\sigma(\rho_o) = \frac{1}{V_c}\left[\sum_{hkl}(\Delta F)^2\right]^{1/2} \tag{7.108}$$

where $\Delta F = |F_o| - |F_c|$ and the sum extends over all symmetry-independent reflections.

(d) The molecular geometry should be chemically sensible, within the limits of current structural knowledge.* Abnormal bond lengths and angles may be correct, but they must be supported by strong evidence of their validity in order to gain acceptance. Normally, a deviation of less than three times the corresponding standard deviation is not considered to be statistically significant.

As a guide to the interpretation of acceptable stereochemistry, we include ionic radii, bond lengths, and bond angles in Tables 7.23, 7.24, and 7.25, respectively.

7.8   Limitations of X-Ray Structure Analysis

There are certain things which X-ray analysis cannot do well, and it is prudent to consider the more important of them.

* See Bibliography.

TABLE 7.24. Selected Bond Lengths (Å)^a

Formal single bonds			
C4—H	1.09	C3—C2	1.45
C3—H	1.08	C3—N3	1.40
C2—H	1.06	C3—N2	1.40
N3—H	1.01	C3—O2	1.36
N2—H	0.99	C2—C2	1.38
O2—H	0.96	C2—N3	1.33
C4—C4	1.54	C2—N2	1.33
C4—C3	1.52	C2—O2	1.36
C4—C2	1.46	N3—N3	1.45
C4—N3	1.47	N3—N2	1.45
C4—N2	1.47	N3—O2	1.36
C4—O2	1.43	N2—N2	1.45
C3—C3	1.46	N2—O2	1.41

Formal double bonds			
C3—C3	1.34	C2—O1	1.16
C3—C2	1.31	N3—O1	1.24
C3—N2	1.32	N2—N2	1.25
C3—O1	1.22	N2—O1	1.22
C2—C2	1.28	O1—O1	1.21
C2—N2	1.32		

Formal triple bonds			
C2—C2	1.20	N1—N1	1.10
C2—N1	1.16		

Aromatic bonds			
C3—C3	1.40	N2—N2	1.35
C2—N2	1.34		

^{a} The notation in the table indicates the connectivity of the atoms.

Liquids and gases lack three-dimensional order, and cannot be used in diffraction experiments in the same way as are crystals. Certain information about the radial distribution of electron density can be obtained, but it lacks the distinctive detail of crystal analysis.

It is not easy to locate light atoms in the presence of heavy atoms. Difference-Fourier maps alleviate the situation to some extent, but the atomic positions are not precise. Least-squares refinement of light-atom parameters is not always successful, because the contributions to the structure factor from the light atoms are relatively small.

TABLE 7.25. Selected Bond Angles

Atom	Geometry	Angle (degrees)
C4	Tetrahedral	109.47
C3	Planar	120
C2	Bent	109.47
C2	Linear	180
N4	Tetrahedral	109.47
N3	Pyramidal	109.47
N3	Planar	120
N2	Bent	109.47
N2	Linear	180
O3	Pyramidal	109.47
O2	Bent	109.47

Hydrogen atoms are particularly difficult to locate with precision because of their small scattering power and the fact that the center of the hydrogen atom does not, in general, coincide with the maximum of its electron density. Terminal hydrogen atoms have a more aspherical electron density distribution than do hydrogen-bonded hydrogen atoms, and their bond distances, from X-ray studies, often appear short when compared with spectroscopic or neutron diffraction values. For similar reasons, refinement of hydrogen atom parameters in a structure analysis may be imprecise, and the standard deviations in their coordinate values may be as much as ten times greater than those for a carbon atom in the same structure. It is, nevertheless, desirable to include hydrogen atom positions in the final structure model. They lead to a best fit, and are useful when comparing the results of X-ray structure determination with those of other techniques, notably nuclear magnetic resonance.

In general, bond lengths determined by X-ray methods represent distances between the centers of gravity of the electron clouds, which may not be the same as the internuclear separations. Internuclear distances can be found from neutron diffraction data, because neutrons are scattered by the atomic nuclei. If, for a given crystal, the synthesized neutron scattering density is subtracted from that of the X-ray scattering density, a much truer picture of the electron density can be obtained—but that is another story.

7.9 Disorder in Single Crystals

A typical small-molecule analysis involves less than about 100 nonhydrogen atoms in the asymmetric unit. With Cu $K\alpha$ radiation and to $\theta \geqslant 70°$, it would be expected to lead to the determination of bond lengths with esd's of about 0.005 Å and of bond angles with esd's of about 0.2°. Isotropic thermal parameters (U_{iso})* for nonhydrogen atoms usually range from 0.050 to 0.090 Å^2 and may have esd's from 0.003 to 0.007 Å^2. However, it is sometimes found that the refined thermal parameters for certain atoms in a structure have atypical values. For example, U_{iso} may increase progressively and significantly toward the end of a chainlike moiety compared to the more rigid areas of the structure. The obvious and reasonable physical interpretation is simply that the atoms near the end of the chain experience greater thermal motion than do the atoms in the bulk of the molecule. For example, the hydroxyethyl side-chain atoms of an azasteroid derivative[†] have the following U_{iso} parameters:

$$>N(0.058)-C(0.073)H_3-C(0.109)H_2-O(0.176)H$$

In this analysis all atoms, including those in the side-chain, were resolved and refined successfully by least-squares. Atoms in solvent of crystallization molecules may exhibit high thermal parameters, and for similar reasons. Exceptions occur from time to time, and in the above example a well resolved solvent water-oxygen atom had a refined U_{iso} value of 0.088 Å^2 and was so well ordered that its hydrogen atoms were clearly located in a ΔF map (Figure 7.19). In this particular case, the clarity of definition in the electron density is associated with the formation of two strong hydrogen bonds donated by each of the water hydrogen atoms holding it firmly in position. However, in the structure of another steroid derivative,[‡] the carbon atoms of the side-chains are so badly disordered that some atoms are not resolved in the difference electron density and appear as diffuse patches (Figures 7.20 and 7.21). Such disorder is probably of a statistical nature, with the atoms taking up slightly different positions from one unit cell to another. The effect can be compensated, albeit somewhat artificially, by the refinement of the isotropic thermal parameters assigned to the atoms concerned. In the example, the U_{iso} values are 3–6 times greater than those of the ordered atoms in the structure.

Disorder may also arise by groups of atoms either in free rotation in the solid state (dynamic disorder) or in more than one position of similar

---

* $U_{iso} = 8\pi^2 \overline{U^2} \sin^2 \theta / \lambda^2$, where $\overline{U^2}$ is the mean square amplitude of vibration of an atom.
† A. I. El-Shora, R. A. Palmer, H. Singh, T. R. Bhardwaj, and D. Paul, *Journal of Crystallographic and Spectroscopic Research* **14**, 89 (1984).
‡ J. Husain, I. J. Tickle, R. A. Palmer, H. Singh, and K. K. Bhutani, *Acta Crystallographica* **B38**, 2845–2851 (1982).

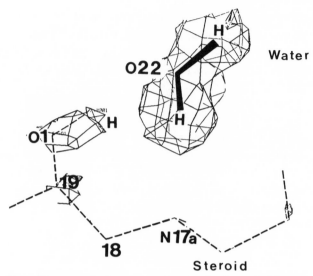

Note: The next six figures are electron density and difference
electron density maps photographed from the screen of an Evans
and Sutherland Picture System 2 cathode ray tube display unit
coupled to a computer that holds the electron density data. The
interactive computer graphics system is programmed such that the
user can simulate a three-dimensional effect by rotating the map
about one or more of three mutually perpendicular axes. The
contouring of the maps encloses the electron density in a cage of
"chicken wire" hoops running in several directions. Unlike the
sectional contour maps used elsewhere in this book, only one
contour level is used, selected so as to optimize the desired features
of electron density.

FIGURE 7.19. Difference electron density map for the azasteroid
HS626, showing the hydrogen atom on O1 (hydroxyl) and the two
water-hydrogen atoms, none of which was included in the structure
factor calculation. The steroid molecule, part of which is shown
by the dashed line, has been subtracted out in the difference
synthesis.

energy (static disorder). Methyl groups in large organic molecules often
show this type of behavior. It may be possible to distinguish between dynamic
and static disorder by a complete reexamination of the structure at a much
reduced temperature.

Protein structures are of particular topical interest, and recent innova-
tions in this field include the development of techniques for refining structures.* The molecules involved in protein analysis are very large, typically
with more than 1000 nonhydrogen atoms in the asymmetric unit. Con-

* D. S. Moss and A. Morffew, *Computational Chemistry* **6**, 1–3 (1981).

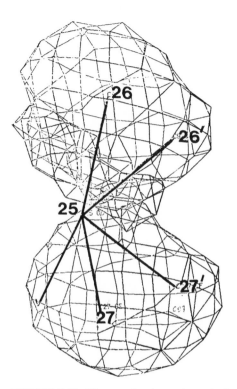

FIGURE 7.20. Electron density at the end of
the cholesteryl side chain of HS650 (molecule
B). The density is smeared out, and at least
two stereochemically sensible positions for

$$-C25\begin{matrix} \nearrow C26 \\ \\ \searrow C27 \end{matrix}$$

can be fitted to the density, as
indicated.

sequently, the crystals have large unit cells, and many possible X-ray
reflections occur within a given θ range compared to small-molecule crystals.
It is customary to limit severely the maximum θ value during the course
of a protein structure analysis, depending on the particular stage reached.
Corresponding to a given maximum θ value (θ_{max}) there is a minimum d
value (d_{min}) (given by $\lambda/2 \sin \theta_{max}$), and it is customary to speak of the d_{min}
of an analysis as the (nominal) resolution. The protein analysis may proceed
through stages of progressively higher resolution—for example, 6 Å, 3 Å,
2.5 Å, 2 Å, and 1.5 Å—the electron density image undergoing gradually
improved mathematical focusing in the process. In addition to the large

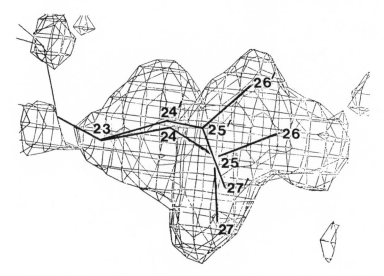

FIGURE 7.21. As for Figure 7.20, but with molecule A of HS650, in which the disorder in the side chain is more extensive and encompasses C23 to C27. In this structure the side chains are loosely held, having little contact with neighboring molecules in the crystal.

quantity of data associated with protein analysis, there is a further technical problem which limits the quality of most studies. During the crystallization process from solution, solvent molecules (typically 40–60% by weight) are trapped in the structure: the protein molecules almost float in a solvated crystalline state (see for example Blundell and Johnson*), and consequently many regions of electron density in the protein structure may be subject to the type of disorder described above. Even in a very high-resolution analysis, protein data rarely extend beyond 1.5 Å, and individual atoms in the protein molecule will not be revealed. A protein structure refinement involves the use of both least-squares analysis and geometrically constrained positioning of groups in order to produce a plausible model (Moss and Morffew, 1981).

We conclude this section with a few examples of electron density determined in the high-resolution X-ray analysis of the enzyme ribonuclease,† a small protein of molecular weight 13,700. In the first example, a tyrosine residue (Figure 7.22) is seen at 1.45 Å resolution as a hollow

* Bibliography, Chapter 6.
† N. Borkakoti, D. S. Moss, M. J. Stanford, and R. A. Palmer, *Journal of Crystallographic and Spectroscopic Research* **14**, 467 (1984).

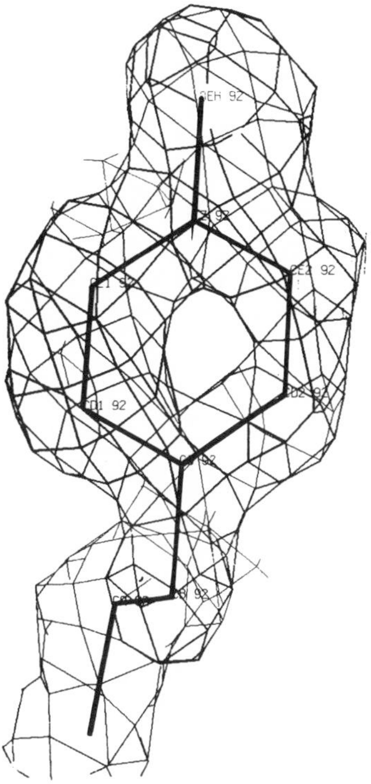

FIGURE 7.22. Figures 7.22–7.24 are extracts from the electron density map of ribonuclease-A at 1.45-Å resolution. The maps are calculated with the coefficients

$$|F_o| + (|F_o| - |F_c|) = 2|F_o| - |F_c|$$

and thus show features of both the electron density and the difference electron density. This figure shows a tyrosyl residue ($HOC_6H_4CH_2CH<$), with the hole of the phenyl ring and the —OH group (on the top) clearly indicated.

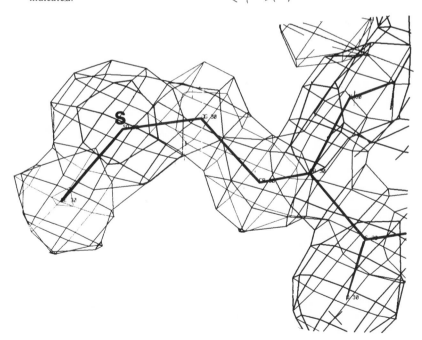

FIGURE 7.23. Methionyl residue ($CH_3SCH_2CH_2CH<$) showing the outstanding electron density region around the sulfur atom.

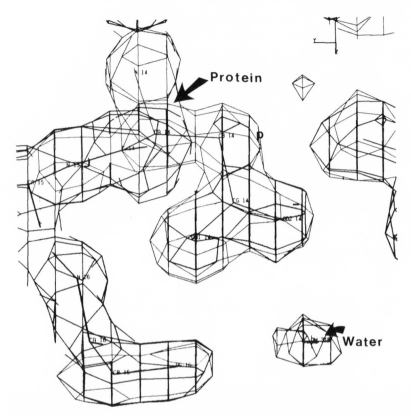

FIGURE 7.24. This electron density portion shows a clearly resolved solvent molecule (water), not included in the structure factor calculation.

ring of density and, although the individual atoms are not resolved, the shape of the density image is strikingly good. As would be expected, the sulfur atom of a methionine residue (Figure 7.23) is quite outstanding, but it does not swamp the rest of this slender aliphatic side-chain. At this resolution, the high quality of the refined analysis is evident in the appearance of resolved solvent (water) molecules, as shown in Figure 7.24.

Bibliography

Crystallographic Computing

AHMED, F. R., HALL, S. R., and HUBER, C. P. (Editors), *Crystallographic Computing*, Copenhagen, Munksgaard (1970).

DIAMOND, R., RAMASESHAN, S., and VENKATESAN, K. (Editors), *Computing in Crystallography*, Bangalore, Indian Academy of Sciences (1980).

PEPINSKY, R., ROBERTSON, J. M., and SPEAKMAN, J. C. (Editors), *Computing Methods and the Phase Problem in X-Ray Crystal Analysis*, Oxford, Pergamon Press (1961).

ROLLETT, J. S. (Editor), *Computing Methods in Crystallography*, Oxford, Pergamon Press (1965).

Direct Methods

GIACOVAZZO, C., *Direct Methods in Crystallography*, New York, Academic Press (1980).

LADD, M. F. C., and PALMER, R. A. (Editors), *Theory and Practice of Direct Methods in Crystallography*, New York, Plenum Press (1980).

STOUT, G. H., and JENSEN, L. H., *X-Ray Structure Determination—A Practical Guide*, New York, Macmillan (1968).

WOOLFSON, M. M., *An Introduction to X-Ray Crystallography*, Cambridge, Cambridge University Press (1970).

Refinement and Molecular Geometry

STOUT, G. H., and JENSEN, L. H., *X-Ray Structure Determination—A Practical Guide*, New York, Macmillan (1968).

WOOLFSON, M. M., *An Introduction to X-Ray Crystallography*, Cambridge, Cambridge University Press (1970).

Chemical Data

SUTTON, L. E. (Editor), *Tables of Interatomic Distances and Configurations in Molecules and Ions*, London, The Chemical Society (1958; supplement, 1965).

Problems

7.1. Choose three of the following reflections to fix an origin in space group $P\bar{1}$, giving reasons for your choice.

hkl	$|E|$	hkl	$|E|$
705	2.2	$6\bar{1}\bar{7}$	3.2
$42\bar{6}$	2.7	203	2.3
$4\bar{3}2$	1.1	$8\bar{1}\bar{4}$	2.1

Are there any triplets which meet the vector requirements of the Σ_2 formula?

7.2. The geometric structure factor formulae for space group $P2_1$ are

$$A = 2 \cos 2\pi(hx + lz + k/4) \cos 2\pi(ky - k/4)$$

$$B = 2 \cos 2\pi(hx + lz + k/4) \sin 2\pi(ky - k/4)$$

Deduce the amplitude symmetry and the phase symmetry for this space group according to the two conditions $k = 2n$ and $k = 2n + 1$.

7.3. In space group $P2_1/c$, two starting sets of reflections for the application of the Σ_2 formula are proposed:

	Origin-fixing	Symbols
(a)	$041, 117, \bar{1}23$	$242, \bar{1}62$
(b)	$223, 012, 13\bar{7}$	$111, 162$

Which starting set would be chosen in practice? Give reasons. What modification would have to be made to the starting set if the space group is $C2/c$?

7.4. The following values of $\log_e[\Sigma_j f_j^2(hkl)/|\overline{F_o(hkl)}|^2]$ and $\overline{(\sin^2\theta)/\lambda^2}$ were obtained from a set of three-dimensional data for a monoclinic crystal. Use the method of least squares to obtain values for the scale K (of $|F_o|$) and temperature factor B by Wilson's method.

$\log_e[\Sigma_j f_j^2(hkl)/|\overline{F_o(hkl)}|^2]$	$\overline{(\sin^2\theta)/\lambda^2}$
4.0	0.10
5.6	0.20
6.5	0.30
7.9	0.40
9.4	0.50

What is the value of the root mean square atomic displacement corresponding to the derived value of B?

7.5. An orthorhombic crystal contains four molecules of a chloro compound in a unit cell of dimensions $a = 7.210(4)$, $b = 10.43(1)$, $c =$

15.22(2) Å. The coordinates of the Cl atoms are

$$\tfrac{1}{4}, y, z; \quad \tfrac{3}{4}, \bar{y}, z; \quad \tfrac{1}{4}, \tfrac{1}{2}+y, \tfrac{1}{2}+z; \quad \tfrac{3}{4}, \tfrac{1}{2}-y, \tfrac{1}{2}+z$$

with $y = 0.140(2)$ and $z = 0.000(2)$. Calculate the shortest Cl $\cdots$ Cl contact distance and its estimated standard deviation.

7.6. The following data give phase indications for the reflection 771 ($|E_h| = 2.2$, $\phi_{\text{calc}} = -14°$) in a crystal of space group $P2_12_12_1$. Determine ϕ_h by both (7.25) and (7.28). For simplicity, let w_h in (7.28) be taken as unity.

k			ϕ_k (deg)	$h-k$			ϕ_{h-k} (deg)	$|E_k||E_{h-k}|$
12	1	0	0	$\bar{5}$	6	1	−37	4.4
7	$\bar{1}$	4	177	0	8	$\bar{3}$	180	5.1
12	0	$\bar{1}$	90	$\bar{5}$	7	2	−114	4.5
12	0	1	90	$\bar{5}$	7	0	−90	3.3
6	1	3	102	1	6	$\bar{2}$	−64	2.7
$\bar{1}$	4	5	−79	8	3	$\bar{4}$	92	3.7

8

Examples of Crystal Structure Analysis

8.1 Introduction

In this final chapter we wish to draw together, by means of two actual examples, material presented earlier in the book. It may be desirable for the reader to refer back to the previous chapters for descriptions of the techniques used, since we shall present here mainly the results obtained at each stage.

Both of the examples selected can be solved by either the heavy-atom method or direct methods. Nowadays, it is quite commonplace to attempt the solving by direct methods of those structures which, at one time, would have been treated by the heavy-atom method. Where a powerful and sophisticated computer package like MULTAN (described in Chapter 7) is available, direct methods frequently provide the most expeditious route to the solution of a structure. However, in order that we may illustrate the methods described, we shall use the heavy-atom technique for the first structure and direct methods for the second.

In the chapter, we list the $|F_o|$ data and the corresponding $|E|$ values derived from them for those projections of the structures that show the maximum resolution of the atoms. We hope that readers will be encouraged to carry through the structure determinations, at least in projection. The sets of two-dimensional $|F|$ data listed are, with other crystal information provided, ideally suited for use with the excellent CALCHEM X-ray teaching package now available.*

---

* Readers interested in this package should write to Dr. Neil Bailey, Department of Chemistry, University of Sheffield, Sheffield S3 7HF, England.

8.2 Crystal Structure of 2-Bromobenzo[b]indeno[1,2-e]pyran (BBIP)*

BBIP is an organic compound which is prepared by heating a solution in ethanol of equimolar amounts of 3-bromo-6-hydroxybenzaldehyde (I) and 2-oxoindane (II) under reflux in the presence of piperidine acetate. The two molecules condense with the elimination of two molecules of water. Upon recrystallization of the product from toluene, it has a m.p. of 176.5–177.0°C. Its molecular formula is $C_{16}H_9BrO$, and its classical structural formula is shown by III.

8.2.1 Preliminary Physical and X-Ray Measurements

The compound was recrystallized slowly as red, acicular (needle-shaped) crystals, with the forms (subsequently named) {100}, {001}, and {011} predominant (Figure 8.1). The red color is characteristic of the chromophoric nature of a conjugated double-bond system.

The density of the crystals was measured by suspending them in aqueous sodium bromide solution in a stoppered measuring cylinder in a thermostat bath at 20°C. Water or concentrated sodium bromide solution, as necessary, was added to the suspension until the crystals neither settled to the bottom of the cylinder nor floated to the surface of the solution. Then, the crystals and solution were of the same density, and the density of the solution was measured with a pyknometer. A convenient variant here is to measure the refractive index of the final solution with an Abbe refractometer, having first prepared a graph of refractive index v density from data in the literature.†

Under a polarizing microscope, the crystals showed straight extinction on (100) and (001), and oblique extinction (about 3° to a crystal edge) on a section cut normal to the needle axis (b). These observations suggested that the crystals were probably monoclinic.

---

* M. F. C. Ladd and D. C. Povey, *Journal of Crystal and Molecular Structure* **2**, 243 (1972).
† *International Critical Tables* **3**, 80 (1928), **7**, 73 (1930); *Mellor's Treatise on Inorganic and Theoretical Chemistry* **2**, 941 (Supplement, 1961).

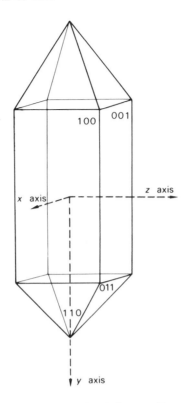

FIGURE 8.1.   Crystal habit of BBIP with the crystallographic axes drawn in.

The crystals chosen for X-ray studies had the approximate dimensions 0.2, 0.4, and 0.3 mm parallel to a, b, and c, respectively. A crystal was mounted on the end of an annealed quartz fiber with "Araldite" or "Eastman 910" adhesive and the fiber was attached to an X-ray goniometer head (arcs) (Figure 8.2) with dental wax. The arcs were affixed to a single-crystal oscillation camera, and the crystal was set with the needle axis accurately parallel to the axis of oscillation, first by eye and then by X-ray methods. Copper $K\alpha$ radiation ($\tilde{\lambda} = 1.5418$ Å) was used throughout the work.

A symmetric oscillation photograph taken about the b axis is shown in Figure 8.3. The horizontal mirror symmetry line indicates that the Laue group of the crystal has an m plane normal to the needle axis. Further X-ray photographs, for example, Figure 8.4, showed that the only axial symmetry was 2 parallel to b, thus confirming the monoclinic system for BBIP.

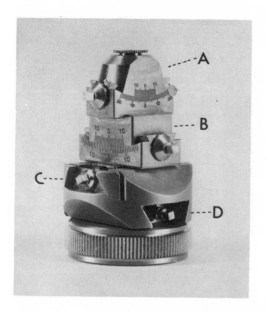

FIGURE 8.2. Standard goniometer head (by courtesy of Stoe et Cie); *A* and *B* are two arcs for angular adjustments; *C* and *D* are two sledges for horizontal adjustments.

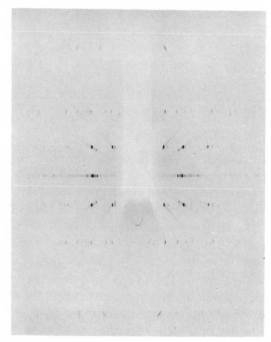

FIGURE 8.3. Symmetric oscillation photograph taken with the X-ray beam normal to *b*; the "shadow" arises from the beam stop and holder.

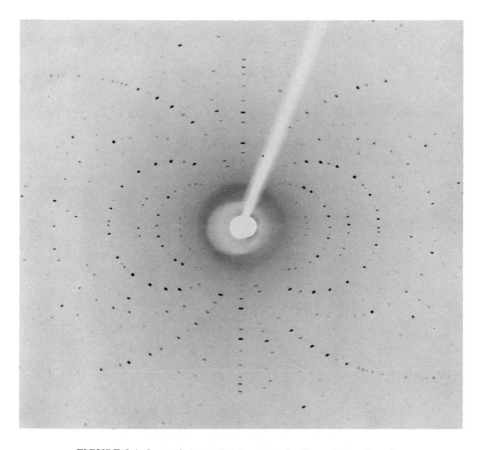

FIGURE 8.4. Laue photograph taken with the X-ray beam along *b*.

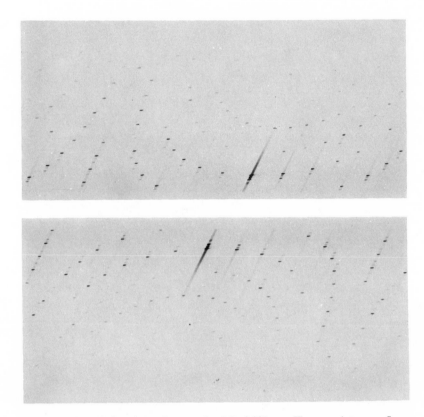

FIGURE 8.5. Weissenberg photograph of the $h0l$ layer. The more intense reflec-
tions show spots arising from both Cu $K\alpha$ ($\lambda = 1.542$ Å) and Cu $K\beta$ ($\lambda = 1.392$ Å)
radiations. In some areas, spots from W $L\alpha$ radiation ($\lambda = 1.48$ Å) arise due to
sputtering of the copper target in the X-ray tube with tungsten from the filament (see
Appendix A.4).

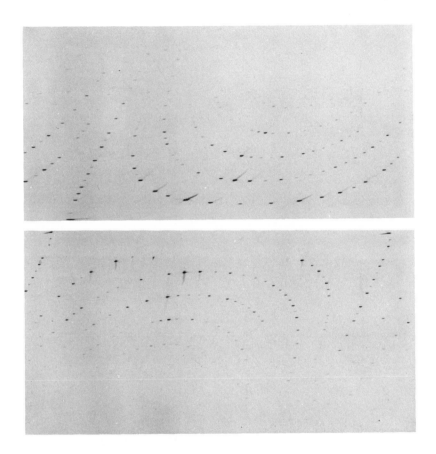

FIGURE 8.6. Weissenberg photograph of the $h1l$ layer. The $01l$ reciprocal lattice row
shows evidence of slight mis-setting of the crystal.

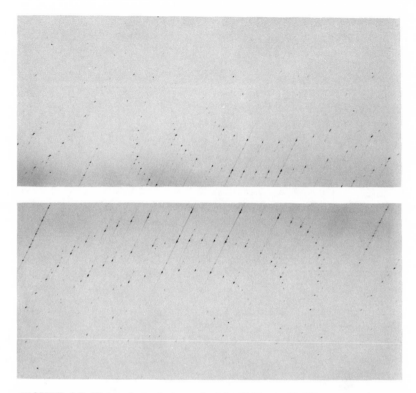

FIGURE 8.7. Weissenberg photograph of the $0kl$ layer; Cu $K\beta$ spots can be seen for the more intense reflections. The $00l$ (z^*) reciprocal lattice row is common to this photograph and that of the $h0l$ layer.

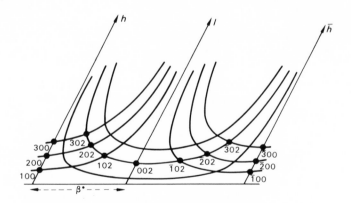

FIGURE 8.8. Sample of indexed reflections on an $h0l$ Weissenberg photograph diagram.

Weissenberg photographs are shown in Figures 8.5–8.7. The indexing of the reflections can be understood with reference to Figures 8.8 and 8.9. There are no systematic absences for the hkl reflections, so that the unit cell is primitive (P), but systematic absences do arise for $h0l$ with l odd and for $0k0$ with k odd. These observations confirm the monoclinic symmetry, and the systematic absences lead unambiguously to space group $P2_1/c$.

Measurements on the X-ray photographs gave the approximate unit cell dimensions $a = 7.51$, $b = 5.96$, $c = 26.2$ Å, and $\beta = 92.5°$.

The Bragg angles θ of 20 high-order reflections of known indices, distributed evenly in reciprocal space, were measured to the nearest $0.01°$ on a four-circle diffractometer. From these data, the unit-cell dimensions were calculated accurately by the method of least squares (page 343). The complete crystal data are listed in Table 8.1. The calculated density D_c is in good agreement with the measured value D_m, which indicates a high degree of self-consistency in the parameters involved.

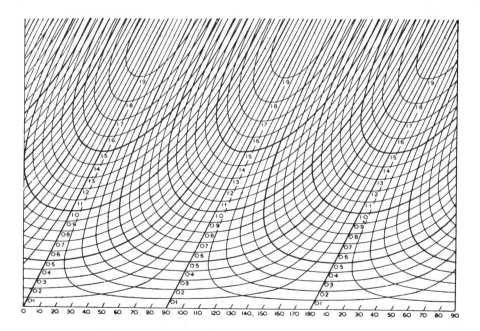

FIGURE 8.9. Weissenberg chart: camera diameter 57.30 mm, 2° rotation per mm travel. (Reproduced with the permission of the Institute of Physics and the Physical Society, London.)

TABLE 8.1. Crystal Data for BBIP at 20°C^a

Molecular formula	$C_{16}H_9BrO$
Molecular weight	297.16
Space group	$P2_1/c$
a, Å	7.508(4)
b, Å	5.959(5)
c, Å	26.172(6)
β, deg	92.55(2)
Unit cell volume, Å^3	1169(2)
D_m, g cm^{-3}	1.68(1)
D_c, g cm^{-3}	1.688(3)
Z	4
$F(000)$	592

^{a} The numbers in parentheses are estimated standard deviations, to be applied to the least significant figure.

8.2.2   Intensity Measurement and Correction

The intensities of about 1700 symmetry-independent reflections with* $(\sin\theta)/\lambda \leqslant 0.56$ Å^{-1} were measured, and corrections were applied to take account of polarization and Lorentz effects, but not absorption (pages 423, 438). Approximate scale (K) and isotropic temperature (B) factors were obtained by Wilson's method (page 257). The straight line was fitted by a least-squares method, and has the equation

$$\log_e\left\{\sum_j f_j^2/\overline{|F_o|^2}\right\} = -1.759 + 3.480\,\overline{\sin^2\theta} \tag{8.1}$$

From the slope $(2B/\lambda^2)$ and intercept $(2\log_e K)$, $B = 4.1$ Å^2 and $K = 0.13$. The graphical Wilson plot is shown in Figure 8.10 and the scaled $|F(h0l)|$ data in Table 8.2. We are now ready to proceed with the structure analysis.

8.2.3   Structure Analysis in the (010) Projection

This projection of the unit cell has the largest area and thus would be expected to show an appreciable resolution of the molecule. It is uncommon for a normal three-dimensional study to be preceded by an analysis in projection. However, from the standpoint of introductory teaching, carefully chosen two-dimensional examples have much to offer.

Using Figure 2.33 (page 97), we can associate the coordinates of the general equivalent positions,

* Large thermal vibrations of the atoms in the structure led to this rather low practicable upper limit for $(\sin\theta)/\lambda$.

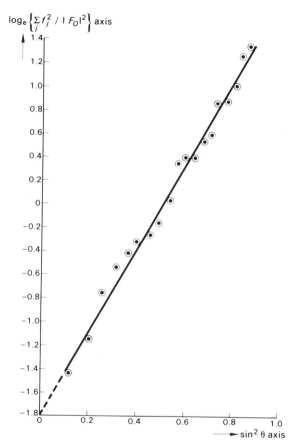

FIGURE 8.10. Wilson plot.

$$\pm\{x, y, z; \quad x, \tfrac{1}{2}-y, \tfrac{1}{2}+z\}$$

with the four bromine atoms in the unit cell. In the (010) projection, these coordinates give rise to two repeats within the length c, so we may consider this projection in terms of plane group $p2$ and compute the Patterson function from 0 to $a/2$ and 0 to $c/2$, which is equivalent to one-half of the unit cell in $p2$. This portion of the projection $P(u, w)$ would be expected to show one Br—Br vector at $(2x, 2z)$, as in Figure 8.11.

The two rows of peaks indicate that, in this projection, the molecules lie closely parallel to the z axis; this conclusion is supported by the large magnitude of $|F(200)|$. This value may be compared with $F(000)$ in Table 8.1, and, more significantly, with $\sum_j f_{j,\theta_{200}}$, which is 474. The peak arising from the Br—Br vector is marked A, and by direct measurement we obtain

TABLE 8.2. $|F_o|$ and $|E|$ Data for the $h0l$ Zone of BBIP^a

h	k	l	$|F_o|$	$|E|$	F_{Br}	h	k	l	$|F_o|$	$|E|$	F_{Br}
0	0	4	34	0.43	121	−5	0	20	12	0.88	19
0	0	6	57	0.79	101	−5	0	22	19	1.66	13
0	0	8	53	0.82	78	−5	0	24	13	1.37	8*
0	0	10	80	1.40	55	−4	0	2	61	1.44	66
0	0	12	52	1.05	34	−4	0	4	63	1.53	57
0	0	16	6	0.17	2*	−4	0	6	53	1.35	46
0	0	20	19	0.76	−14	−4	0	8	45	1.23	33
0	0	22	68	3.29	−18	−4	0	10	36	1.04	20
0	0	24	7	0.42	−19	−4	0	12	28	0.93	8*
1	0	0	25	0.31	8*	−4	0	16	5	0.21	−8*
2	0	0	303	4.40	−114	−4	0	18	10	0.50	−14
3	0	0	4	0.07	−17	−4	0	20	42	2.46	−17
4	0	0	61	1.43	71	−4	0	22	7	0.49	−18
5	0	0	15	0.47	17	−4	0	24	22	1.87	−17
6	0	0	17	0.74	−38	−3	0	2	48	0.87	−34
7	0	0	18	1.12	−12	−3	0	4	33	0.62	−49
8	0	0	11	1.01	17	−3	0	6	52	1.04	−59
−8	0	2	10	0.92	15	−3	0	8	94	2.03	−65
−8	0	4	6	0.56	12	−3	0	10	59	1.41	−66
−8	0	6	10	0.96	8*	−3	0	12	47	1.26	−63
−8	0	10	6	0.65	1*	−3	0	14	60	1.84	−57
−7	0	4	20	1.26	−21	−3	0	16	43	1.52	−49
−7	0	6	15	0.98	−23	−3	0	18	45	1.86	−40
−7	0	8	12	0.82	−24	−3	0	20	37	1.82	−31
−7	0	10	20	1.47	−24	−3	0	22	37	2.18	−22
−7	0	12	8	0.64	−22	−3	0	24	18	1.29	−15
−7	0	14	15	1.33	−19	−2	0	2	129	1.89	−108
−7	0	16	26	2.60	−16	−2	0	4	90	1.38	−95
−6	0	2	9	0.39	−34	−2	0	6	54	0.88	−78
−6	0	4	13	0.58	−29	−2	0	8	50	0.90	−59
−6	0	6	30	1.38	−22	−2	0	10	6	0.12	−39
−6	0	8	23	1.12	−15	−2	0	12	58	1.33	−22
−6	0	10	16	0.89	−7*	−2	0	14	8	0.21	−6
−6	0	12	10	0.57	−1*	−2	0	18	15	0.55	13
−6	0	14	11	0.70	4*	−2	0	20	23	1.00	18
−6	0	18	7	0.58	11	−2	0	22	37	1.94	20
−6	0	20	34	3.28	12	−2	0	24	30	1.92	21
−5	0	2	16	0.51	27	−1	0	2	53	0.67	32
−5	0	4	36	1.16	35	−1	0	4	51	0.68	53
−5	0	6	18	0.61	40	−1	0	6	87	1.25	68
−5	0	8	43	1.54	43	−1	0	8	86	1.37	77
−5	0	10	50	1.94	43	−1	0	10	79	1.43	80
−5	0	12	18	0.77	41	−1	0	12	57	1.18	77
−5	0	14	40	1.91	37	−1	0	14	54	1.30	70
−5	0	16	34	1.85	31	−1	0	16	76	2.16	61
−5	0	18	19	1.19	25	−1	0	18	71	2.39	51

^{a} The scale factor K is 1.11 and $B = 4.13$ Å.

TABLE 8.2.—*cont.*

h	k	l	$|F_o|$	$|E|$	F_{Br}	h	k	l	$|F_o|$	$|E|$	F_{Br}
-1	0	20	63	2.55	40	3	0	18	30	1.34	39
-1	0	22	37	1.81	30	3	0	20	49	2.62	33
-1	0	24	17	1.02	21	3	0	22	6	0.32	26
1	0	2	25	0.32	-16	3	0	24	21	1.66	19
1	0	4	62	0.84	-38	4	0	2	42	1.01	72
1	0	6	57	0.83	-55	4	0	4	50	1.25	69
1	0	8	53	0.86	-66	4	0	6	22	0.58	62
1	0	10	76	0.52	-71	4	0	8	46	1.32	53
1	0	12	28	0.59	-71	4	0	10	42	1.33	42
1	0	14	67	1.66	-67	4	0	12	44	1.57	31
1	0	16	89	2.59	-59	4	0	14	3	0.12	20
1	0	18	58	2.01	-51	4	0	16	13	0.61	11
1	0	20	67	2.79	-41	4	0	18	4	0.22	3*
1	0	22	21	1.06	-31	4	0	22	13	1.02	-6*
1	0	24	23	1.43	-22	4	0	24	10	0.26	-8*
2	0	2	51	0.76	-112	5	0	2	11	0.35	7*
2	0	4	60	0.93	-104	5	0	4	6	0.20	-3*
2	0	6	76	1.28	-90	5	0	6	11	0.36	-12
2	0	8	60	1.11	-73	5	0	8	17	0.65	-20
2	0	10	80	1.67	-55	5	0	12	29	1.35	-28
2	0	12	41	0.98	-38	5	0	14	26	1.37	-28
2	0	14	6	0.17	-22	5	0	16	13	0.79	-27
2	0	16	19	0.62	-9*	5	0	18	18	1.27	-25
2	0	20	9	0.41	8*	5	0	20	21	1.76	-21
2	0	22	47	2.62	12	5	0	22	6	0.61	-17
2	0	24	16	1.09	14	6	0	2	49	2.17	-39
3	0	2	3	0.06	0*	6	0	4	9	0.41	-39
3	0	4	30	0.58	16	6	0	6	14	0.68	-36
3	0	6	20	0.41	30	6	0	8	28	1.46	-32
3	0	8	35	0.79	41	6	0	10	17	0.97	-26
3	0	10	30	0.75	47	6	0	12	23	1.45	-20
3	0	12	29	0.83	50	6	0	14	9	0.64	-14
3	0	14	63	2.06	49	6	0	18	3	0.29	-4*
3	0	16	50	1.90	45						

---

[a] F_{Br} is the calculated contribution of the bromine atoms to $F(h0l)$, assuming a B value of 4.1 $Å^2$. Values marked with asterisks have small amplitudes and their signs may not be correct for the corresponding $F(h0l)$ reflections.

the fractional coordinates $x = 0.25$, $z = 0.015$ for the Br atom in the asymmetric unit.

An electron density map, in this projection, was calculated using the signs given by F_{Br} with the experimental values of $|F(h0l)|$ provided that $|F_{Br}|$ was not less than about $|F_o|/3$, for which reflections the signs would be likely to be the more uncertain at this stage. Figure 8.12 shows the

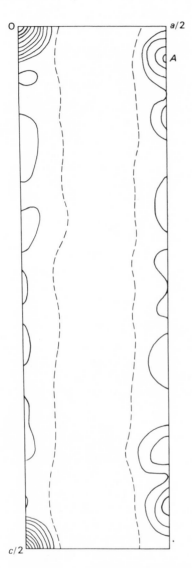

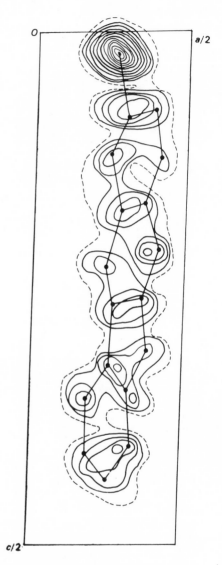

FIGURE 8.11. Asymmetric unit of $P(u, w)$. Since we are concerned here mainly with the Br—Br vector, the slight distortion arising from drawing β as 90° is inconsequential.

FIGURE 8.12. Asymmetric unit of $\rho(x, z)$ phased on the bromine atoms; the probable atomic positions are marked in.

electron density map with the molecule, fitted with the aid of a model, marked in. The resolution is moderately good, and we can see that we are working along the right lines. From the shapes of the rings, it is apparent that the molecule is inclined to the plane of this projection, and there will be a limit to the possible improvement of the resolution attainable in projection. In consequence, we shall begin three-dimensional studies.

8.2.4 Three-Dimensional Structure Determination

In order to obtain values for all spatial coordinates, we proceeded first to a three-dimensional Patterson map $P(u, v, w)$, calculated section by section normal to the b axis.

The coordinates of the general positions show that the Br—Br vectors in the asymmetric unit will be found at $2x, 2y, 2z$ (single-weight peak, B), $0, \frac{1}{2} - 2y, \frac{1}{2}$ (double-weight peak, C), and $2\bar{x}, \frac{1}{2}, \frac{1}{2} + 2z$ (double-weight peak, D). Hence, we must study the Patterson map carefully, particularly the line $[0, v, \frac{1}{2}]$ and the Harker section $(u, \frac{1}{2}, w)$. Figures 8.13 and 8.14 show these two regions of Patterson space, and Figure 8.15 illustrates the general

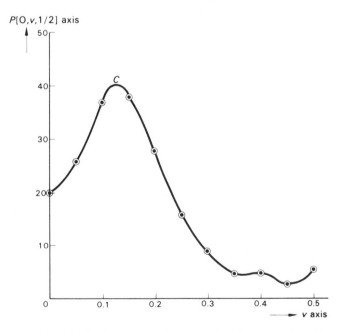

FIGURE 8.13. Patterson function along the Harker line $[0, v, \frac{1}{2}]$.

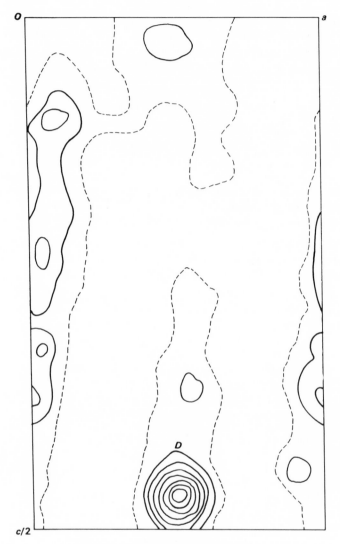

FIGURE 8.14. Patterson section $(u, \frac{1}{2}, w)$.

section containing the single-weight peak B. From the peaks B, C, and D, the coordinates for the bromine atom in the asymmetric unit were found to be 0.248, 0.188, 0.016. Repeating the phasing procedure, but now for *hkl* reflections, and calculating a three-dimensional electron density map produced a good resolution of the complete structure, with the exception

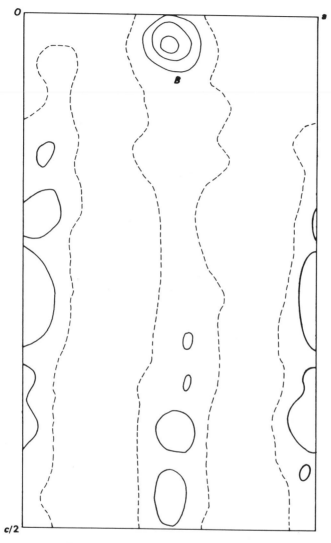

FIGURE 8.15. Patterson section $(u, 0.375, w)$.

of the hydrogen atoms. Figure 8.16 illustrates a composite electron density map, which consists of superposed sections calculated at intervals along a.

The scattering of X-rays by hydrogen is small in magnitude, and these atoms are not resolved by the direct summation of the electron density. They may often be located by means of a difference-Fourier synthesis (page

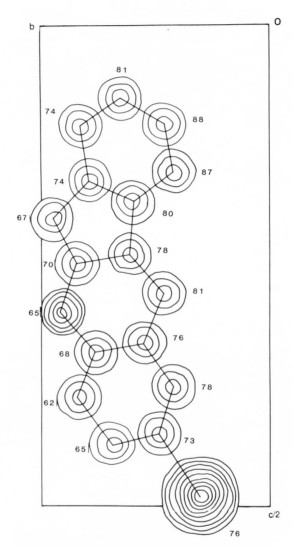

FIGURE 8.16. Composite three-dimensional electron
density map with the molecule (excluding H atoms)
marked in, as seen along a. The contour of zero electron
density is not shown, and the numbers represent $100x$
for each atom.

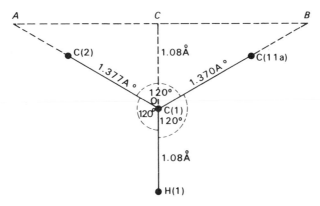

FIGURE 8.17. Calculation of the coordinates of H(1); the configuration of atoms is planar.

264), which may be carried out at this stage of the structure analysis. We can calculate the probable positions for the hydrogen atoms from the geometry of the structure, for comparing with or supplementing those obtained from the difference-Fourier synthesis. The principle of a simple calculation is shown by Figure 8.17.

Consider the hydrogen atom H(1) attached to C(1), and assume a C—H bond length* of 1.0 Å, with trigonal geometry at C(1). This atom is transferred, temporarily, to the origin, and the transformed fractional coordinates converted to absolute values in Å. Next, the atoms are, for convenience, referred to rectangular axes (Figure 8.18). Triangle OAB is isosceles, with $OC = C(1)$—H(1), and the coordinates of point C are the averages of those at A and B. The coordinates of H(1) are now obtained by inverting those of C through O and then transforming back to fractional values in the monoclinic unit cell.

Finally, we arrive at the complete structure of BBIP, shown in Figure 8.19 with a convenient atom numbering scheme.

8.2.5 Refinement

During the final stages of refinement of the structure, the hydrogen atoms were included in the evaluation of the structure factors F_c, but no attempt was made to refine the parameters of the hydrogen atoms because

---

* The idealized value (see page 359) is 1.08 Å, but the experimental X-ray value is closer to 1 Å.

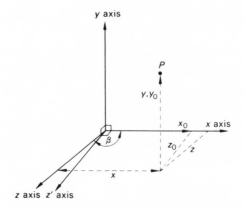

FIGURE 8.18. Orthogonal coordinates in
the monoclinic system. The point P has coordinates X, Y, Z (Å) on the monoclinic axes
and X_0, Y_0, Z_0 on orthogonal axes, where
$X_0 = X + Z \cos \beta$, $Y_0 = Y$, and $Z_0 = Z \sin \beta$.

the main interest in the problem lay in determining the molecular conformation. The final adjustments of the structural parameters of the Br, O, and
C atoms (x, y, z, and anisotropic temperature factors) and the scale factor
were carried out by the method of least squares (page 344). The refinement
converged with an R-factor of 0.070, and a final difference-Fourier synthesis
showed no fluctuations in density greater than about twice $\sigma(\rho_o)$. The
analysis was considered to be satisfactory, and the refinement was terminated
at this stage.

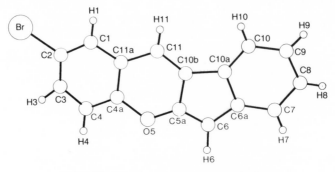

FIGURE 8.19. Structural formula for BBIP.

8.2.6 Molecular Geometry

It remained to determine the bond lengths, bond angles, and other features of the geometry of the molecule and its relationship with other molecules in the unit cell.

From the coordinates of the atomic positions (Table 8.3) and using equations such as (7.89) and (7.91), bond lengths and angles were calculated. They are shown on the drawings of the molecule in Figures 8.20 and 8.21. Figure 8.22 illustrates the packing of the molecules in the unit cell, as seen along the b axis; the average intermolecular contact distance is 3.7 Å, which

TABLE 8.3. Fractional Atomic Coordinates in BBIP, with esd's in Parentheses[a]

Atom	x	y	z
Br[b]	0.7602(2)	0.3152(3)	0.4848(0)
C(1)	0.7820(16)	0.4187(22)	0.3789(4)
C(2)	0.7310(16)	0.4951(24)	0.4252(4)
C(3)	0.6524(16)	0.7075(25)	0.4297(5)
C(4)	0.6214(16)	0.8413(23)	0.3871(5)
C(4a)	0.6794(16)	0.7619(22)	0.3406(4)
O(5)	0.6520(11)	0.9051(14)	0.2995(3)
C(5a)	0.6973(14)	0.8329(21)	0.2526(4)
C(6)	0.6714(14)	0.9397(19)	0.2077(5)
C(6a)	0.7384(15)	0.7990(19)	0.1678(4)
C(7)	0.7401(17)	0.8230(24)	0.1150(4)
C(8)	0.8078(18)	0.6526(24)	0.0858(4)
C(9)	0.8766(17)	0.4574(24)	0.1079(5)
C(10)	0.8731(16)	0.4268(21)	0.1605(4)
C(10a)	0.8035(16)	0.5954(20)	0.1908(4)
C(10b)	0.7767(14)	0.6076(21)	0.2454(5)
C(11)	0.8064(15)	0.4734(21)	0.2850(4)
C(11a)	0.7593(14)	0.5475(20)	0.3359(5)
H(1)	0.809	0.239	0.375
H(3)	0.622	0.789	0.460
H(4)	0.565	0.999	0.389
H(6)	0.630	0.121	0.208
H(7)	0.674	0.944	0.097
H(8)	0.804	0.667	0.043
H(9)	0.886	0.361	0.076
H(10)	0.809	0.253	0.375
H(11)	0.870	0.311	0.285

[a] There are no esd's for the hydrogen atom coordinates because these parameters were not included in the least-squares refinement.

[b] A symmetry-related position to that in Figure 8.12 has been chosen to ensure right-handed reference axes.

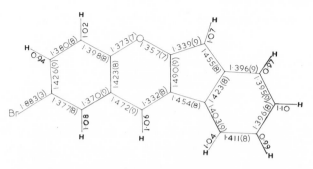

FIGURE 8.20. Bond lengths in BBIP, with their estimated
standard deviations in parentheses.

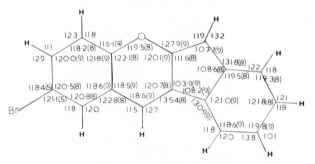

FIGURE 8.21. Bond angles in BBIP, with their estimated
standard deviations in parentheses.

is typical of the van der Waals forces which link molecules in this type of
compound.

In a molecule of this nature, the planarity of the ring system is of
stereochemical interest. The equation of a plane, $Ax + By + Cz = D$, can be
solved by three triplets x, y, z. Hence, the best molecular plane is obtained
by a least-squares procedure that minimizes the sum of the squares of the
deviations d of all of the atoms from the plane. The results are listed in
Table 8.4. It can be seen that the deviations of the atoms from the best
plane are not significant, and it is possible to conclude, therefore, that the

FIGURE 8.22. Stereoview of molecular packing in the structure of BBIP, as seen along a.

TABLE 8.4. Deviations of Atoms from the Least-Squares
Plane through the Molecule[a]

Atom	Deviation (Å)	Atom	Deviation (Å)
Br	0.03	C(6a)	0.02
C(1)	0.01	C(7)	−0.06
C(2)	−0.04	C(8)	−0.08
C(3)	−0.05	C(9)	−0.03
C(4)	−0.06	C(10)	0.03
C(4a)	0.01	C(10a)	0.04
O(5)	0.05	C(10b)	0.04
C(5a)	0.06	C(11)	0.03
C(6)	0.02	C(11a)	0.03

[a] The mean estimated standard deviation of the deviations is 0.02, so that
hardly any atoms deviate significantly from the best plane at a 3σ level.

introduction of the heteroatom has but little effect on the planarity of the
benzofluorene moiety.

8.3 Crystal Structure of Potassium 2-Hydroxy-3,4-dioxocyclobut-1-ene-1-olate Monohydrate (KHSQ)*

1,2-Dihydroxy-3,4-dioxocyclobut-1-ene (IV) may be prepared by
acid-catalyzed hydrolysis of 1,2-diethoxy-3,3,4,4-tetrafluorocyclobut-1-
ene (V). On recrystallization from water, it has a melting point of 293°C, at
which temperature it decomposes.

It has been called by the trivial name, squaric acid; the hydrogen atoms
in the hydroxyl groups are acidic, and can be replaced by a metal. The
potassium hydrogen salt monohydrate (VI), which is the subject of this
example, can be obtained by mixing hot, concentrated, equimolar, aqueous
solutions of potassium hydroxide and squaric acid and then cooling the
reaction mixture.

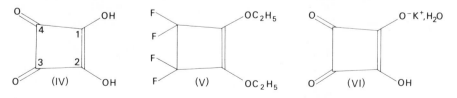

*R. J. Bull, M. F. C. Ladd, D. C. Povey, and R. Shirley, *Crystal Structure Communications*
2, 625 (1973).

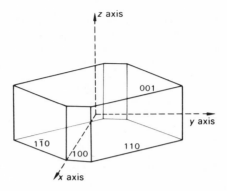

FIGURE 8.23. Crystal habit of KHSQ
with the crystallographic axes drawn in.

8.3.1 Preliminary X-Ray and Physical Measurements

The compound was recrystallized from water as colorless, prismatic crystals with the forms {001}, {110}, and {100} predominant (Figure 8.23). Under a polarizing microscope, straight extinction was observed on {001} and {100}, and an extinction angle of about 2° was obtained on a section cut normal to b. These results suggest strongly that the crystals belong to the monoclinic system.

The crystal specimen chosen for X-ray work had the dimensions 0.5, 0.5, and 0.3 mm parallel to a, b, and c, respectively. The details of the preliminary measurements are similar to those described for the previous example, and we list the crystal data immediately (Table 8.5). Copper $K\alpha$ radiation ($\bar{\lambda} = 1.5418$ Å) was used throughout this work.

TABLE 8.5. Crystal Data for KHSQ at 20°C

Molecular formula	$C_4HO_4^-$, K^+, H_2O
Molecular weight	170.17
Space group	$P2_1/c$
a, Å	8.641(1)
b, Å	10.909(1)
c, Å	6.563(2)
β, deg	99.81(1)
Unit-cell volume, Å³	609.6(2)
D_m, g cm^{-3}	1.839(7)
D_c, g cm^{-3}	1.854(1)
Z	4
$F(000)$	296

TABLE 8.6. $|F_o|$ and $|E|$ Data for the $hk0$ Zone of KHSQ[a]

h	k	l	$|F_o|$	$|E|$	h	k	l	$|F_o|$	$|E|$
1	0	0	0.2	0.00	4	1	0	21.2	0.63
2	0	0	20.2	0.30	4	2	0	9.9	0.32
3	0	0	90.5	1.75	4	3	0	32.5	1.09
4	0	0	13.3	0.29	4	4	0	27.5	1.06
5	0	0	32.1	0.93	4	5	0	2.3	0.09
6	0	0	30.6	1.02	4	6	0	2.6	0.13
7	0	0	50.5	2.26	4	7	0	9.6	0.51
8	0	0	42.4	2.16	4	8	0	10.0	0.64
9	0	0	14.9	1.02	4	9	0	18.3	1.28
0	2	0	13.8	0.19	4	10	0	7.1	0.60
0	4	0	59.5	1.08	4	11	0	16.2	1.51
0	6	0	14.4	0.36	5	1	0	30.2	1.13
0	8	0	23.3	0.81	5	2	0	11.7	0.51
0	10	0	18.9	0.93	5	3	0	6.2	0.26
0	12	0	3.5	0.24	5	4	0	38.6	1.92
1	1	0	9.3	0.17	5	5	0	18.8	0.93
1	2	0	15.2	0.34	5	6	0	8.9	0.55
1	3	0	26.7	0.60	5	7	0	50.3	3.12
1	4	0	31.1	0.90	5	8	0	10.8	0.85
1	5	0	14.9	0.45	5	9	0	5.4	0.43
1	6	0	9.6	0.38	5	10	0	13.1	1.33
1	7	0	10.9	0.45	6	1	0	3.3	0.15
1	8	0	0.8	0.04	6	2	0	32.3	1.58
1	9	0	20.2	1.17	6	3	0	12.2	0.60
1	10	0	16.6	1.27	6	4	0	22.3	1.23
1	11	0	12.8	1.03	6	5	0	16.7	0.96
1	12	0	5.3	0.56	6	6	0	29.1	1.93
2	1	0	32.2	0.66	6	7	0	36.0	2.54
2	2	0	0.2	0.00	6	8	0	5.4	0.45
2	3	0	6.8	0.17	6	9	0	3.3	0.29
2	4	0	34.8	1.01	7	1	0	9.2	0.53
2	5	0	20.0	0.63	7	2	0	10.2	0.67
2	6	0	19.4	0.75	7	3	0	31.7	1.96
2	7	0	42.9	1.84	7	4	0	11.2	0.82
2	8	0	17.4	0.92	7	5	0	4.6	0.32
2	9	0	0.2	0.01	7	6	0	1.6	0.14
2	10	0	16.1	1.18	7	7	0	17.2	1.45
2	11	0	16.8	1.37	7	8	0	3.8	0.40
3	1	0	35.7	0.89	8	1	0	3.2	0.22
3	2	0	30.7	0.90	8	2	0	4.6	0.34
3	3	0	5.2	0.15	8	3	0	4.4	0.32
3	4	0	51.9	1.85	8	4	0	21.9	1.79
3	5	0	21.8	0.79	8	5	0	12.2	1.01
3	6	0	17.2	0.80	8	6	0	10.4	0.97
3	7	0	29.5	1.42	9	1	0	5.7	0.50
3	8	0	6.5	0.41	9	2	0	10.9	1.08
3	9	0	9.4	0.61	9	3	0	2.2	0.20
3	10	0	4.3	0.36	9	4	0	12.8	1.37
3	11	0	9.1	0.80					

[a] The scale factor K is 1.000 and $B = 1.89$ Å^2.

8.3.2 Intensity Measurement and Correction

Nine hundred symmetry-independent intensities with $(\sin\theta)/\lambda \leqslant$ $0.57\,\text{Å}^{-1}$ were measured. Corrections were applied for polarization and Lorentz effects, but not for absorption. Scale (K) and isotropic temperature (B) factors were deduced by Wilson's method (page 257) and the $|F_o|$ data were converted to $|E|$ values through (7.1). Table 8.6 lists $|F_o|$ and $|E|$ values for the [001] zone in this compound.

In using direct methods, a structure analysis often begins with those reflections for which $|E|$ is greater than some value $|E_{\min}|$, which, typically is chosen to be about 1.5. For a centrosymmetric crystal, these reflections usually number about 12–15% of the total data. There were 142 of such reflections for KHSQ, representing 15.8% of the experimental reflection data, and their statistics are shown in Table 8.7. The agreement with the theoretical values for a centric distribution of $|E|$ values is very close, in accord with the centrosymmeric space group (Table 8.5).

8.3.3 Σ_2 Listing

The next stage is the preparation of a Σ_2 listing (page 305). Symmetry-related reflections now become very important in generating triplet relationships: 300 and $30\bar{4}$ can lead to both 004 and $60\bar{4}$, the latter by replacing $30\bar{4}$ by $\bar{3}04$, taking note of the phase symmetry. We recall the relevant phase symmetry for space group $P2_1/c$ (see page 305), which may be summarized as follows:

$$s(hkl) = s(\bar{h}\bar{k}\bar{l}) \qquad (8.2)$$
$$s(hkl) = s(h\bar{k}l)(-1)^{k+l} \qquad (8.3)$$

TABLE 8.7. Statistics of $|E|$ Values in KHSQ

	Acentric	Centric	This structure
$|E|^2$	1.00	1.00	1.00
$|E|$	0.89	0.80	0.81
$|E|^2-1$	0.74	0.97	0.95
% ≥ 1.0	36.8	31.7	33.9
% ≥ 1.5	10.5	13.4	14.6
% ≥ 1.75	4.7	8.0	8.4
% ≥ 2.0	1.8	4.6	4.9
% ≥ 2.5	0.2	1.2	1.1

TABLE 8.8. Part of the Σ_2 Listing for KHSQ

k	$|E(\mathbf{k})|$	h − k	$|E(\mathbf{h}-\mathbf{k})|$	h	$|E(\mathbf{h})|$	$|E(\mathbf{h})|\,|E(\mathbf{k})|\,|E(\mathbf{h}-\mathbf{k})|$
53$\bar{1}$	2.6	010,4	2.8	573	2.6	18.9
(37)		041	2.2	57$\bar{2}$	3.3	17.2
		041	2.0	570	2.7	14.0
		114	2.3	62$\bar{5}$	1.7	10.2
		032	1.7	56$\bar{3}$	2.0	8.8
11$\bar{4}$	2.3	57$\bar{2}$	3.3	482	1.9	14.4
(45)		66$\bar{4}$	1.8	570	2.7	11.2
		68$\bar{1}$	1.5	573	2.6	9.0
		56$\bar{3}$	2.0	451	1.5	6.9
		454	1.6	540	1.5	5.5
032	1.7	53$\bar{1}$	2.6	56$\bar{3}$	2.0	8.8
(54)		57$\bar{2}$	3.3	540	1.5	8.4
		482	1.9	454	1.6	5.2
		451	1.5	48$\bar{1}$	2.0	5.1
11$\bar{2}$	2.5	57$\bar{2}$	3.3	66$\bar{4}$	1.8	14.9
(39)		482	1.9	570	2.7	12.8
		11$\bar{4}$	2.3	002	1.9	10.2
		571	1.7	68$\bar{1}$	1.5	6.4
010,4	2.8	33$\bar{2}$	2.2	372	1.9	11.7
(35)		62$\bar{5}$	1.7	68$\bar{1}$	1.5	7.1
33$\bar{2}$	2.2	33$\bar{2}$	2.2	66$\bar{4}$	1.8	8.7
(46)		11$\bar{4}$	2.3	242	1.7	8.6
		31$\bar{3}$	1.8	041	2.0	7.9
		625	1.7	313	1.8	6.7
002	1.9	041	2.0	041	2.0	7.4
(25)		11$\bar{4}$	2.3	11$\bar{6}$	1.5	6.6
		68$\bar{1}$	1.5	681	1.6	4.6

A portion of the Σ_2 listing is shown in Table 8.8. The numbers in parentheses under each **h** are the total numbers of Σ_2 triplets for each of the reflections* **h**, and **k**, and **h** − **k** represent vectors forming a triplet with **h**.

8.3.4 Specifying the Origin

Following the procedure described on pages 298ff and using the reflections in Table 8.8, three reflections were chosen and allocated positive signs, in order to fix the origin at 0, 0, 0. The symmetry relationships in the space group of this compound allowed, in all, 12 signs in the origin set (Table

* We use now the convenient notation **h** for the reflection hkl, **k** for $h'k'l'$, and **h** − **k** for $h - h'$, $k - k'$, $l - l'$.

TABLE 8.9. Origin-Fixing Reflections
and Their Symmetry Equivalents

hkl	Sign	$\lvert E \rvert$	No. of Σ_2 triplets
$53\bar{1}$	+	2.6	37
$\bar{5}31$	+		
$5\bar{3}\bar{1}$	+		
$\bar{5}31$	+		
$11\bar{4}$	+	2.3	45
$\bar{1}\bar{1}4$	+		
$1\bar{1}\bar{4}$	−		
$\bar{1}14$	−		
032	+	1.7	54
$0\bar{3}\bar{2}$	+		
$0\bar{3}2$	−		
$03\bar{2}$	−		

8.9). The reader should check the signs, starting from the first one in each group of four, with equations (8.2) and (8.3).

8.3.5   Sign Determination

The Σ_2 listing is examined with a view to generating new signs, using (7.9), which may be given in relation to Table 8.8 by

$$s[E(\mathbf{h})] \approx s\left[\sum_{\mathbf{k}} E(\mathbf{k})E(\mathbf{h}-\mathbf{k})\right] \tag{8.4}$$

where the sum is taken over the several $\mathbf{k}$ triplets all involved with the given $\mathbf{h}$. The probability of (8.4) is given by (7.10). If only a single Σ_2 interaction is considered, (8.4) becomes [cf. (7.7)]

$$s[E(\mathbf{h})] \approx s[E(\mathbf{k})]s[E(\mathbf{h}-\mathbf{k})] \tag{8.5}$$

and the probability calculation omits the summation in (7.11).

We shall assume that the values of $P_+(\mathbf{h})$ are sufficiently high for the sign to be accepted as correct; very small or zero values of $P_+(\mathbf{h})$ indicate strongly a negative sign for $\mathbf{h}$. Some examples of the application of (8.5) are given in Table 8.10.*

* It does not matter which reflections in a triplet are labeled $\mathbf{h}$, $\mathbf{k}$.

TABLE 8.10. Sign Determination
Starting from the "Origin Set"

k	h−k	h	Indication for s (h)
$53\bar{1}+$	$1\bar{1}\bar{4}-$	$62\bar{5}$	−
$53\bar{1}+$	$03\bar{2}-$	$56\bar{3}$	−
$56\bar{3}-$	$\bar{1}\bar{1}4+$	451	−
$56\bar{3}-$	$032+$	$59\bar{1}$	−
$451-$	$03\bar{2}-$	$48\bar{1}$	+

Use of Sign Symbols

The above process of sign determination was applied to the entire Σ_2 listing, which, although it contained 1276 triple products, was exhausted after only 24 signs had been found. To enable further progress to be made, three reflections were assigned the symbols A, B, and C, where each symbol represented either a plus or minus sign. Twelve symbolic signs (Table 8.11) were added to the set, and the sign determination was continued, now in terms of both signs and symbols. It may be noted that although the symbols are given to reflections with large $|E|$ values and large numbers of Σ_2 interactions, there are not, necessarily, any restrictions on either parity groups or the use of structure invariants.

TABLE 8.11. Symbolic Signs

h	Sign	$|E|$	No. of Σ_2 relationships
$11\bar{2}$	A	2.5	39
$\bar{1}12$	A		
$1\bar{1}\bar{2}$	$-A$		
$\bar{1}12$	$-A$		
$010,4$	B	2.8	35
$0\bar{1}0,\bar{4}$	B		
$0\bar{1}0,4$	B		
$010,\bar{4}$	B		
$33\bar{2}$	C	2.2	46
$\bar{3}\bar{3}2$	C		
$3\bar{3}\bar{2}$	$-C$		
$\bar{3}32$	$-C$		

Some examples of this stage of the process are given in Table 8.12. The values of **h** and **k** are taken from either Tables 8.9 and 8.11, which constitute the "starting set," or as determined through (8.4). The reader is invited to follow through the stages in Table 8.12, working out the correct symmetry-equivalent signs from (8.2) and (8.3) as necessary.

From Table 8.12, we see that six more reflections have been allocated signs, and another 17 are determined in terms of A, B, and C. Multiple indications can now be seen. For example, there are two indications that $s(573) = B$, two indications that $s(570) = -$, and two indications that $s(540) = A$. Three indications for 041 suggest that both $s(041) = -$ and $A = -$.

Continuing in this manner, it was found possible to allot signs and symbols to all 142 $|E|$ values greater than 1.5. The symbols A, B, and C were involved in 65, 72, and 55 relationships, respectively. Consistent

TABLE 8.12. Further Sign Determinations[a]

k	$s(\mathbf{k})$	$\mathbf{h-k}$	$s(\mathbf{h-k})$	**h**	Sign indication, $s(\mathbf{h})$
010,4	B	$62\bar{5}$	$-$	$68\bar{1}$	$s(68\bar{1}) = B$
010,4	B	$33\bar{2}$	C	372	$s(372) = -BC$
$11\bar{2}$	A	$68\bar{1}$	B	571	$s(571) = AB$
$53\bar{1}$	$+$	010,4	B	573	$s(573) = B$
$11\bar{4}$	$+$	$68\bar{1}$	B	573	$s(573) = B$
$33\bar{2}$	C	$33\bar{2}$	C	$66\bar{4}$	$s(66\bar{4}) = CC = +$
$11\bar{4}$	$+$	$66\bar{4}$	$+$	570	$s(570) = -$
$11\bar{2}$	A	570	$-$	482	$s(482) = A$
$11\bar{2}$	A	$66\bar{4}$	$+$	$57\bar{2}$	$s(57\bar{2}) = -A$
$11\bar{4}$	$+$	$57\bar{2}$	$-A$	482	$s(482) = A$
032	$+$	$57\bar{2}$	$-A$	540	$s(540) = A$
032	$+$	482	A	454	$s(454) = -A$
$11\bar{4}$	$+$	454	$-A$	540	$s(540) = A$
$33\bar{2}$	C	$11\bar{4}$	$+$	242	$s(242) = -C$
$11\bar{2}$	A	$11\bar{4}$	$+$	002	$s(002) = A$
$11\bar{2}$	A	482	A	570	$s(570) = -AA = -$
570	$-$	$53\bar{1}$	$+$	041	$s(041) = -$
$62\bar{5}$	$-$	$33\bar{2}$	C	$31\bar{3}$	$s(31\bar{3}) = C$
$31\bar{3}$	C	$33\bar{2}$	C	041	$s(041) = -CC = -$
$53\bar{1}$	$+$	041	$-$	$57\bar{2}$	$s(57\bar{2}) = +$
002	A	041	$-$	041	$s(041) = A$
002	A	$68\bar{1}$	B	681	$s(681) = AB$
002	A	$11\bar{4}$	$+$	$11\bar{6}$	$s(11\bar{6}) = A$

[a] Symmetry relations should be employed as necessary.

indications, such as those mentioned above for $s(041)$, led finally to the sign relationships $A = AC = B = -$, from which it follows that $C = +$. It does not always turn out that the signs represented by symbols can be allocated from the analysis in this complete and satisfactory manner. If there are n undetermined symbols, then there will be 2^n sets of signs to be examined. In this case, figures of merit, such as those discussed on page 328, can be used to indicate that set of signs most likely to be correct. It does not follow always that the indicated set *is* correct, and some trials may be needed at this stage in order to elicit the correct result.

8.3.6 The E Map

The signs of the 142 $|E|$ values used in this procedure were obtained with a high probability, and an electron density map was computed using the signed $|E|$ values as coefficients. The sections of this map $\rho(x, y, z)$ at $z = 0.15, 0.20, 0.25$, and 0.30 are shown in Figures 8.24–8.27. They reveal the K^+ ion and the $C_4O_4^-$ ring system clearly; the oxygen atom O_w of the water molecule was not indicated convincingly at this stage of the analysis. A tilt of the plane of the molecule with respect to (001) can be inferred from Figures 8.25–8.27. Some spurious peaks S may be seen. This is a common feature of E maps. We must remember that a limited data set (142 out of 900) is being used, and that the $|E|$ values are sharpened coefficients corresponding to an approximate point-atom model. The data set is therefore terminated while the coefficients for the Fourier series are relatively large, a procedure which can lead to spurious maxima (page 217).

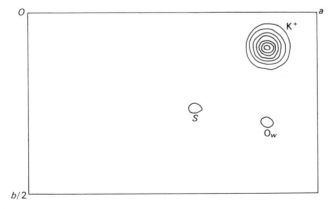

FIGURE 8.24. E map at $z = 0.15$.

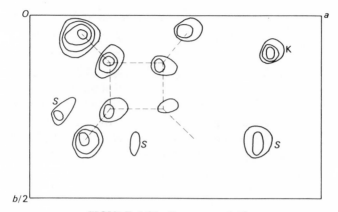

FIGURE 8.25. E map at $z = 0.20$.

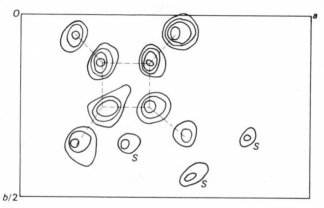

FIGURE 8.26. E map at $z = 0.25$.

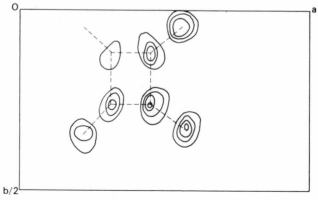

FIGURE 8.27. E map at $z = 0.30$.

However, such peaks are often of smaller weight than those that correspond to atomic positions.

8.3.7 Completion and Refinement of the Structure

Sometimes, all atomic positions are not contained among the peaks in an E map. Those peaks that do correspond to atomic positions may be used to form a trial structure for calculation of structure factors and an $|F_o|$ electron density map (page 260ff). A certain amount of subjective judgement may be required to decide upon the best peaks for the trial structure at such a stage.

This situation was obtained for KHSQ, although it was not difficult to pick out a good trial structure. Coordinates were obtained for all non-hydrogen atoms except the oxygen atom of the water molecule. The R-factor for this trial structure was 0.30, and the composite three-dimensional electron density map obtained is shown in Figure 8.28, which now reveals O_w clearly. It may be noted in passing that the small peak labeled O_w in Figure 8.24 corresponds to the position of this atom, but this fact could not be determined conclusively at that stage of the analysis.

Further refinement was carried out by the method of least squares, and an R-factor of 0.078 was obtained. Figure 8.29 shows a composite three-dimensional difference-Fourier map (page 264) of KHSQ. Peaks numerically greater than 0.5 electron per $\mathring{A}^3$, representing about twice $\sigma(\rho_0)$, are significant, and have been contoured. Some of these peaks indicate areas

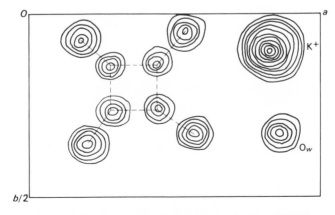

FIGURE 8.28. Composite electron density map for KHSQ (excluding H atoms); the atomic coordinates are listed in Table 8.13.

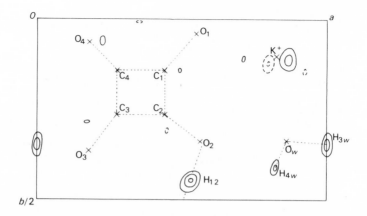

FIGURE 8.29. Composite difference electron density map for KHSQ: Positive contours are solid lines and negative contours are broken lines. Bonds in the squarate ring and those involving hydrogen atoms are shown in dotted lines. Some spurious, small peaks (unlabeled) are shown by this synthesis.

of small disagreement between the true structure and the model. Three positive peaks, however, are in positions expected for hydrogen atoms. Inclusion of these atoms in the structure factor calculations in the final cycles of least-squares refinement had a small effect on the R-factor, bringing it to its final value of 0.077. The fractional atomic coordinates for the atoms in the asymmetric unit are listed in Table 8.13.

TABLE 8.13. Fractional Atomic Coordinates for KHSQ

	x	y	z
K^+	0.8249(2)	0.1040(2)	0.1295(3)
C(1)	0.4353(9)	0.1295(7)	0.2572(12)
C(2)	0.4495(9)	0.2597(7)	0.2714(12)
C(3)	0.2795(9)	0.2714(8)	0.2462(11)
C(4)	0.2659(9)	0.1345(7)	0.2305(12)
O(1)	0.5399(6)	0.0450(5)	0.2649(10)
O(2)	0.5649(6)	0.3346(5)	0.2920(10)
O(3)	0.1874(7)	0.3582(6)	0.2386(10)
O(4)	0.1578(6)	0.0605(5)	0.2022(10)
O_w	0.8789(7)	0.3429(6)	0.0424(10)
H(12)	0.522	0.413	0.246
H(3w)	1.000	0.346	0.075
H(4w)	0.826	0.400	0.100

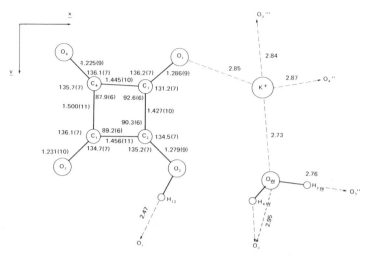

FIGURE 8.30. Bond lengths and bond angles in the crystal-chemical unit of KHSQ; the O—H · · · O distances refer to the overall O · · · O separations. Primes on atom symbols indicate neighboring crystal-chemical units; this diagram should be studied in conjunction with Figure 8.31.

Interatomic distances and angles are shown in Figure 8.30, and a molecular packing diagram, as seen along c, is given in Figure 8.31. From the analysis, we find that intermolecular hydrogen bonds exist between O(2) and O(1)′ [2.47(1) Å], between O(3)″ and O_w [2.76(1) Å], and between O(4)′ and O_w[2.95(1) Å]; they are largely responsible for the cohesion between molecules in the solid state.* We cannot say at this stage how the negative charge on the HSQ⁻ ion is distributed; a more detailed investigation would be needed in order to answer this question.

8.4 Concluding Remarks

No description of crystal structure analysis can be as complete or as satisfying as a practical involvement in the subject. In teaching crystallography, extended projects in structure analysis are becoming an important part of the work. Those readers who have access to appropriate computing facilities may want to follow through these structure determinations with the three-dimensional data.†

---

* Single and double primes indicate different neighboring molecules.

† Available from one of the authors (M.F.C.L.) as lists of both $|F_o|$ and $|E|$ values, at nominal cost to cover postage.

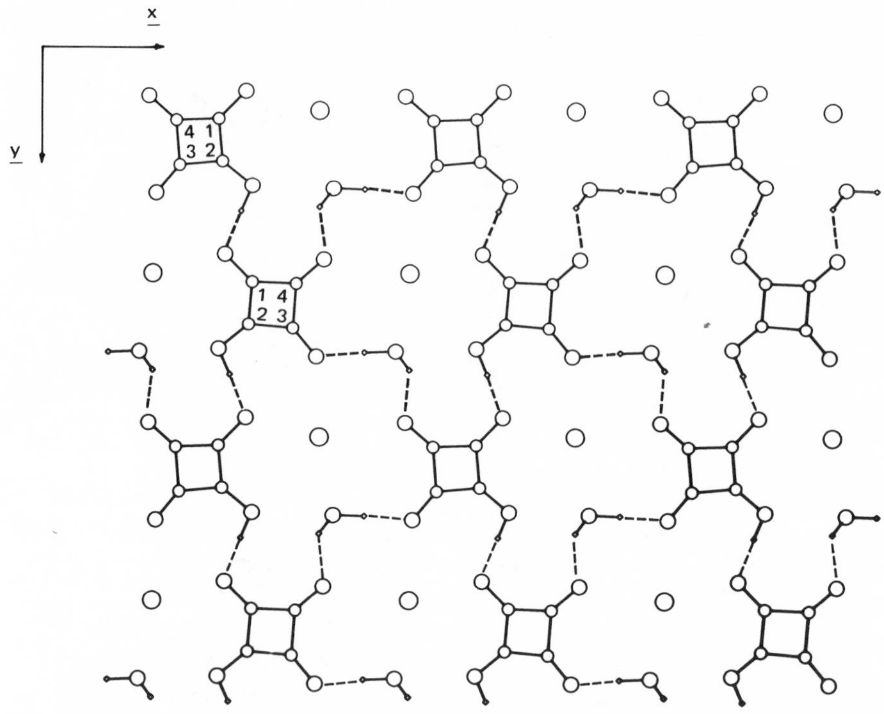

FIGURE 8.31. Molecular packing diagram of one layer as seen along c. The circles in order of decreasing size represent K, O, C, and H. The hydrogen-bond network is shown by dashed lines.

Bibliography

Published Structure Analyses

Acta Crystallographica (the early issues contain most detail for the beginner).

Journal of Crystallographic and Spectroscopic Research (formerly *Journal of Crystal and Molecular Structure*).

Zeitschrift für Kristallographie.

General Structural Data

WYCKOFF, R. W. G., *Crystal Structures*, Vols. 1–6 (1963–1966, 1968, 1971, New York, Wiley.

KENNARD, O., *et al.* (Editors), *Molecular Structures and Dimensions*, Published annually since 1970 by the International Union of Crystallography.

Problems

8.1. The unit cell of euphenyl iodoacetate, $C_{32}H_{53}IO_2$, has the dimensions $a = 7.26$, $b = 11.55$, $c = 19.22$ Å, and $\beta = 94.07°$. The space group is $P2_1$ and $Z = 2$. Figure P8.1 is the Patterson section $(u, \frac{1}{2}, w)$.

(a) Determine the x and z coordinates for the iodine atoms in the unit cell.

(b) Atomic scattering factor data for iodine are tabulated below; temperature factor corrections may be ignored.

$(\sin \theta)/\lambda$	0.00	0.05	0.10	0.15	0.20	0.25	0.30	0.35	0.40
f_I	53.0	51.7	48.6	45.0	41.6	38.7	36.1	33.7	31.5

Determine probable signs for the reflections 001 ($|F_o| = 40$), 0014 ($|F_o| = 37$), 106 ($|F_o| = 33$), and 300 ($|F_o| = 35$). Comment upon the likelihood of the correctness of the signs which you have determined.

(c) Calculate the length of the shortest iodine–iodine vector in the structure.

FIGURE P8.1. Sharpened Harker section, $P(u, \frac{1}{2}, w)$, for euphenyl iodoacetate.

8.2. The following two-dimensional $|E|$ values were determined for the [100] zone of a crystal of space group $P2_1/a$. Prepare a Σ_2 listing, assign an origin, and determine signs for as many reflections as possible, and give reasons for each step that you carry out. In this projection, two reflections for which the indices are not both even may be used to specify the origin.

0kl	$|E|$	0kl	$|E|$
0018	2.4	0310	1.9
011	1.0	0312	0.1
021	0.1	059	1.9
024	2.8	081	2.2
026	0.3	0817	1.8
035	1.8	011,7	1.3
038	2.1	011,9	2.2

8.3. The chart in Fig. P8.2 shows some $|E|$ values taken from the $hk0$ data for potassium hydrogen squarate. Take an origin at the center

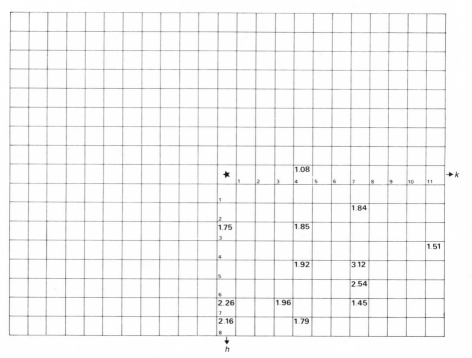

FIGURE P8.2.

of a sheet of centimeter graph paper and copy the $|E|$ values onto it, using the top left portion of each appropriate square. For each $|E|$ value plotted, add the corresponding symmetry-related $|E|$ values to the other three portions of the graphical reciprocal space representation. Remember to change the signs of $|E|$ in accordance with the space group. Next, draw an identical chart on transparent paper, but with the $|E|$ values in the bottom right portion of each appropriate square.

(a) Obtain a Σ_2 listing: take each plotted $|E|$ value in turn on the original chart and superimpose the transparency, with the origin of the transparency over the chosen $|E|$ value and keeping the two sets of h,k axes in register. Look for any superimposed $|E|$ values. A Σ_2 triplet is given by the $|E|$ value on the original chart under the origin of the transparency, together with the superimposed values, with h,k read one from the original and the other from the transparency. Thus, with the origin of the transparency on the original $|E(300)|$, we read 840 on the original and 540 on the transparency. Set up the Σ_2 listing as follows:

$\mathbf{h}$	$|E_\mathbf{h}|$	$\mathbf{k}$	$|E_\mathbf{k}|$	$\mathbf{h-k}$	$|E_{\mathbf{h-k}}|$	$|E_\mathbf{h}||E_\mathbf{k}||E_{\mathbf{h-k}}|$
300	1.75	840	1.79	$\overline{5}40$	1.92	6.0
⋮						
700	2.26...					

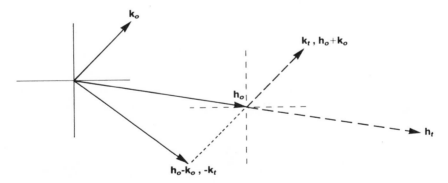

FIGURE P8.3. Σ_2 relationships: subscript o refers to the original chart, and subscript t refers to the transparency. If the origin of the transparency is placed over a chosen $\mathbf{h}_o$, it may be seen that coincidences of $|E|$ values given by $+\mathbf{k}_t$, $\mathbf{h}_o + \mathbf{k}_o$, or by $-\mathbf{k}_t$, $\mathbf{h}_o - \mathbf{k}_o$ will represent Σ_2 triplets. It should be noted that this technique applies only to centrosymmetric projections of space groups.

The rationale for the graphical procedure may be seen from Figure P8.3.

(b) Assign an origin in accordance with the rules discussed in the previous chapter and allocate signs to as many reflections as possible; use symbols if necessary. It may be assumed that the products $|E_h||E_k||E_{h-k}|$ are all sufficiently large for the indications to be accepted.

Appendix

A1   Stereoviews and Crystal Models

A1.1   Stereoviews

The representation of crystal and molecular structures by stereoscopic pairs of drawings has become commonplace in recent years. Indeed, some very sophisticated computer programs have been written which draw stereoviews from crystallographic data. Two diagrams of a given object are necessary, and they must correspond to the views seen by the eyes in normal vision. Correct viewing requires that each eye sees only the appropriate drawing, and there are several ways in which it can be accomplished.

1. A stereoviewer can be purchased for a modest sum. Two suppliers are:

(a) C. F. Casella and Company Limited, Regent House, Britannia Walk, London N1 7ND, England. This maker supplies two grades of stereoscope.

(b) Taylor-Merchant Corporation, 25 West 45th Street, New York, N.Y. 10036, U.S.A.

Stereoscopic pairs of drawings may then be viewed directly.

2. The unaided eyes can be trained to defocus, so that each eye sees only the appropriate diagram. The eyes must be relaxed, and look straight ahead. This process may be aided by placing a white card edgeways between the drawings so as to act as an optical barrier. When viewed correctly, a third (stereoscopic) image is seen in the center of the given two views.

3. An inexpensive stereoviewer can be constructed with comparative ease. A pair of planoconvex or biconvex lenses each of focal length about

411

10 cm and diameter 2–3 cm are mounted in a framework of opaque material so that the centers of the lenses are about 60–65 mm apart. The frame must be so shaped that the lenses can be held close to the eyes. Two pieces of cardboard shaped as shown in Figure A1.1 and glued together with the lenses in position represents the simplest construction. This basic stereoviewer can be refined in various ways.

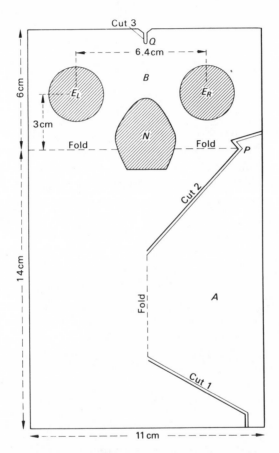

FIGURE A1.1. Simple stereoviewer. Cut out two pieces of card as shown and discard the shaded portions. Make cuts along the double lines. Glue the two cards together with the lenses E_L and E_R in position, fold the portions A and B backward, and fix P into the cut at Q. View from the side marked B. (A similar stereoviewer is marketed by the Taylor–Merchant Corporation, New York.)

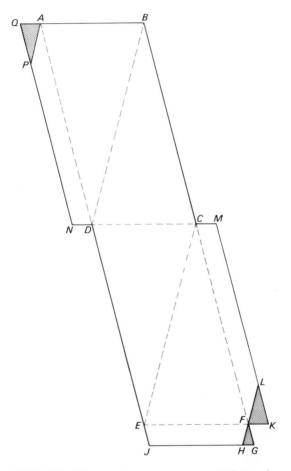

FIGURE A1.2. Construction of a tetragonal crystal
of point group $\bar{4}2m$:

$NQ = AD = BD = BC = DE = CE = CF = KM$
  $= 10$ cm;

$AB = CD = EF = GJ = 5$ cm;

$AP = PQ = FL = KL = 2$ cm;

$AQ = DN = CM = FK = FG = JH = EJ = 1$ cm.

A1.2   Model of a Tetragonal Crystal

The crystal model illustrated in Figure 1.30 can be constructed easily. This particular model has been chosen because it exhibits a $\bar{4}$ axis, which is one of the more difficult symmetry elements to appreciate from plane drawings.

A good quality paper or thin card should be used for the model. The card should be marked out in accordance with Figure A1.2 and then cut out along the solid lines, discarding the shaded portions. Folds are made in the same sense along all dotted lines, the flaps *ADNP* and *CFLM* are glued internally, and the flap *EFHJ* is glued externally. The resultant model belongs to crystal class $\bar{4}2m$.

A1.3   Stereoscopic Space-Group Drawings

A valuable teaching aid has been developed* in respect of the triclinic, monoclinic, and orthorhombic space groups. In the stereoscopic illustrations, the standard setting has been used for the monoclinic system (Fig. A1.3), and the diagrams may be used in conjunction with the *International Tables for X-Ray Crystallography*.† A copy of the complete set can be obtained from the address below (a nominal handling fee will be charged).

The pattern motif used in the illustrations has four different sizes of atoms, the minimum required for it to have only trivial symmetry in itself. The diagrams are suitable for stereoprojection‡ in a lecture theatre.

A2   Crystallographic Point-Group Study and Recognition Scheme

The first step in this scheme is a search for the center of symmetry and mirror plane; they are probably the easiest to recognize. If a model with a center of symmetry is placed on a flat surface, it will have a similar face uppermost and parallel to the supporting surface. For the *m* plane, a search is made for the left-hand/right-hand relationship in the crystal.

* J. E. Quinn, Crystallography Department, Birkbeck College, Malet Street, London WC1E 7HX, England.
† Bibliography, Chapter 1.
‡ A suitable projector may be obtained from Albion Instrument Company, 2 Albion Road, Folkestone, Kent CT19 5SE, England.

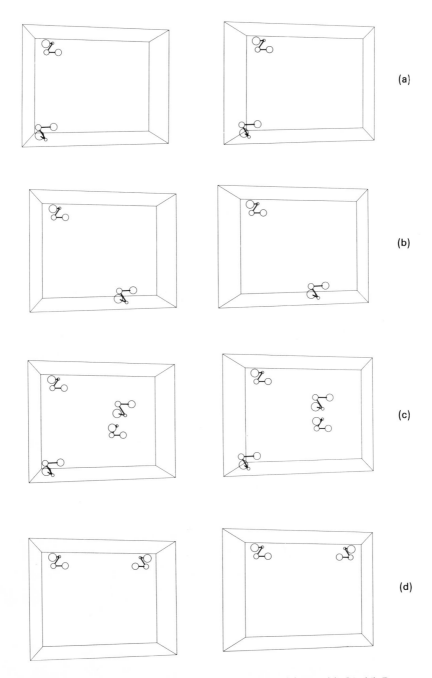

FIGURE A1.3. Stereoscopic space-group diagrams: (a) $P2$, (b) $P2_1$, (c) $C2$, (d) Pm.

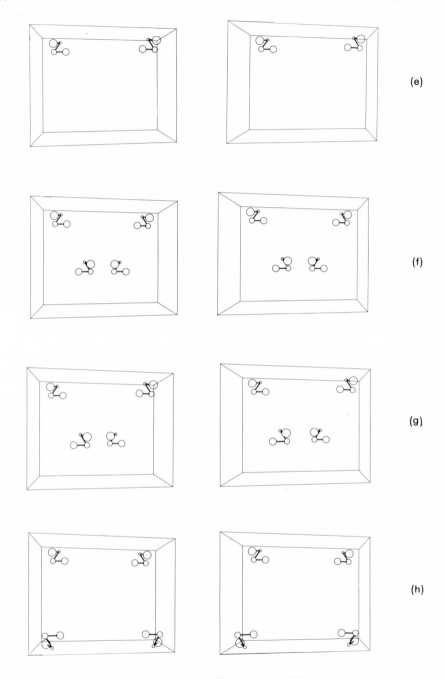

FIGURE A1.3.—cont. (e) Pc, (f) Cm, (g) Cc, (h) P2/m.

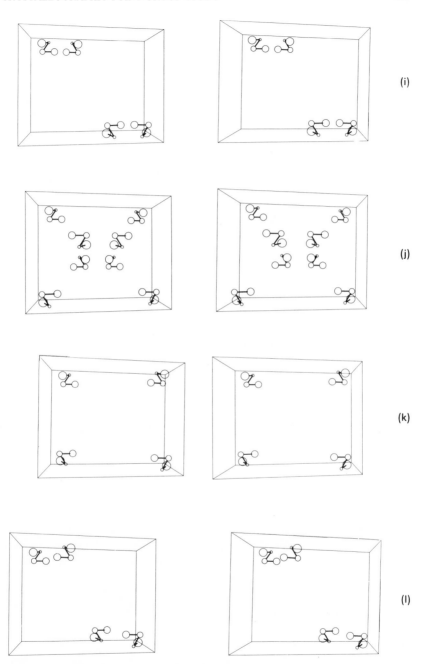

FIGURE A1.3.—*cont.* (i) $P2_1/m$, (j) $C2/m$, (k) $P2/c$, (l) $P2_1/c$.

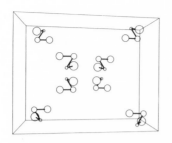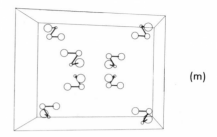

(m)

FIGURE A1.3.—*cont.* (m) $C2/c$.

The point groups may be classified into four sections:

(I)   No m and no $\bar{1}$:
      1, 2, 222, 3, 32, 4, $\bar{4}$, 422, 6, 622, 23, 432
(II) m present but no $\bar{1}$:
      $m, mm2, 3m, 4mm, \bar{4}2m, \bar{6}, 6mm, \bar{6}m2, \bar{4}3m$
(III) $\bar{1}$ present but no m:
      $\bar{1}, \bar{3}$
(IV) m and $\bar{1}$ both present:

$$2/m,\ mmm,\ \bar{3}m,\ 4/m,\ \frac{4}{m}mm,\ 6/m,\ \frac{6}{m}mm,\ m3,\ m3m$$

The further systematic identification is illustrated by means of the block diagram in Figure A2.1. Here R refers to the maximum degree of *rotational* symmetry in a crystal, or crystal model, and N is the number of such axes. Questions are given in ovals, point groups in squares, and error paths in diamonds. It may be noted that in sections I, II, and IV, the first three questions (with a small difference in II) are similar. The cubic point groups evolve from question 2 in I, II, and IV.

Readers familiar with computer programming may liken Figure A2.1 to a flow diagram. Indeed, this scheme is ideally suited to a computer-aided self-study enhancement of a lecture course on crystal symmetry, and some success with the method has been obtained.*

A3   Schoenflies' Symmetry Notation

Theoretical chemists and spectroscopists use the Schoenflies notation for describing point-group symmetry, which is a little unfortunate, because

---

* M. F. C. Ladd, *International Journal of Mathematical Education in Science and Technology,* **7**, 395–400 (1976).

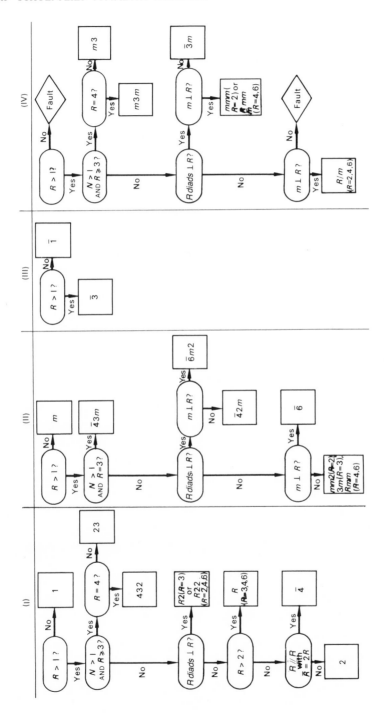

FIGURE A2.1. Flow diagram for point-group recognition.

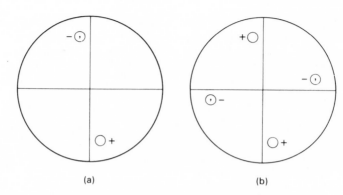

(a)   (b)

FIGURE A3.1. Stereograms of point groups: (a) S_2, (b) S_4.

although the crystallographic (Hermann–Mauguin) and Schoenflies notations are adequate for point groups, only the Hermann–Mauguin system is
satisfactory for space groups.

The Schoenflies notation uses the rotation axis and mirror plane
symmetry elements with which we are now familiar, but introduces the
alternating axis of symmetry in place of the inversion axis.

A3.1 Alternating Axis of Symmetry

A crystal is said to have an alternating axis of symmetry S_n of degree n,
if it can be brought into apparent self-coincidence by the combined operation of rotation through $(360/n)$ degrees and reflection across a plane
normal to the axis. It must be stressed that this plane is *not* necessarily a
mirror plane.* Operations S_n are nonperformable. Figure A3.1 shows
stereograms of S_2 and S_4; we recognize them as $\bar{1}$ and $\bar{4}$, respectively. The
reader should consider which point groups are obtained if, additionally, the
plane of the diagram *is* a mirror plane.

A3.2 Notation

Rotation axes are symbolized by C_n, where n takes the meaning of R in
the Hermann–Mauguin system. Mirror planes are indicated by subscripts v,

---

* The usual Schoenflies symbol for $\bar{6}$ is $C_{3h}(3/m)$. The reason that $3/m$ is not used in the
  Hermann–Mauguin system is that point groups containing the element $\bar{6}$ describe crystals that
  belong to the hexagonal system rather than to the trigonal system; $\bar{6}$ cannot operate on a
  rhombohedral lattice.

TABLE A3.1. Schoenflies and Hermann–Mauguin Point-Group Symbols

Schoenflies	Hermann–Mauguin[a]	Schoenflies	Hermann–Mauguin[a]
C_1	1	D_4	422
C_2	2	D_6	622
C_3	3	D_{2h}	mmm
C_4	4	D_{3h}	$\bar{6}m2$
C_6	6		
C_i, S_2	$\bar{1}$	D_{4h}	$\frac{4}{m}mm$
C_s, S_1	$m\,(\bar{2})$		
S_6	$\bar{3}$		
S_4	$\bar{4}$	D_{6h}	$\frac{6}{m}mm$
C_{3h}	$\bar{6}$		
C_{2h}	$2/m$	D_{2d}	$\bar{4}2m$
C_{4h}	$4/m$	D_{3d}	$\bar{3}m$
C_{6h}	$6/m$	T	23
C_{2v}	$mm2$	T_h	$m3$
C_{3v}	$3m$	O	432
C_{4v}	$4mm$	T_d	$\bar{4}3m$
C_{6v}	$6mm$	O_h	$m3m$
D_2	222	$C_{\infty v}$	∞
D_3	32	$D_{\infty h}$	$\infty/m\,(\overline{\infty})$

[a] $2/m$ is an acceptable way of writing $\frac{2}{m}$, but $4/mmm$ is not as satisfactory as $\frac{4}{m}mm$.

d, and h; v and d refer to mirror planes containing the principal axis, and h indicates a mirror plane normal to that axis. The symbol D_n is introduced for point groups in which there are n twofold axes in a plane normal to the principal axis of degree n. The cubic point groups are represented through the special symbols T and O. Table A3.1 compares the Schoenflies and Hermann–Mauguin symmetry notations.

A4   Generation and Properties of X-Rays

A4.1   X-Rays and White Radiation

X-rays are electromagnetic radiations of short wavelength, and are produced by the sudden deceleration of rapidly moving electrons at a target material. If an electron falls through a potential difference of V volts, it acquires an energy of eV electron-volts. If this energy were converted entirely into a quantum $h\nu$ of X-rays, the wavelength λ would be given by

$$\lambda = hc/eV \qquad\qquad (A4.1)$$

where h is Planck's constant, c is the velocity of light, and e is the charge on the electron. Substitution of numerical values in (A4.1) leads to the equation

$$\lambda = 12.4/V \qquad (A4.2)$$

where V is measured in kilovolts (kV).

Generally, an electron does not lose all its energy in this way. It enters into multiple collisions with the atoms of the target material, increasing their vibrations and so generating heat in the target. Thus, (A4.2) gives the minimum value of wavelength for a given accelerating voltage. Longer wavelengths are more probable, but very long wavelengths have a small probability and the upper limit is indeterminate. Figure A4.1 is a schematic diagram of an X-ray tube, and Figure A4.2 shows typical intensity vs. wavelength curves for X-rays. Because of the continuous nature of the spectrum from an X-ray tube, it is often referred to as "white" radiation. The generation of X-rays is a very uneconomical process. Most of the incident electron energy appears as heat in the target, which must be thoroughly water-cooled; about 0.1% of the energy is usefully converted for crystallographic purposes.

A4.2 Characteristic X-Rays

If the accelerating voltage applied to an X-ray tube is sufficiently large, the impinging electrons excite inner electrons in the target atoms, which may

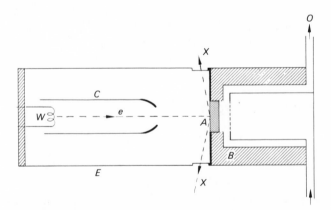

FIGURE A4.1. Schematic diagram of an X-ray tube: W, heated tungsten filament; E, evacuated glass envelope; C, accelerating cathode; e, electron beam; A, target anode; X, X-rays (about 6° angle to target surface); B, anode supporting block of material of high thermal conductivity; I, cooling water in; and O, cooling water out.

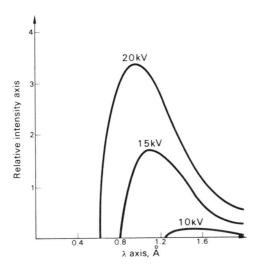

FIGURE A4.2. Variation of X-ray intensity
with wavelength λ.

be expelled from the atoms. Then, other electrons, from higher energy
levels, fall back to the inner levels and their transition is accompanied by the
emission of X-rays. In this case, the X-rays have a wavelength dependent
upon the energies of the two levels involved. If this energy difference is ΔE,
we may write

$$\lambda = hc/\Delta E \qquad (A4.3)$$

This wavelength is characteristic of the target material. The white radiation
distribution now has sharp lines of very high intensity superimposed on it
(Figure A4.3). In the case of a copper target, very commonly used in X-ray
crystallography, the characteristic spectrum consists of $K\alpha$ ($\lambda = 1.542$ Å)
and $K\beta$ ($\lambda = 1.392$ Å); $K\alpha$ and $K\beta$ are always produced together.

A4.3   Absorption of X-Rays

All materials absorb X-rays according to an exponential law:

$$I = I_0 \exp(-\mu t) \qquad (A4.4)$$

where I and I_0 are, respectively, the transmitted and incident intensities, μ is
the linear absorption coefficient, and t is the path length through the

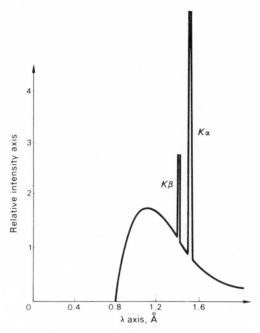

FIGURE A4.3. Characteristic K spectrum superposed on the "white" radiation continuum.

material. The absorption of X-rays increases with increase in the atomic
number of the elements in the material.

The variation of μ with λ is represented by the curve of Figure A4.4;
μ decreases approximately as λ^3. At a value which is specific to a given
atom in the material, the absorption rises sharply. This wavelength corresponds to a resonance level in the atom: a process similar to that involved
in the production of the characteristic X-rays occurs, with the exciting
species being the incident X-rays themselves. The particular wavelength is
called the absorption edge; for metallic nickel it is 1.487 Å.

A4.4 Filtered Radiation

If we superimpose Figures A4.3 and A4.4, we see that the absorption
edge of nickel lies between the $K\alpha$ and $K\beta$ characteristic lines of copper
(Figure A4.5). Thus, the effect of passing X-rays from a copper target
through a thin (0.018 mm) nickel foil is that the $K\beta$ radiation is selectively
almost completely absorbed. The intensities of both $K\alpha$ and the white
radiation are also reduced, but the overall effect is a spectrum in which the

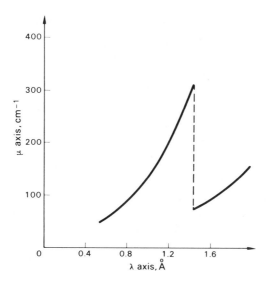

FIGURE A4.4. Variation of μ (Ni) with wavelength λ of X-radiation.

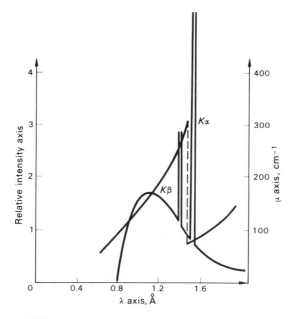

FIGURE A4.5. Superposition of Figures A4.3 and A4.4 to show diagrammatically the production of "filtered" radiation.

most intense part is the $K\alpha$ line; we speak of filtered radiation, to indicate the production of effectively monochromatic radiation by this process. The copper $K\alpha$ line ($\bar{\lambda} = 1.542\,\text{Å}$) actually consists of a doublet, α_1 ($\lambda = 1.5405\,\text{Å}$) and α_2 ($\lambda = 1.5443\,\text{Å}$); the doublet is resolved on photographs at high θ values, but we shall not be concerned here with that feature. The value of $1.542\,\text{Å}$ is a weighted mean $(2\alpha_1 + \alpha_2)/3$, the weights being derived from the relative intensities (2:1) of the α_1 and α_2 lines.

The absorption effect is important also in considering the radiation to be used for different materials. We have mentioned that Cu $K\alpha$ is very commonly used, but it would be unsatisfactory for materials containing a high percentage of iron (absorption edge $1.742\,\text{Å}$) since radiation of this wavelength is highly absorbed by iron atoms and re-emitted as characteristic Fe K spectrum. In this case, Mo $K\alpha$ ($\lambda = 0.7107\,\text{Å}$) is a satisfactory alternative.

A5   Crystal Perfection and Intensity Measurement

A5.1   Crystal Perfection

In the development of the Bragg equation (3.16), we assumed geometric perfection of the crystal, with all unit cells in the crystal stacked side by side in a completely regular manner. Few, if any, crystals exhibit this high degree of perfection. Figure A5.1 shows a family of planes, all in exactly the same orientation with respect to the X-ray beam, at the correct angle for a Bragg reflection. It is clear that the first reflected ray BC is in the correct

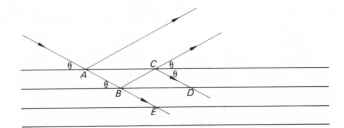

FIGURE A5.1. Primary extinction: The phase changes on reflection at B and C are each $\pi/2$, so that between the directions BE and CD there is a total phase difference of π. Hence, some attenuation of the intensity occurs for the beam incident upon planes deeper in the crystal.

FIGURE A5.2. "Mosaic" character in a crystal; the angular misalignment between blocks may vary from 2' to about 30' of arc.

position for a second reflection *CD*, and so on. Since there is a phase change of $\pi/2$ on reflection,* the doubly reflected ray has π phase difference with respect to the incident ray (*BE*). In general, rays reflected n and $n-2$ times differ in phase by π, and the net result is a reduction in the intensity of the X-ray beam passing through the crystal. This effect is termed primary extinction, and is a feature of geometric perfection of a crystal. In the *ideally perfect* crystal, $I \propto |F|$.

Most crystals, however, are composed of an array of slightly misoriented crystal blocks (mosaic character) (Figure A5.2). The ranges of geometric perfection are quite small. Even crystals that show some primary extinction exhibit mosaic character to some degree, and we may write

$$I \propto |F|^m \tag{A5.1}$$

Generally, the mosaic blocks are small, and m is effectively 2.

Another process which leads to attenuation of the X-ray beam by a crystal set at the Bragg angle is known as secondary extinction. It may be encountered in single-crystal X-ray studies, and the magnitude of the effect can be appreciable. Consider a situation in which the first planes encountered by the X-ray beam reflect a high proportion of the incident beam. Parallel planes further in the crystal receive less incident intensity, and, hence, reflect less than might be expected. The effect is most noticeable with large crystals and intense (usually low-order) reflections. Crystals in which the mosaic blocks are highly misaligned have negligible secondary extinction, because only a small number of planes are in the reflecting position at a given time. Such crystals are termed *ideally imperfect*; this condition can be developed by subjecting the crystals to the thermal shock of dipping them in

* This $\pi/2$ phase change is usually neglected since it arises for all reflections.

liquid air. The effect of secondary extinction on the intensity of a reflection can be brought into the least-squares refinement (page 344) as an additional variable, the extinction parameter ζ. The quantity minimized in the refinement of the atomic parameters is then

$$\sum_{hkl} w \left(|F_o| - \frac{1}{K\zeta} |F_c| \right)^2 \qquad (A5.2)$$

A5.2   Intensity of Reflected Beam

The real or imperfect crystal will reflect X-rays over a small angular range centered on a Bragg angle θ. We need to determine the total energy of a diffracted beam $\mathscr{E}(hkl)$ as the crystal, which is completely bathed in an X-ray beam of incident intensity I_0, passes through the reflecting range.

At a given angle θ, let the power of the reflected beam be $d\mathscr{E}(hkl)/dt$. The greater the value of I_0, the greater the power. Hence,

$$d\mathscr{E}(hkl)/dt = R(\theta)I_0 \qquad (A5.3)$$

where $R(\theta)$ is the reflecting power. Figure A5.3 shows a typical curve of $R(\theta)$ against θ. The area under the curve is called the integrated reflection $J(hkl)$:

$$J(hkl) = \int_{-\delta\theta_0}^{\delta\theta_0} R(\theta) \, d\theta \qquad (A5.4)$$

Using (A5.3), we obtain

$$J(hkl) = (1/I_o) \int_{-\delta\theta_0}^{\delta\theta_0} [d\mathscr{E}(hkl)/dt] \, d\theta \qquad (A5.5)$$

If the crystal is rotating with angular velocity ω ($= d\theta/dt$),

$$J(hkl) = \omega\mathscr{E}(hkl)/I_0 \qquad (A5.6)$$

where $\mathscr{E}(hkl)$ is the total energy of the diffracted beam for one pass of the crystal through the reflecting range, $\pm\delta\theta_0$. Since intensity is a measure of

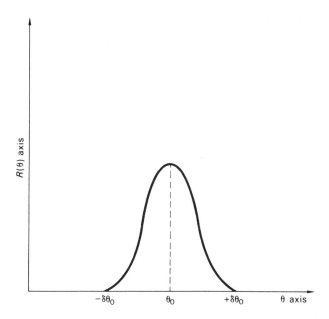

FIGURE A5.3. Variation of reflecting power $R(\theta)$ with θ arising from "mosaic" character: θ_0 is the ideal Bragg angle, and $\pm\delta\theta_0$ represent the limits of reflection.

energy per unit time, we have

$$\mathscr{E}(hkl) = I_0(hkl)t \qquad (A5.7)$$

and, from (4.57), we obtain

$$\mathscr{E}(hkl) \propto K^2 C(hkl)|F_o(hkl)|^2 \qquad (A5.8)$$

where $C(hkl)$ includes correcting factors for absorption and extinction, and for the Lorentz and polarization effects (page 437). Because of the proportionality between energy and intensity (A5.7), although we are actually measuring the energy of the diffracted beam, we usually speak of the corresponding intensity.

A5.3 Intensity Measurements

X-ray intensities are measured either from the blackening of photographic film emulsion or by direct quantum counting.

A5.3.1   Film Measurements

The optical density D of a uniformly blackened area of an X-ray diffraction spot on a photographic film is given by

$$D = \log(i_0/i) \tag{A5.9}$$

where i_0 is the intensity of light hitting the spot and i is the intensity of light transmitted by it. D is proportional to the intensity of the X-ray beam I_0 for values of D less than about 1. In practice, this means spots which are just visible to those of a medium-dark grey on the film.

An intensity scale can be prepared by allowing a reflected beam from a crystal to strike a film for different numbers of times and according each spot a value in proportion to this number; Figure A5.4 shows one such scale. Intensities may be measured by visual comparison with the scale, and, with care, the average deviation of intensity from the true value would be about 15%.

In place of the scale and the human eye, a photometric device may be used to estimate the blackening. In this method, the background intensity is measured and subtracted from the peak intensity. This process is carried out automatically in the visual method. Carefully photometered intensities would have an average deviation of less than 10%.

The accuracy of film measurements can be enhanced if an integrating mechanism is used in conjunction with either a Weissenberg or a precession camera in recording intensities. In this method, a diffraction spot (Figure A5.5a) is allowed to strike the film successively over a grid of points (Figure A5.5b). Each point acts as a center for building up the spot. The results of this process are a central plateau of uniform intensity in each spot and a series of spots of similar, regular shape: Figure A5.6 illustrates, diagrammatically, the building up of the plateau, and Figure A5.7 shows a Weissenberg photograph comparing the normal and integrating methods with the same crystal.

The average deviation in intensity measurements from carefully photometered, integrated Weissenberg photographs is about 5%. The general

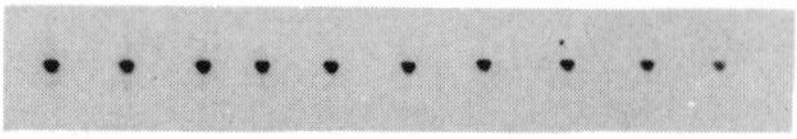

FIGURE A5.4. Sketch of a crystal-intensity scale.

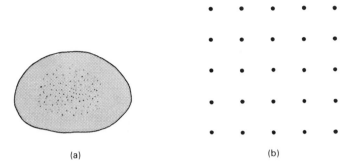

(a)   (b)

FIGURE A5.5. Spot integration: (a) typical diffraction spot, (b) 5×5 grid of points.

subject of accuracy in photographic measurements has been discussed exhaustively by Jeffery.*

A5.3.2 Diffractometer Geometry and Data Collection

The principle of the four-circle single-crystal diffractometer is shown in Figure A5.8. It consists of (1) an X-ray source, (2) an Eulerian cradle to set the crystal into any desired orientation, and (3) a movable scintillation counter as an X-ray detector, D. The diffractometer operates in normal-beam equatorial geometry. The zero position of the detector on the 2θ circle is defined to be on the incident beam in the $-y$ direction. The plane of the detector and the source is the equatorial (xy) plane. The plane of the χ-circle is normal to the X-ray beam at $\omega = 0$. At $\chi = 0$, the ϕ axis carrying the goniometer head is along z; the new position of ϕ is defined arbitrarily.

S is a general reciprocal lattice vector defined by

$$\mathbf{S} = \mathbf{M} \cdot \mathbf{h} = (\mathbf{U} \cdot \mathbf{B}) \cdot \mathbf{h} \qquad (A5.10)$$

where **B** is a matrix which orthogonalizes the reciprocal lattice coordinates referred to the crystal axes and **U** is a matrix which rotates the orthogonalized crystal reciprocal lattice coordinates into the diffractometer reference frame, with all circles set at their respective zero positions.

The general vector **S** with components (S_x, S_y, S_z) is moved into the diffracting condition as follows:

(a) **S** is moved into the χ plane (xz) by rotation of ϕ by $\tan^{-1}(-S_y/S_x)$; ϕ is arbitrary if $S_x = S_y = 0$.

* See Bibliography, Chapter 3.

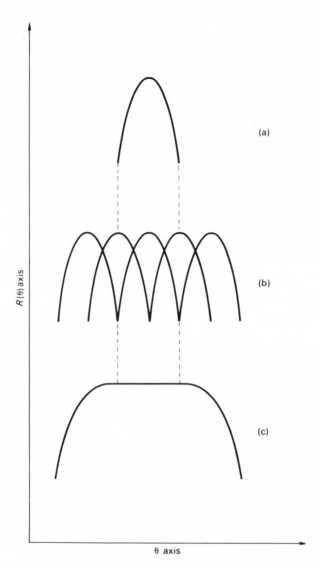

FIGURE A5.6. Spot integration: (a) ideal peak profile,
(b) superposition, by translation, of five profiles, (c)
integrated profile showing a central plateau.

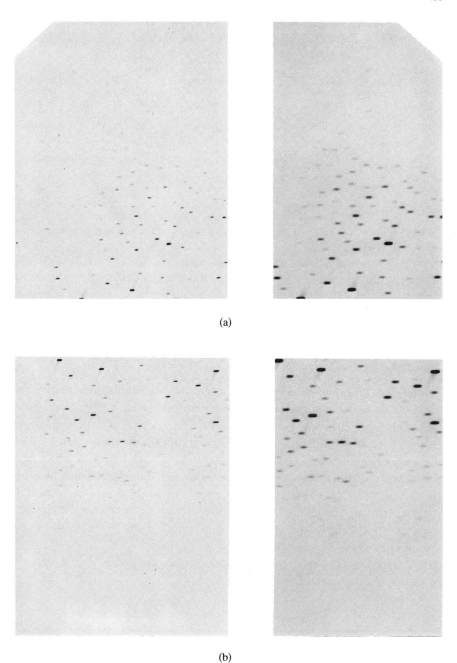

(a)

(b)

FIGURE A5.7. Weissenberg photographs: (a) normal, (b) integrated.

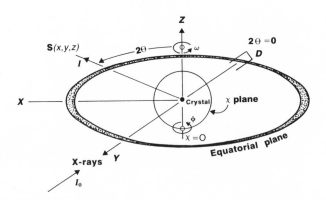

FIGURE A5.8. Geometry of a four-circle diffractometer.

(b) Next χ is rotated by $\tan^{-1}[S_z/(S_x^2+S_y^2)^{1/2}]$ so that $\mathbf{S}$ moves into the equatorial plane (xy); at this point, $\mathbf{S}$ is perpendicular to the incident X-ray beam I_0.

(c) The ω circle is moved by an angle $\omega = \theta$ (from Bragg's equation) so that the reciprocal lattice point intersects the Ewald sphere.

(d) Finally the detector is rotated in the same direction by 2θ to receive the diffracted beam, I.

The condition $\omega = \theta$ is imposed since only two independent angles are required to define the direction of the scattering vector. Rotation about that vector, which would (unless $\chi = \pm 90°$) require the $\omega = \theta$ condition to be relaxed, has no effect on the geometry of diffraction, other than to alter the path of the incident and diffracted rays in the crystal. This effect may be utilized to estimate absorption corrections. The $\omega = \theta$ condition ensures, by making the χ plane bisect I_0 and I, that the possibility of obstruction of the incident or diffracted X-rays by the mechanical parts of the diffractometer—particularly the χ circle, the base of the ϕ circle, and the goniometer head—is minimized.

The problem of the general 3-circle setting ω, χ, ϕ is best dealt with in terms of Eulerian angles.

Eulerian Angles. The Eulerian angles are defined as the three successive angles of rotation by which one can carry out the transformation from a given Cartesian coordinate system to another. The sequence will be started by rotating an initial system of axes, *xyz*, by an angle ϕ counterclockwise about the *z* axis, and the axes in the resultant coordinate system will be

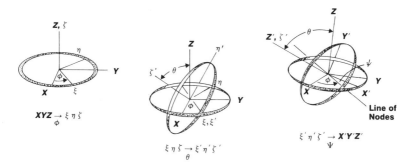

FIGURE A5.9. Eulerian angles.

labeled $\xi\eta\zeta$. In the second stage the intermediate axes, $\xi\eta\zeta$, are rotated about the ξ axis counterclockwise by an angle θ to produce another intermediate axes set, the $\xi'\eta'\zeta'$ axes. The ξ' axis is at the intersection of the xy and $\xi'\eta'$ planes and is called the line of nodes.

Finally the $\xi'\eta'\zeta'$ axes are rotated counterclockwise by an angle ψ about the ζ' axis to produce the desired $x'y'z'$ system of axes (Figure A5.9).

In matrix terms the complete transformation $\mathbf{A}$ can be written as the triple product of the separate rotations. Thus the initial rotation about z can be described by a matrix $\mathbf{D}$.

$$\xi = \mathbf{D}\mathbf{X} \tag{5.11}$$

The transformation from $\xi\eta\zeta$ to $\xi'\eta'\zeta'$ is given by

$$\xi' = \mathbf{C}\xi \tag{5.12}$$

and the final rotation as

$$\mathbf{X}' = \mathbf{B}\xi' \tag{5.13}$$

The complete rotation from $\mathbf{X}$ to $\mathbf{X}'$ is then given by

$$\mathbf{X}' = \mathbf{A}\mathbf{X} \tag{5.14}$$

where $\mathbf{A}$ is the product of the successive matrices:

$$\mathbf{A} = \mathbf{B}\mathbf{C}\mathbf{D} \tag{5.15}$$

where

$$D = \begin{pmatrix} \cos\theta & \sin\theta & 0 \\ -\sin\theta & \cos\theta & 0 \\ 0 & 0 & 1 \end{pmatrix}$$

$$C = \begin{pmatrix} 1 & 0 & 0 \\ 0 & \cos\theta & \sin\theta \\ 0 & -\sin\theta & \cos\theta \end{pmatrix} \tag{5.16}$$

$$B = \begin{pmatrix} \cos\psi & \sin\psi & 0 \\ -\sin\psi & \cos\psi & 0 \\ 0 & 0 & 1 \end{pmatrix}$$

Choice of Collimator Sizes and Scan Width. The incident beam collimator must be such that the crystal is completely bathed with a uniform intensity of radiation. Usually a 1-mm diameter is suitable. The diffracted beam collimator must subtend an angle at the crystal of at least $S + 2C + 2M$ for an ω/θ scan, where S is the angle subtended by the source at the crystal (typically 5×10^{-3} radian), C is the angle subtended by the crystal at the source (typically 2.5×10^{-3} radian), and M is the angular dispersion of the mosaic blocks (typically 1.5 to 5×10^{-3} radian). The latter parameter is usually quite anisotropic, and in practice the collimator size is estimated empirically by observing the effect of varying it on the signal/background ratio of several low-angle reflections.

The scan-width required for peak integration must be at least $a + b \tan\theta$ where $a = S + C + M$ and $b = \delta\lambda/\lambda = 0.142$ for Cu $K\bar{\alpha}$. This scan-width must be multiplied by a factor ($\sim$1.5–2) to allow for background scanning on each side of the peak. The peak is either scanned at a fixed position relative to the calculated setting angles and the background measured at equidistant points on each side ("B-P-B" method), or its window is allowed to "float" in a combined peak–background scan and the actual peak position determined experimentally for each scan ("moving-window" method). The latter is useful in cases where the crystal suffers small random movements (as always happens to some extent), and the scan width can be reduced to a minimum with no danger of losing peak counts to the background portion of the scan.

A5.4   Data Processing

A5.4.1   Introduction

From (A5.8) and (A5.9), we see that certain corrections are necessary in order to convert measured intensities into values of $|F|^2$. We shall write

$$I_0(hkl) \propto ALp|F_o(hkl)|^2_{\text{rel}} \qquad (A5.17)$$

and

$$|F_o(hkl)| = K|F(hkl)|_{\text{rel}} \qquad (A5.18)$$

where A is an absorption factor (including extinction for the purpose of this discussion), L is the Lorentz factor, p is the polarization factor, and K is the scale factor which places $|F|$ values onto an absolute scale, represented by $|F_o|$; it includes, implicitly, the proportionality constant of (A5.8). The Lorentz factor expresses the fact that, for a constant angular velocity of rotation of the crystal, different reciprocal lattice points pass through the sphere of reflection at different rates and thus have different times-of-reflection opportunity. The form of the L factor depends upon the experimental arrangement. For both zero-level photographs taken with the X-ray beam normal to the rotation axis and four-circle diffractometer measurements, L has the simple form of $1/\sin 2\theta$.

The radiation from a normal X-ray tube is unpolarized, but after reflection from a crystal the beam is polarized. The fraction of energy lost in this process is dependent only on the Bragg angle:

$$p = (1 + \cos^2 2\theta)/2 \qquad (A5.19)$$

Application of the L and p factors, where absorption and secondary extinction are negligible, is essential in order to bring the $|F|^2$ data onto a correct relative scale. The scale factor K can be determined approximately by Wilson's method (page 257) and refined as a parameter in a least-squares analysis.

A5.4.2   Standard Deviation of Intensity

The net integrated intensity I and background B are measured, most conveniently in diffractometry, with a step-scan moving-window method.*
The standard deviation in I arising only from statistical counting fluctuations

---

* I. J. Tickle, *Acta Crystallographica* **B31**, 329 (1975).

is given by

$$\sigma(I) = (I + rB + r^2B)^{1/2} \tag{A5.20}$$

where r is the ratio of the time spent in measuring I to that spent in measuring B, typically 1.5.

A5.4.3 Absorption Corrections

The absorption of X-rays by matter is governed by the equation

$$I = I_0 \exp(-\mu\tau) \tag{A5.21}$$

where I is the diffracted beam intensity, I_0 is the incident beam intensity, μ is the linear absorption coefficient, and τ is the thickness of specimen. Hence the transmission of the X-ray beam through a crystal is given by

$$T = I/I_0 = \exp[-\mu(\tau_i + \tau_d)] \tag{A5.22}$$

where τ_i and τ_d are the incident and diffracted beam path lengths. If the shape of the crystal is known exactly, then it is possible to correct for absorption by calculating

$$T = V^{-1} \int^V \exp[-\mu(\tau_i + \tau_d)]\, \mathbf{d}V \tag{A5.23}$$

where $\mathbf{d}V$ is an infinitesimal volume of crystal (Busing and Levy).*

Frequently, however, the crystal faces are not well defined and it is necessary to resort to empirical methods for estimating the transmission factor.

Empirical Absorption Correction. The incident and diffracted X-rays for a general reflection with $\phi = \phi_0$ will intersect the transmission profile at $\phi_0 - \delta$ and $\phi_0 + \delta$, where

$$\delta = \tan^{-1}(\tan\theta\cos\lambda) \tag{A5.24}$$

Hence, $\delta = 0$ at $\chi = \pm 90°$. The transmission profile used is that with θ nearest to the equi-inclination angle ν, where

$$\nu = \sin^{-1}(\sin\theta\sin\chi) \tag{A5.25}$$

---

*W. R. Busing and H. A. Levy, *Acta Crystallographica* **10**, 180 (1957).

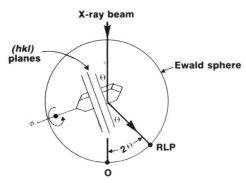

FIGURE A5.10. Geometry of absorption
correction.

The transmission T is given either as the arithmetic mean or as the
geometric mean of the estimated incident and reflected ray transmissions:

$$T = [T_\nu(\phi - \delta) + T_\nu(\phi + \delta)]/2 \qquad (A5.26)$$

or

$$T = [T_\nu(\phi - \delta) \times T_\nu(\phi + \delta)]^{1/2} \qquad (A5.27)$$

Transmission Profiles. The transmission is measured for axial reflections $(\chi \approx \pm 90°)$ as a function of ϕ (Figure A5.10). The transmission is
given by

$$T_\theta(\phi) = I_\theta(\phi)/I_\theta(\text{max}) \qquad (A5.28)$$

The variation of T with θ is neglected as it has the same effect as a small
isotropic temperature factor.

A set of profiles of T as a function of ϕ are obtained for different
values of θ and applied in data processing as detailed above.

A5.4.4   Scaling

Fluctuations in the incident X-ray beam intensity and possible radiation
damage to the crystal may be monitored by measuring four standard
reflections of moderate intensity at a regular interval, say, hourly. Two of
these reflections should have χ at about 0° and two at about 90°, with each

pair about $90°$ apart in ϕ. The average of these intensities relative to the average of their starting values is smoothed and used to rescale the raw intensity data. If S is this scale factor, then the total correction applied is now

$$I_{rel} = (Lp)^{-1} T^{-1} S^{-1} I_{raw} \tag{A5.29}$$

$$\sigma I_{rel} = (Lp)^{-1} T^{-1} S^{-1} \sigma I_{raw} \tag{A5.30}$$

A5.4.5 Merging Equivalent Reflections

Where more (n) than one symmetry equivalent of a given reflection is measured, the weighted mean is calculated:

$$\bar{I} = \sum_{j=1}^{n} w_j I_j \bigg/ \sum_{j=1}^{n} w_j \tag{A5.31}$$

where

$$w_j = \sigma_j^{-2} \tag{A5.32}$$

A chi-square test may be used to detect equivalents which may have a systematic error:

$$\chi^2 = \sum_{j=1}^{n} [(I_j - \bar{I}_j)/\sigma_j]^2 \tag{A5.33}$$

where there are $n-1$ degrees of freedom. If χ^2 exceeds χ^2_{n-1} ($\alpha = 0.001$), then the equivalent with the greatest weighted deviation from the mean, $w_j|I_j - \bar{I}_j|$, is rejected and the test repeated on the remaining equivalents. If $n = 2$, then the smaller intensity value is rejected.

The merging R value is defined by

$$R_m = \frac{\sum_{hkl} [\sum_j |I_j - \bar{I}_j|]}{\sum_{hkl} [\sum_j I_j]} \tag{A5.34}$$

A5.5 Synchrotron Radiation

Synchrotron radiation is emitted when an electron or a positron experiences radial acceleration. It is focused by a relativistic effect in the plane of its trajectory, and is polarized in the same plane. The spectrum consists of white radiation, the energy of which increases as E^4/R, where E is the electron energy and R is the radius of the trajectory. The radiation is characterized by its high intensity, fine focus, maximum energy in the region of 4 keV, and high stability.

Synchrotron sources have been developed over the past decade to a point where they are now available for crystallographic research. The intensity of the storage ring source can vary from 10 to 10^8 times that of conventional X-ray tubes, and the X-ray photon energy (and wavelength) can be treated as a variable parameter, enabling one to use the large changes in $\Delta f'$ and $\Delta f''$ that occur as λ varies to good purpose.

Phase information in X-ray diffraction may be said to be lost through the unavoidable application of the Friedel law. The wavelength tunability of synchrotron radiation permits diffraction experiments to be carried out at different wavelengths near the absorption edge of an atom in the structure, thereby modifying $\Delta f'$ and $\Delta f''$. Changes in $\Delta f'$ are equivalent to isomorphous replacement. For example, small wavelength changes in the neighborhood of the L_{III} absorption edge of platinum at 1.072 Å produce effects that are equivalent to a change in f of between 10 and 12 electrons. Changes in $\Delta f''$ break the Friedel law (anomalous scattering, q.v.) so as to lead to phase information. These effects are important developments, particularly with large structures (proteins and nucleic acids), and the next decade will surely see their wide use.

A6 Transformations

The main purpose of this appendix is to obtain a relationship between the indices of a given plane referred to two different unit cells in one and the same lattice. However, several other useful equations will emerge in the discussion.

In Figure A6.1, a centered unit cell $(\mathbf{A}, \mathbf{B})$ and a primitive unit cell $(\mathbf{a}, \mathbf{b})$ are shown; for simplicity, only two dimensions are considered. From the geometry of the diagram,

$$\mathbf{A} = \mathbf{a} - \mathbf{b} \qquad \qquad (A6.1)$$

$$\mathbf{B} = \mathbf{a} + \mathbf{b} \qquad \qquad (A6.2)$$

$$\mathbf{a} = \mathbf{A}/2 + \mathbf{B}/2 \qquad \qquad (A6.3)$$

$$\mathbf{b} = -\mathbf{A}/2 + \mathbf{B}/2 \qquad \qquad (A6.4)$$

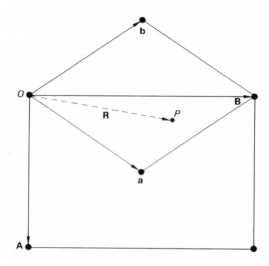

FIGURE A6.1. Unit-cell transformations.

We have encountered this type of transformation before, in our study of lattices (page 81).

The point P may be represented by fractional coordinates X, Y in the centered unit cell and by x, y in the primitive cell. Since $\mathbf{OP}$ is invariant under unit cell transformation,

$$\mathbf{R} = X\mathbf{A} + Y\mathbf{B} = x\mathbf{a} + y\mathbf{b} \qquad (A6.5)$$

Substituting for $\mathbf{A}$ and $\mathbf{B}$ from (A6.1) and (A6.2), we obtain

$$(X + Y)\mathbf{a} + (-X + Y)\mathbf{b} = x\mathbf{a} + y\mathbf{b} \qquad (A6.6)$$

whence

$$x = X + Y \qquad (A6.7)$$

$$y = -X + Y \qquad (A6.8)$$

Similarly, it may be shown that

$$X = x/2 - y/2 \qquad (A6.9)$$

$$Y = x/2 + y/2 \qquad (A6.10)$$

The vector to the reciprocal lattice point hk is given, from (2.15), by

$$\mathbf{d}^*(hk) = h\mathbf{a}^* + k\mathbf{b}^* \tag{A6.11}$$

and that to the same point, but represented by HK, is

$$\mathbf{d}^*(HK) = H\mathbf{a}^* + K\mathbf{b}^* \tag{A6.12}$$

The scalar $\mathbf{d}^* \cdot \mathbf{R}$ is invariant with respect to unit cell transformation, since it represents the path difference between that point and the origin† (see page 154). Hence, evaluating $\mathbf{d}^* \cdot \mathbf{R}$ with respect to both unit cells and using the properties of the reciprocal lattice discussed on pages 70–74, we obtain

$$hx + ky = HX + KY \tag{A6.13}$$

Substituting for x and y from (A6.7) and (A6.8), we find

$$(h-k)X + (h+k)Y = HX + KY \tag{A6.14}$$

Hence,

$$H = h - k \tag{A6.15}$$

$$K = h + k \tag{A6.16}$$

which is the same form of transformation as that for the unit cell, given by (A6.1) and (A6.2). Generalization of this treatment to three dimensions and oblique unit cells is straightforward, if a little time consuming.

A7   Comments on Some Orthorhombic and Monoclinic Space Groups

A7.1   Orthorhombic Space Groups

In Chapter 2, we looked briefly at the problem of choosing the positions of the symmetry planes in the space groups of class $mmm \left(\dfrac{2}{m} \dfrac{2}{m} \dfrac{2}{m} \right)$ with

---

† The full significance of this statement can be appreciated after studying Chapter 4.

respect to a center of symmetry at the origin of the unit cell. We give now some simple rules whereby this task can be accomplished readily, while still making use implicitly of the ideas already discussed, including the relative orientations of the symmetry elements given by the space-group symbol itself (see Tables 1.5 and 2.5).

Half-Translation Rule

Location of Symmetry Planes. Consider space group *Pnna*; the translations associated with the three symmetry planes are $(b+c)/2$, $(c+a)/2$, and $a/2$, respectively. If they are summed, the result (T) is $(a+b/2+c)$. We disregard the *whole* translations a and c, because they refer to neighboring unit cells; thus, T becomes $b/2$, and the center of symmetry is displaced by $T/2$, or $b/4$, from the point of intersection of the three symmetry planes n, n and a. As a second example, consider *Pmma*. The only translation is $a/2$; thus, $T = a/2$, and the center of symmetry is displaced by $a/4$ from *mma*.

Space group *Imma* may be formed from *Pmma* by introducing the body-centering translation $\frac{1}{2}, \frac{1}{2}, \frac{1}{2}$ (Fig. 6.18b). Alternatively, the half-translation rule may be applied to the complete space-group symbol. In all, *Imma* contains the translations $(a+b+c)/2$ and $a/2$, and $T = a+(b+c)/2$, or $(b+c)/2$; hence, the center of symmetry is displaced by $(b+c)/4$ from *mma*. This center of symmetry is one of a second set of eight introduced, by the body-centering translation, at $\frac{1}{4}, \frac{1}{4}, \frac{1}{4}$ (half the I translation) from a *Pmma* center of symmetry. This alternative setting is given in the *International Tables for X-Ray Crystallography**; it corresponds to that in Figure 6.18b with the origin shifted to the center of symmetry at $\frac{1}{4}, \frac{1}{4}, \frac{1}{4}$. Space groups based on A, B, C, and F unit cells similarly introduce additional sets of centers of symmetry. The reader may care to apply these rules to space group *Pnma* and then check the result with Figure 2.37.

Type and Location of Symmetry Axes. The quantity T, reduced as above to contain half-translations only, readily gives the types of twofold axes parallel to a, b, and c. Thus, if T contains an $a/2$ component, then 2_x (parallel to a) $\equiv 2_1$, otherwise $2_x \equiv 2$. Similarly for 2_y and 2_z, with reference to the $b/2$ and $c/2$ components. Thus, in *Pnna*, $T = b/2$, and so $2_x \equiv 2$, $2_y \equiv 2_1$, and $2_z \equiv 2$. In *Pmma*, $T = a/2$; hence, $2_x \equiv 2_1$, $2_y \equiv 2$, and $2_z \equiv 2$.

* See Bibliography, Chapter 1.

The location of each twofold axis may be obtained from the symbol of the symmetry plane perpendicular to it, being displaced by half the corresponding glide translation (if any). Thus, in *Pnna*, we find 2 along $[x, \frac{1}{4}, \frac{1}{4}]$, 2_1 along $[\frac{1}{4}, y, \frac{1}{4}]$, and another 2 along $[\frac{1}{4}, 0, z]$. In *Pmma*, 2_1 is along $[x, 0, 0]$, 2 is along $[0, y, 0]$, and another 2 is along $[\frac{1}{4}, 0, z]$. The reader may care to continue the study of *Pnma*, and then check the result, again against Figure 2.37.

General Equivalent Positions

Once we know the positions of the symmetry elements in a space-group pattern, the coordinates of the general equivalent positions in the unit cell follow readily.

Consider again *Pmma*. From the above analysis, we may write

$$\bar{1} \text{ at } 0, 0, 0 \text{ (choice of origin)}$$
$$m_x \parallel (\tfrac{1}{4}, y, z), \qquad m_y \parallel (x, 0, z), \qquad a \parallel (x, y, 0)$$

Taking a point x, y, z across the three symmetry planes in turn, we have (from Figure 2.34)

$$x, y, z \xrightarrow{\;m_x\;} \tfrac{1}{2} - x, y, z$$

$$\xrightarrow{\;m_y\;} x, \bar{y}, z$$

$$\xrightarrow{\;a\;} \tfrac{1}{2} + x, y, \bar{z}$$

These four points are now operated on by $\bar{1}$ to give the total of eight equivalent positions for *Pmma*:

$$\pm\{x, y, z; \quad \tfrac{1}{2} - x, y, z; \quad x, \bar{y}, z; \quad \tfrac{1}{2} + x, y, \bar{z}\}$$

The reader may now like to complete the example of *Pnma*.

A similar analysis may be carried out for the space groups in the *mm*2 class, with respect to origins on 2 or 2_1 (consider, for example, Figure 4.16), although we have not discussed specifically these space groups in this book.

A7.2   Monoclinic Space Groups

In the monoclinic space groups of class $2/m$, a 2_1 axis, with a translational component of $b/2$, shifts the center of symmetry by $b/4$ with respect to
the point of intersection of 2_1 with m (Figure S5.4b). In $P2/c$, the center
of symmetry is shifted by $c/4$ with respect to $2/c$, and in $P2_1/c$ the
corresponding shift is $(b+c)/4$ (Figure 2.33).

A8   Vector Algebraic Relationships in Reciprocal Space

A8.1   Introduction

The reciprocal lattice was introduced earlier in a geometrical manner,
as we find that treatment suitable for the beginner. With practice and
familiarity in reciprocal space concepts, a vector algebraic approach has the
appeal of conciseness and elegance, and we introduce this method here.

A8.2   Reciprocal Lattice

In considering the stereographic projection, we showed that the morphology of a crystal can be represented by a bundle of lines, drawn from
a point, normal to the faces of the crystal. This description, though angle-
true, lacks linear dimensions. The representation may be extended by giving
each normal a length that is inversely proportional to the corresponding
interplanar spacing in the real lattice, so forming a reciprocal lattice.

The non-coplanar vectors **a**, **b**, and **c** have been used to delineate a
unit cell in the real (Bravais) lattice (see page 57ff). The corresponding
vectors for the unit cell of a reciprocal lattice, $\mathbf{a}^*$, $\mathbf{b}^*$, and $\mathbf{c}^*$, will be defined
by

$$\mathbf{a}^* = \frac{\mathbf{b} \times \mathbf{c}}{V}, \qquad \mathbf{b}^* = \frac{\mathbf{c} \times \mathbf{a}}{V}, \qquad \mathbf{c}^* = \frac{\mathbf{a} \times \mathbf{b}}{V} \qquad \text{(A8.1)}$$

where the unit-cell volume V is given by

$$V = \mathbf{a} \cdot \mathbf{b} \times \mathbf{c} \qquad \text{(A8.2)}$$

In Section 2.4, particularly equation (2.11), we included a constant K. In
this appendix, we take the value of K as unity, so that the reciprocal lattice

has the dimensions of length^{-1} and is independent of the wavelength of X-radiation. In practical applications, K is nearly always chosen as λ; the size of the reciprocal lattice depends on the value of λ.

The magnitude of $\mathbf{a}^*$ is given by (see Figure A8.1)

$$a^* = \frac{\text{area } OBGC}{V} = 1/OP \qquad (A8.3)$$

where OP is the perpendicular from the origin O of the Bravais unit cell to the plane $ADFE$ which contains the point P. Similar relationships may be written for b^* and c^*, in terms of OQ and OR respectively. Hence, the reciprocal lattice vectors $\mathbf{a}^*$, $\mathbf{b}^*$, and $\mathbf{c}^*$ are normal to the planes bc, ca, and ab respectively in the Bravais unit cell. From Figure A8.1, it is now easy to see that [cf. (2.13) et seq.]

$$\mathbf{a} \cdot \mathbf{a}^* = \mathbf{b} \cdot \mathbf{b}^* = \mathbf{c} \cdot \mathbf{c}^* = 1 \qquad (A8.4)$$

and

$$\mathbf{a} \cdot \mathbf{b}^* = \mathbf{a} \cdot \mathbf{c}^* = \mathbf{a}^* \cdot \mathbf{b} = \text{etc.} = 0 \qquad (A8.5)$$

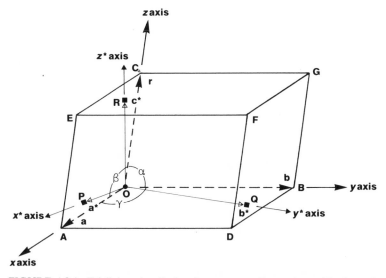

FIGURE A8.1. Triclinic unit cell, showing corresponding reciprocal lattice unit-cell vectors.

A8.2.1   Interplanar Spacings

Any vector $\mathbf{d}^*(hkl)$ from the origin to the point hkl in a reciprocal lattice can be written as

$$\mathbf{d}^*(hkl) = h\mathbf{a}^* + k\mathbf{b}^* + l\mathbf{c}^* \tag{A8.6}$$

[In some sections we have used $\mathbf{h}$ for $\mathbf{d}^*(hkl)$ for convenience.] The vector $\mathbf{r}$ from the origin to a point x,y,z in the unit cell of a Bravais lattice is given by

$$\mathbf{r} = x\mathbf{a} + y\mathbf{b} + z\mathbf{c} \tag{A8.7}$$

The scalar product $\mathbf{d}^*(hkl) \cdot \mathbf{r}$ leads to the equation of a plane in the Bravais lattice, normal to the direction of $\mathbf{d}^*(hkl)$ (see Figure A8.2). From (A8.6) and (A8.7), the scalar product gives

$$hx + ky + lz = \text{a constant, } \kappa \tag{A8.8}$$

For $\kappa = 0$, the plane passes through the origin; for $\kappa = 1$, it is the first plane from the origin of the family of (hkl) planes. These two cases are expressed also in equations (1.11) and (1.12), since $x = X/a$ and etc.

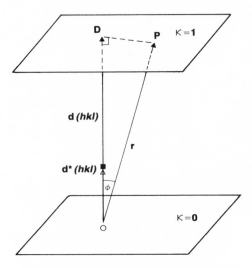

FIGURE A8.2. Planes for $\kappa = 0$ and $\kappa = 1$ in a Bravais lattice, showing the corresponding $\mathbf{d}$ and $\mathbf{d}^*$ vectors and the vector $\mathbf{r}$ to $P(x, y, z)$.

Since D, the foot of the perpendicular from O, lies in the same plane as P, the termination of $\mathbf{r}$ from the same point O, it follows that

$$\mathbf{d}^*(hkl) \cdot \mathbf{d}(hkl) = 1 \qquad (A8.9)$$

or

$$d^*(hkl) = \frac{1}{d(hkl)} \qquad (A8.10)$$

which may be compared with equation (2.11), the starting point of the geometrical treatment. Hence, any point hkl in the reciprocal lattice may be said to correspond to a family of planes (hkl) in the Bravais lattice, with interplanar spacing $d(hkl)$.

It may be noted that if P is a Bravais lattice point, then

$$\mathbf{r} = U\mathbf{a} + V\mathbf{b} + W\mathbf{c} \qquad (A8.11)$$

where U, V, W are the coordinates of the Bravais lattice point—cf. equation (2.1). The scalar product $\mathbf{d}^*(hkl) \cdot \mathbf{r}$ may now be written as

$$hU + kV - lW = \kappa \qquad (A8.12)$$

from which it follows that κ is an integer, as defined above. The coordinates U, V, W describe the direction, or directed line, $[UVW]$, as discussed on page 57.

We can now appreciate a fundamental difference between the stereographic projection and the reciprocal lattice constructions. Equation (A8.6) places no restrictions on the values of h, k, and l. Implicitly, they are prime to one another, but this limitation is not essential. Suppose that h, k, and l may be written as mh', mk', and ml', where h', k', and l' are prime to one another and m is an integer. If we carry out the same analysis as before, equation (A8.10) becomes

$$\frac{1}{m} d^*(h'k'l') = \frac{1}{d(h'k'l')} \qquad (A8.13)$$

or

$$d^*(h'k'l') = \frac{1}{d(h'k'l')/m} \qquad (A8.14)$$

But $d(h'k'l')/m$ is $d(hkl)$ and, since the definition of Miller indices identifies each family of lattice planes uniquely (see pages 68 and 69), equations (A8.6)–(A8.10) apply to all (hkl). In the stereographic projection, h, k, and l are prime to one another. It is clear that the normals to mh, mk, and ml ($m = 1, 2, 3, \ldots$) would all intercept the sphere (Figure 1.20) in the same point.

A8.2.2 Reciprocity of Unit Cell Volumes

Following (A8.2), we may write

$$V^* = \mathbf{a}^* \times \mathbf{b}^* \cdot \mathbf{c}^*$$

and from (A8.1)

$$V^* = \frac{1}{V^3}\{(\mathbf{b} \times \mathbf{c}) \times (\mathbf{c} \times \mathbf{a}) \cdot (\mathbf{a} \times \mathbf{b})\}$$

$$= \frac{1}{V^3}\{(\mathbf{b} \times \mathbf{c} \cdot \mathbf{a})\mathbf{c} - (\mathbf{b} \times \mathbf{c} \cdot \mathbf{c})\mathbf{a}\} \cdot (\mathbf{a} \times \mathbf{b})$$

$$= \frac{1}{V^3}\{V\mathbf{c} - 0\} \cdot (\mathbf{a} \times \mathbf{b})$$

$$= \frac{1}{V}$$

Hence

$$V^* V = 1 \tag{A8.15}$$

A8.2.3 Angle between Bravais Lattice Planes

The angle ϕ in Figure A8.3 between planes $(h_1 k_1 l_1)$ and $(h_2 k_2 l_2)$ may be most readily obtained from the supplement of the angle between the corresponding reciprocal lattice vectors $\mathbf{d}^*(h_1 k_1 l_1)$ and $\mathbf{d}^*(h_2 k_2 l_2)$. We have

$$\mathbf{d}^*(h_1 k_1 l_1) \cdot \mathbf{d}^*(h_2 k_2 l_2) = -d^*(h_1 k_1 l_1) d^*(h_2 k_2 l_2) \cos \phi \tag{A8.16}$$

Hence, for the orthorhombic system,

$$-\cos \phi = \frac{h_1 h_2/a^2 + k_1 k_2/b^2 + l_1 l_2/c^2}{(h_1^2/a^2 + k_1^2/b^2 + l_1^2/c^2)^{1/2}(h_2^2/a^2 + k_2^2/b^2 + l_2^2/c^2)^{1/2}} \tag{A8.17}$$

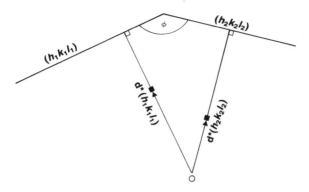

FIGURE A8.3. The interplanar angle ϕ, calculated in terms of the reciprocal lattice vectors to the two planes concerned.

From (A8.17), we see that the angle between (111) and (11$\bar{1}$) in the cubic system is $\cos^{-1}(-1/3)$, or 109.47°. The equation (A8.17) can be generalized by using the full form for $\mathbf{d}_1^* \cdot \mathbf{d}_2^*$, which incorporates the cross-terms such as $(k_1 l_2 + l_1 k_2)b^* c^* \cos \alpha^*$.

A8.2.4 Reciprocity of F and I Unit Cells

In Figure A8.4, we select a primitive unit cell by means of the transformation

$$\mathbf{a}_P = \mathbf{b}_F/2 + \mathbf{c}_F/2$$
$$\mathbf{b}_P = \mathbf{c}_F/2 + \mathbf{a}_F/2 \qquad (A8.18)$$
$$\mathbf{c}_P = \mathbf{a}_F/2 + \mathbf{b}_F/2$$

From (A8.1) and (A8.18), we have

$$\mathbf{a}_P^* = \frac{\mathbf{b}_P \times \mathbf{c}_P}{V_P} = \frac{1}{V_P}[(\mathbf{c}_F/2 + \mathbf{a}_F/2) \times (\mathbf{a}_F/2 + \mathbf{b}_F/2)]$$

or

$$\mathbf{a}_P^* = \frac{1}{V_F}[(\mathbf{c}_F \times \mathbf{b}_F) + (\mathbf{c}_F \times \mathbf{a}_F) + (\mathbf{a}_F \times \mathbf{b}_F)] \qquad (A8.19)$$

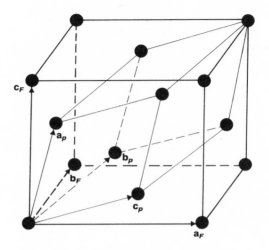

FIGURE A8.4. An F unit cell, with the related P
unit cell outlined within it.

since $V_F = 4V_P$. Hence,

$$\mathbf{a}_P^* = -\mathbf{a}_F^* + \mathbf{b}_F^* + \mathbf{c}_F^* \qquad (A8.20)$$

The negative sign before $\mathbf{a}_F^*$ is needed to preserve right-handed axes from
the product $(\mathbf{c}_F \times \mathbf{b}_F)$. Similar equations can be deduced for $\mathbf{b}_P^*$ and $\mathbf{c}_P^*$.
Turning next to a body-centered unit cell, the equations similar to (A8.18)
are

$$\mathbf{a}_P = -\mathbf{a}_I/2 + \mathbf{b}_I/2 + \mathbf{c}_I/2 \qquad \text{etc.} \qquad (A8.21)$$

Writing (A8.20) as

$$\mathbf{a}_P^* = -2\mathbf{a}_F^*/2 + 2\mathbf{b}_F^*/2 + 2\mathbf{c}_F^*/2 \qquad (A8.22)$$

we see that an F unit cell in a Bravais lattice reciprocates into an I unit
cell in the corresponding reciprocal lattice, where the I unit cell is defined
by the vectors $2\mathbf{a}_F^*$, $2\mathbf{b}_F^*$, and $2\mathbf{c}_F^*$. If, as is customary in practice, we define
the reciprocal of F by vectors $\mathbf{a}_F^*$, $\mathbf{b}_F^*$, and $\mathbf{c}_F^*$, then only those reciprocal
lattice points for which $h+k$, $k+l$, and $l+h$ are each integral belong to
the reciprocal of F. In other words, Bragg reflections from an F unit cell
have indices of the same parity.

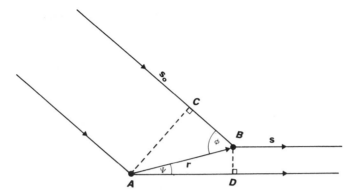

FIGURE A8.5. Scattering of X-rays at two centers A and B.

A8.3 X-Ray Diffraction and the Reciprocal Lattice

In Figure A8.5, let s_0 be a vector of magnitude $1/\lambda$ in the direction of the parallel incident beam and s the corresponding vector in the scattered beam. Let A and B be two scattering centres, situated at lattice points, separated by the vector $\mathbf{r}$; AC and BD indicate respectively the incident and diffracted wavefronts. The path difference δ between the scattered rays is given by

$$\delta = AD - BC \tag{A8.23}$$

Now

$$\mathbf{r} \cdot \mathbf{s} = \frac{1}{\lambda} r \cos \psi = AD/\lambda \tag{A8.24}$$

Similarly

$$\mathbf{r} \cdot \mathbf{s}_0 = \frac{1}{\lambda} r \cos \phi = BC/\lambda \tag{A8.25}$$

Hence

$$\delta = \lambda \mathbf{r} \cdot \mathbf{S} \tag{A8.26}$$

where $\mathbf{S} = \mathbf{s} - \mathbf{s}_0$, and may be called the scattering vector. If the waves scattered at A and B are to be in phase, $\mathbf{r} \cdot \mathbf{S}$ must be integral. Thus,

$$(U\mathbf{a} + V\mathbf{b} + W\mathbf{c}) \cdot \mathbf{S} = n \qquad (A8.27)$$

where U, V, and W are the coordinates of the lattice point at the end of $\mathbf{r}$ and n is an integer. This equation holds for any integral change in U and/or V and/or W. Hence,

$$\mathbf{a} \cdot \mathbf{S} = h$$
$$\mathbf{b} \cdot \mathbf{S} = k \qquad (A8.28)$$
$$\mathbf{c} \cdot \mathbf{S} = l$$

where h, k, and l are integers: (A8.28) is a vectorial expression of the Laue equations (q.v.).

In Figure A8.6, three planes normal to the x axis are shown. For the plane $h = 0$, $\mathbf{a} \cdot \mathbf{S} = 0$, which means that the projection of $\mathbf{S}$ on $\mathbf{a}$ is zero: it may be compared with the zero-layer of an oscillation photograph taken with the crystal rotating about a (see page 128ff). For $\mathbf{a} \cdot \mathbf{S} = 1$, we have a similar plane making an intercept of $1/a$ along the x axis. Hence, $\mathbf{a} \cdot \mathbf{S} = h$ represents a family of parallel equidistant planes normal to a. When the

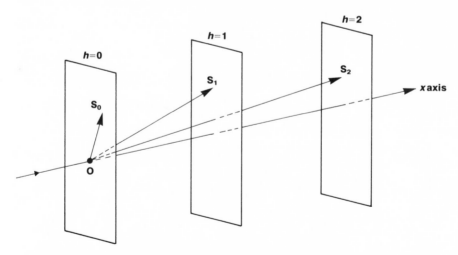

FIGURE A8.6. Planes for h values of 0, 1, and 2 in a Bravais lattice, with the corresponding scattering vectors.

equations (A8.28) are satisfied simultaneously, scattering in the *hkl* spectrum, or from the (*hkl*) family of planes, occurs.

A8.3.1   Bragg's Equation

The equations (A8.27) may be written as

$$(\mathbf{a}/h) \cdot \mathbf{S} = 1$$
$$(\mathbf{b}/k) \cdot \mathbf{S} = 1 \tag{A8.29}$$
$$(\mathbf{c}/l) \cdot \mathbf{S} = 1$$

Hence,

$$(\mathbf{a}/h - \mathbf{b}/k) \cdot \mathbf{S} = 0 \tag{A8.30}$$

which means that **S** is normal to the vector $(\mathbf{a}/h - \mathbf{b}/k)$. From Figure A8.7, it follows that this vector lies in the (*hkl*) plane. Also **S** is normal to

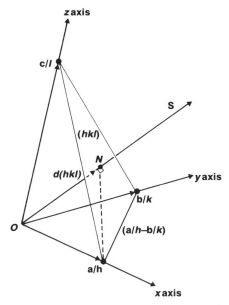

FIGURE A8.7. An (*hkl*) plane in a Bravais lattice; N is the foot of the perpendicular from the origin O to the plane.

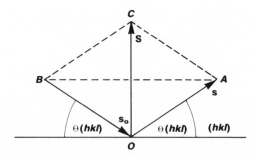

FIGURE A8.8. Relationship of the scattering vector **S** to the corresponding (*hkl*) plane.

($\mathbf{a}/h - \mathbf{c}/l$) and to ($\mathbf{b}/k - \mathbf{c}/l$); hence, **S** is normal to the plane (*hkl*). This result can be seen in another way.

In Figure A8.8, **S** is shown as the bisector of the angle between **s** and $\mathbf{s}_0$, normal to (*hkl*). The magnitudes of *OA* and *OB* are each $1/\lambda$, and the angles *AOC* and *BOC* are each equal to $\pi/2 - \theta(hkl)$. Hence, $S/2 = (1/\lambda) \sin \theta(hkl)$, or

$$S = \frac{2}{\lambda} \sin \theta(hkl) \tag{A8.31}$$

The interplanar spacing $d(hkl)$, shown in Figure A8.7, represents the projection of $\mathbf{a}/h$ (or $\mathbf{b}/k$ or $\mathbf{c}/l$) onto **S**. Hence,

$$d(hkl) = (\mathbf{a}/h) \cdot \mathbf{S}/S \tag{A8.32}$$

Using (A8.28) with (A8.31) and (A8.32), we have

$$(\mathbf{a}/h) \cdot \mathbf{S} = 1$$

$$d(hkl) = 1/S = \lambda/[2 \sin \theta(hkl)]$$

or

$$2d(hkl) \sin \theta(hkl) = \lambda \tag{A8.33}$$

which is Bragg's equation. Figure A8.8 illustrates the idea of "reflection" from a plane, but subject to (A8.33). The reader is invited to redraw the Ewald sphere (Figure 3.25) with a radius of $1/\lambda$ ($\lambda = 1.54$ Å, say, with an appropriate scale) inside the limiting sphere, center O, radius $2/\lambda$, and with $CO = -\mathbf{s}_0$ and $CP = \mathbf{s}$ to show that OP is identified with $\mathbf{S}$.

A8.4 Laue Photographs

A single crystal irradiated with an X-ray beam of wavelength λ may not be in a position to give rise to a reflection: no reciprocal lattice point

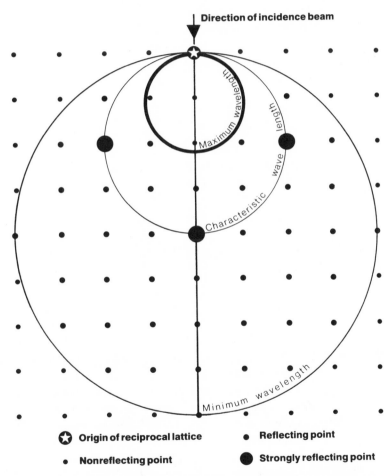

FIGURE A8.9. Reciprocal lattice illustration of the formation of a Laue photograph.

need lie on the Ewald sphere for that wavelength under the static conditions of the experiment (see Section 3.5.1). However, the beam from an X-ray tube consists of a range of wavelengths including, if the excitation voltage is large enough, characteristic radiation of high intensity (see Appendix A4). The wavelength spread is from a sharply defined minimum to an ill-defined maximum. Figure A8.9 illustrates a layer of the reciprocal lattice and the corresponding sections of Ewald spheres for different wavelengths. All reciprocal lattice points lying between the circles for maximum and minimum wavelength can give rise to reflections. Reciprocal lattice points that happen to lie on the circle for the characteristic wavelength will reflect strongly. Each reflection is developed according to the Ewald construction: the crystal selects its own wavelength for each given d and θ, according to (A8.33). The symmetry of the reciprocal lattice (and crystal) is conveyed to the diffraction pattern.

A8.5 Crystal Setting

We shall consider the problem of bringing a given reciprocal lattice plane (equatorial plane) to a position normal to the crystal rotation axis, prior to taking an oscillation photograph. We shall assume that our crystal has a well developed morphology, such as a prismatic (needle-shaped) habit (Figure 3.4).

The crystal is set up on a goniometer head, with its prism axis along the axis of rotation. Two arc adjustments, A and B, and two sledges, C and D (Figure 8.2) enable the crystal to be set, initially to better than $5°$, and arranged so as to rotate within its own volume.

A8.5.1   Setting Technique

A method of Weisz and Cole, as modified by Davis, will be considered: it has the advantage that each arc can be treated independently of the other.

A $15°$ oscillation photograph is taken with the arcs A and B at $45°$ to the X-ray beam (Figure A8.10) at the midpoint of the oscillation range, using unfiltered radiation and an exposure time sufficient to produce intense reflections. The goniometer head is then turned through exactly $180°$ and a second oscillation photograph taken on the same film, but with an exposure time of about one-third that of the first. The form of the double oscillation zero layer-line curve is shown in Figure A8.10.

From Figure A8.10a, the distance of the curve above (P_L) or below the true zero layer-line position at $\theta = 45°$ is $\tan^{-1}(\Delta_L/D)$, where D is the diameter of the film. This value is also that of the ζ reciprocal lattice

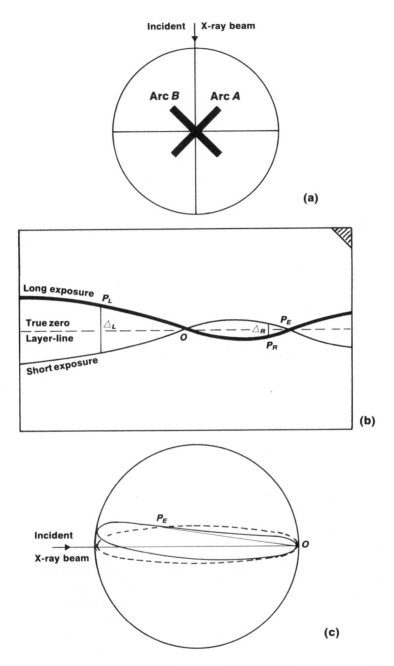

FIGURE A8.10. Crystal setting. (a) Goniometer arcs set at 45° to the X-ray beam. (b) Appearance of a double oscillation photograph—the traces are outlined by the spots from the $K\alpha$ and $K\beta$ radiation and the Laue "streaks"; the top right-hand corner (X-ray beam coming toward the observer) is clipped for identification. (c) Ewald sphere—the dashed line is the true equatorial circle, and the full line represents the longer exposure, related to the equatorial circle by rotation about the line OP_E.

coordinate of a possible reflection at P_L. Thus, the angle of elevation of the reciprocal lattice vector OP_L is $\delta_L = \tan^{-1}(\zeta/OP_L) = \tan^{-1}(\Delta_L/D\sqrt{2})$, taking $K = \lambda$, as in Section 3.5.3. Similarly, $\delta_R = \tan^{-1}(\Delta_R/D\sqrt{2})$. For values of $\delta < 4°$, $\tan\delta = \delta$ to 0.1%. Hence, for $D = 57.3$ mm, $\delta = 0.707\Delta°$, with Δ expressed in mm.

In order to apply the corrections, we return the goniometer head to a reading at or near the center of the range for the longer exposure. With the photograph marked as shown in Figure A8.10, consider the more intense curve. The correction δ_R is applied to arc A, lying in the NE-SW direction (N toward the X-ray source). The direction of movement of the arc is such that a reciprocal lattice vector at $\theta = 45°$, imagined to be protruding from the crystal in the NE direction (the reciprocal lattice origin is transferred to the crystal at this point) will be brought to the equatorial plane. The correction δ_L is applied to arc B in a similar manner.

A9   Intensity Statistics

A9.1   Weighted Reciprocal Lattice

The weights (intensities) associated with the reciprocal lattice points show about five different characteristics; we shall consider them in turn.

A9.1.1   Laue Symmetry

We discussed on page 45 the fact that X-ray diffraction patterns are centrosymmetric. Hence, the symmetry of this pattern, in terms of both position and intensity, corresponds to one of the eleven centrosymmetric (Laue) point groups. Where the crystal *is* centrosymmetric, the Laue symmetry of the diffraction pattern is also the point group of the crystal. Where the crystal is noncentrosymmetric, the diffraction pattern is centrosymmetric only insofar as the Friedel law applies, that is, the effects of anomalous scattering are negligible. This condition is easily satisfied with, say, a compound of C, H, N, and O irradiated with Mo $K\alpha$, but not necessarily with, say, a compound containing C, H, N, O, and Br irradiated with Cu $K\alpha$.

A9.1.2   Systematic Absences

Certain groups of reciprocal lattice points have zero weight because of the space group symmetry, irrespective of the contents of the unit cell. This topic was discussed sufficiently for our purposes in Chapter 2.

A9.1.3 Accidental Absences

A small proportion of possible reflections, although not of zero intensity, is often sufficiently weak not to be recorded with significance in the X-ray diffraction pattern. This effect depends upon both the nature of the atoms present and their relative positions in space. These reflections are often called "unobserved," because they do not produce visible blackening of an X-ray photograph. With diffractometer data, another criterion has to be adopted. Reflections may be so classified if, typically, $I < 3\sigma(I)$, where $\sigma(I)$ is the standard deviation, from counting statistics, of the intensity I. The "unobserved" data are often omitted, without powerful reason, from the structure analysis. In a good structure determination, the unobserved data should, at least, be checked against the corresponding $|F_c|$ to ensure that no significant reflections have been classified erroneously.

A9.1.4 Enhanced Averages

Some groups of reflections have enhanced average intensities. We touched upon this topic on pages 258 and 296, where we saw that such reflections were dependent upon the crystal class. We shall now look at this effect in more detail and, in particular, show that the enhancement factor ε is the same for the space groups $P2/m$, $P2/c$, $P2_1/m$, and $P2_1/c$, all of crystal class $2/m$.

The structure factor equation for these space groups may be written as

$$F(hkl) = \sum_{j=1}^{N/4} 4g_j \cos 2\pi(hx_j + lz_j + n/4) \cos 2\pi(ky_j - n/4) \quad (A9.1)$$

where N is the number of atoms in the unit cell and n is an integer (0, l, k and $k+l$, respectively, for the four space groups being considered). The average value of $F(hkl)$ depends upon the averages $\overline{\cos 2\pi(hx_j + lz_j + n/4)}$ and $\overline{\cos 2\pi(ky_j - n/4)}$. If we assume a random distribution of atomic positions and provided that N is not small, then since $-1 \leqslant \cos(\text{angle}) \leqslant 1$, these averages are zero and, thus, $\overline{F(hkl)}$ is zero. From (A9.1),

$$|F(hkl)|^2 = \sum_{j=1}^{N/4} 16g_j^2 \cos^2 2\pi(hx_j + lz_j + n/4) \cos^2 2\pi(ky_j - n/4)$$

$$+ \sum_{\substack{i=1 \\ i \neq j}}^{N/4} \sum_{j=1}^{N/4} 16g_i g_j \cos 2\pi(hx_i + lz_i + n/4) \cos 2\pi(ky_i - n/4)$$

$$\times \cos 2\pi(hx_j + lz_j + n/4) \cos 2\pi(ky_j - n/4) \quad (A9.2)$$

The terms under the double summation will take both positive and negative values, and for a sufficiently large number N of similar, randomly distributed atoms this sum will tend to zero. Now we have

$$\overline{|F(hkl)|^2} = \sum_{j=1}^{N/4} 16g_j^2 \overline{\cos^2 2\pi(hx_j + lz_j + n/4)} \, \overline{\cos^2 2\pi(ky_j - n/4)} \qquad (A9.3)$$

From reasoning similar to that used before, the average value of $\cos^2(\text{angle}) = 1/2$. Hence,

$$\overline{|F(hkl)|^2} = \sum_{j=1}^{N/4} 4g_j^2 = \sum_{j=1}^{N} g_j^2 = S \qquad (A9.4)$$

which is the Wilson average (page 258). If we consider the $h0l$ reflections,

$$|F(h0l)|^2 = \sum_{j=1}^{N/4} 8g_j^2 = 2 \sum_{j=1}^{N} g_j^2 = 2S \qquad (A9.5)$$

Similarly, we can show that the average value $\overline{|F(0k0)|^2}$ is also $2S$, but that for any other class of reflection, equation (A9.3) gives the value S. Thus, we have shown that the enhancement (epsilon) factors for crystal class $2/m$ are $\varepsilon_{h0l} = \varepsilon_{h00} = \varepsilon_{00l} = \varepsilon_{0k0} = 2$, otherwise $\varepsilon = 1$.

A graphical derivation of these results is afforded by the stereogram of crystal class $2/m$, Figure 1.39. If we consider that the four points shown are $\mathbf{g}_j$ vectors in the stucture factor equation, it is clear that projection onto the m plane ($h0l$ data) or onto the 2 axis ($0k0$ data) leads to superposition of the $\mathbf{g}_j$ vectors: each behaves in projection as though it has doubled weight ($\varepsilon = 2$). Results for different crystal classes may be obtained in either of the ways shown (Table 7.1).

A9.1.5 Special Distributions

The weighted reciprocal lattice, in entirety or in special planes or rows, may approximate to one or more of certain distinctive distributions. Two of them depend upon the presence or absence of a center of symmetry in the crystal. This information can be very important in the determination of the space group, since information from systematic absences may be inconclusive.

The structure factor equation can be written conveniently as

$$F_{\mathbf{h}} = \sum_{j=1}^{N/2} 2g_j \cos 2\pi(\mathbf{h} \cdot \mathbf{r}_j) \qquad (A9.6)$$

for a centrosymmetric crystal. The terms $\mathbf{h}$ and $\mathbf{r}_j$ are the vectors $h\mathbf{a}^* + k\mathbf{b}^* + l\mathbf{c}^*$ and $x_j\mathbf{a} + y_j\mathbf{b} + z_j\mathbf{c}$, so that $\mathbf{h} \cdot \mathbf{r}_j = hx_j + ky_j + lz_j$. We have seen that the mean value $\overline{2g_j \cos 2\pi(\mathbf{h} \cdot \mathbf{r}_j)}$ is zero. The variance of this term may be given as

$$V_j = 4g_j^2[\overline{\cos^2 2\pi(\mathbf{h} \cdot \mathbf{r}_j)} - \overline{\cos 2\pi(\mathbf{h} \cdot \mathbf{r}_j)^2}] = 2g_j^2 \qquad (A9.7)$$

Cramer's central limit theorem in statistics states that the sum of a large number of independent random variables has a normal probability distribution, a mean equal to the sum of the means of the independent variables, and a variance equal to the sum of their variances. We have shown in Section A9.1.4 that $\overline{F(hkl)} = 0$; this result follows directly from Cramer's theorem.

A normal probability distribution of F is given by

$$P(F) = (2\pi\sigma^2)^{-1/2} \exp[-(F - \bar{F})^2/2\sigma^2] \qquad (A9.8)$$

Since $\bar{F} = 0$, and σ^2 is given by

$$\sigma^2 = \sum_j V_j = \sum_{j=1}^{N/2} 2g_j^2 = \sum_{j=1}^{N} g_j^2 = S \qquad (A9.9)$$

the centric probability distribution function is given by

$$P_{\bar{1}}(F) = (2\pi S)^{-1/2} \exp(-F^2/2S) \qquad (A9.10)$$

For noncentrosymmetric crystals, the corresponding acentric distribution function is

$$P_1(|F|) = (2/S)|F| \exp(-|F|^2/S) \qquad (A9.11)$$

From the relationship

$$|E|^2 = |F|^2/S \qquad (A9.12)$$

where S, as well as containing a temperature factor will *here* be assumed to include ε, and with $|F|^2$ on an absolute scale, it follows that

$$P_{\bar{1}}(E) = (2\pi)^{-1/2} \exp(-E^2/2) \tag{A9.13}$$

and

$$P_1(|E|) = 2|E| \exp(-|E|^2) \tag{A9.14}$$

Equations (A9.13) and (A9.14) are independent of the complexity of the structure: they may be used for deducing a number of useful statistical parameters, such as those in Table 7.2.

As an example, we will determine the value of $\overline{|E|}$ for a noncentrosymmetric crystal. The mean, or expectation, value of a variable X, distributed according to a probability function $P(X)$, is given generally by

$$\bar{X} = \int XP(X)\mathbf{d}X / \int P(X)\mathbf{d}X \tag{A9.15}$$

where the integration is carried out over the limits of the variable. Hence,

$$\overline{|E|} = \frac{\int_0^\infty |E|^2 \exp(-|E|^2)\mathbf{d}|E|}{\int_0^\infty |E| \exp(-|E|^2)\mathbf{d}|E|} \tag{A9.16}$$

To solve these integrals easily, let $|E|^2 = t$, so that $\mathbf{d}|E| = \mathbf{d}t/2t^{1/2}$. Thus, the numerator becomes

$$\int_0^\infty t^{3/2-1} \exp(-t)\mathbf{d}t \tag{A9.17}$$

This integral is the gamma function $\Gamma(3/2) = 1/2\Gamma(1/2) = 1/2\sqrt{\pi}$, or 0.89. It is easy to show that the value of the denominator is unity; hence 0.89 is the value of $\overline{|E|}$, as shown in Table 7.2.

The cumulative values in the same table can be obtained from the same distribution equations. Let the fraction of $|E|$ values less than or equal to some value p be $N(p)$, given by

$$N(p) = \int_0^{|E|} P(|E|)\mathbf{d}|E| \tag{A9.18}$$

For the noncentrosymmetric crystal, we have

$$N(p) = 2 \int_0^{|E|} |E| \exp(-|E|^2)\mathbf{d}|E| \qquad (A9.19)$$

or

$$N(p) = [-\exp(-|E|^2)]_0^{|E|} \qquad (A9.20)$$

For $p = 1.5$, $N(1.5) = 0.895$. Hence, the number of $|E|$ values *greater* than 1.5 in the acentric distribution is 0.105, or 10.5%, as shown in Table 7.2. Many other useful results can be obtained quite simply by means of the two distribution equations.

Integrals of the type in (A9.16) are easily evaluated through the properties of the gamma function $\Gamma(n)$. We define

$$\Gamma(n) = \int_0^\infty t^{n-1} \exp(-t)\mathbf{d}t$$

The following results may be used directly:

$$\Gamma(n) = (n-1)! \qquad \text{for } n \text{ integral and greater than zero}$$
$$\Gamma(n) = (n-1)\Gamma(n-1)$$
$$\Gamma(\tfrac{1}{2}) = \sqrt{\pi}$$

We may question the upper limit of ∞ in equation (A9.16). It is easy to show that the maximum $|E|$, $E(000)$, is given by $\sqrt{N/\varepsilon}$, for a unit cell of N similar atoms. If we have a molecule of 25 similar atoms in general positions of space group $P2_1$, $E(000)$ is 5.00: ε here is the symmetry number (number of general equivalent positions) of 2 for the space group. From (A9.20), we see that the fraction of $|E|$ values greater than 5.00 is 1.4×10^{-11}. Hence, the error in taking the upper limit of ∞ is totally negligible, but the convenience is considerable.

A10   Enantiomorph Selection

In those noncentrosymmetric space groups, such as $P2_1$ and $P2_12_12_1$, that contain no inversion symmetry (enantiomorphous space groups), it is always possible to specify two enantiomorphic arrangements of the atoms in the structure that will lead to the same values of $|F|$. For example, in the structure in Figure 1.3, which has two molecules per unit cell in space group $P2_1$, the two arrangements would be related by inversion through the origin, and will be referred to as the structure (S) and its inverse (I).

From the structure factor theory discussed earlier, we can write

$$\mathbf{F}(\mathbf{h})_S = A(\mathbf{h})_S + iB(\mathbf{h})_S \qquad (A10.1)$$

for the structure, and

$$\mathbf{F}(\mathbf{h})_I = A(\mathbf{h})_I + iB(\mathbf{h})_I \qquad (A10.2)$$

for its inverse. From the inversion relationship, we know that $\mathbf{F}(\mathbf{h})_S$ and $\mathbf{F}(\mathbf{h})_I$ are complex conjugates; hence,

$$A(\mathbf{h})_S = A(\mathbf{h})_I \qquad (A10.3)$$

and

$$B(\mathbf{h})_S = -B(\mathbf{h})_I \qquad (A10.4)$$

For either the structure or its inverse, we can choose $B(\mathbf{h})$ to be positive, so that the corresponding phase angle $\phi(\mathbf{h})$ lies in the range $0 \leq \phi(h) \leq \pi$. This procedure was followed in the structure analysis, of tubercidin (Section 7.2.9), where the phase of symbolic reflection a $(13\bar{8})$ was restricted to a value between 0 and π, specifically $3\pi/4$.

In $P2_12_12_1$, another noncentrosymmetric space group of frequent occurrence in practice, the zonal reflections $0kl$, $h0l$, and $hk0$ are centric, and may be given phases equal to $m\pi/2$. The value of m takes the same parity as the index following zero, working in a cyclic manner. Thus, an origin and an enantiomorph could be specified in this space group by the selection

$$
\left.
\begin{array}{rrrl}
5 & 2 & 0 & +\pi/2 \\
0 & 1 & 1 & +\pi/2 \\
11 & 3 & 0 & +\pi/2
\end{array}
\right\} \quad \text{Origin}
$$

$$
\begin{array}{rrrl}
11 & 0 & 0 & +\pi/2 \quad \text{Enantiomorph}
\end{array}
$$

A detailed practical treatment on the origin and enantiomorph for all space groups has been given by Rogers.*

* D. Rogers, in *Theory and Practice of Direct Methods in Crystallography*, New York, Plenum Press (1980).

It is important not to confuse the specifying of the enantiomorph with the selection of the absolute configuration of a structure: in both cases, the same type of space group is involved. Selection of the enantiomorph is essential to a correct application of direct methods to a structure with an enantiomorphous space group. However, the solution of the structure may correspond to either the absolute configuration or its inverse. This dilemma has to be resolved by further tests, usually involving anomalous scattering (see page 282).

Solutions

Chapter 1

1.1. (1, 3.366).

1.2. (a) $(1\bar{2}0)$. (b) (164). (c) $(00\bar{1})$. (d) $(3\bar{3}4)$. (e) $(0\bar{4}3)$. (f) $(\bar{4}2\bar{3})$.

1.3. (a) $[\bar{5}11]$. (b) $[3\bar{5}2]$. (c) $[111]$. (d) $[110]$.

1.4. $(52\bar{3})$; $(52\bar{3})$ and $(\bar{5}2\bar{3})$ are parallel, and $[UVW]$ and $[\bar{U}\bar{V}\bar{W}]$ are coincident.

1.5. (a) See Figure S1.1.
(b) $c/a = \cot \widehat{100\ 10\bar{1}} = \cot 29.4° = 1.775$.
(c) In this example, the zone circles may be sketched in carefully, and the stereogram indexed without using a Wulff's net. Draw on the procedures used in Problems 1.3 and 1.4. (The center of the stereogram corresponds to 001, even though this face is not present on the crystal.) By making use of the axial ratio, the points of intersection of the

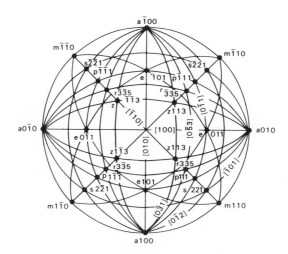

FIGURE S1.1

469

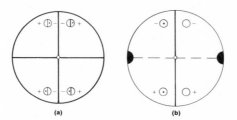

FIGURE S1.2

zone circles with the Y axis may be indexed, even though they do not all represent faces present. Reading from center to right, they are 001, 013, 035, 011, 021, and 010 (letter symbols indicate faces actually present). Hence, the zone symbols and poles may be deduced. Confirm the assignments of indices by means of the Weiss zone law.

1.6. (a) *mmm*. (b) $2/m$. (c) 1.

1.7. See Figure S1.2. (a) *mmm*   (b) $2/m$
                        $m.m.m \equiv \bar{1}$. $2.m \equiv \bar{1}$.

1.8.

	{010}	{$\bar{1}$10}	{11$\bar{3}$}
$2/m$	2	4	4
$\bar{4}2m$	4	4	8
$m3$	6	12	24

1.9. (a) 1. (b) *m*. (c) 2. (d) *m*. (e) 1. (f) 2. (g) 6. (h) 6*mm*. (i) 3. (j) 2*mm*. (Did you remember to use the Laue group in each case?)

1.10. (a) 2. (b) *m*.

    (a) $\bar{6}m2$ D_{3h} {10$\bar{1}$0} or {01$\bar{1}$0}

    (b) $\dfrac{4}{m}mm$ D_{4h} {100} or {110}

    (c) $m3m$ O_h {100}

    (d) $\bar{4}3m$ T_d {111} or {1$\bar{1}$1}

    (e) $3m$ C_{3v}

    (f) $\bar{1}$ C_1

    (g) $\dfrac{6}{m}mm$ D_{6h}

    (h) $mm2$ C_{2v}

    (i) mmm D_{2h}

    (j) m C_s

Chapter 2

2.1. (a) (i) 4*mm*, (ii) 6*mm*. (b) (i) Square, (ii) hexagonal. (c) (i) Another square can be drawn as the conventional (p) unit cell. (ii) The symmetry at each point is degraded to 2*mm*. A rectangular net is produced, and may be described by a p unit cell. The transformation equations for both examples are

$$\mathbf{a}' = \mathbf{a}/2 + \mathbf{b} + 2, \qquad \mathbf{b}' = -\mathbf{a}/2 + \mathbf{b}/2$$

Note. A regular hexagon of points with another point at its center is not a centered hexagonal unit cell; it represents three adjacent *p* hexagonal unit cells in different orientations.

2.2. The *C* unit cell may be obtained by the transformation $\mathbf{a}' = \mathbf{a}$, $\mathbf{b}' = \mathbf{b}$, $\mathbf{c}' = -\mathbf{a}/2 + \mathbf{c}/2$. The new dimensions are $c' = 5.763$ Å and $\beta' = 139.28°$; a' and b' remain as a and b, respectively. $V_c(C \text{ cell}) = V_c(F \text{ cell})/2$.

2.3. (a) The symmetry is no longer tetragonal, although it represents a lattice (orthorhombic).
(b) The tetragonal symmetry is apparently restored, but the unit cell no longer represents a lattice because the points do not all have the same environment.
(c) A tetragonal *F* unit cell is obtained, which is equivalent to *I* under the transformation $\mathbf{a}' = \mathbf{a}/2 + \mathbf{b}/2$, $\mathbf{b}' = -\mathbf{a}/2 + \mathbf{b}/2$, $\mathbf{c}' = \mathbf{c}$.

2.4. 28.74 Å (*F* cell); 28.64 Å. Students familiar with matrices may note that the second result can be obtained by transforming $[31\bar{2}]$ in the second cell to the corresponding direction in the first cell, $[41\bar{1}]$.

2.5. It is not an eighth system because the symmetry of the unit cell is not higher than $\bar{1}$. It represents a special case of the triclinic system with $\gamma = 90°$.

2.6. (a) Plane group *c2mm* (see Figure (S2.1)).

$$(0, 0; \tfrac{1}{2}, \tfrac{1}{2}) + \qquad\qquad \text{Limiting conditions}$$

8	(f)	1	x, y; $x, \bar{y}$; $\bar{x}, y$; $\bar{x}, \bar{y}$	$hk: h + k = 2n$
4	(e)	*m*	$0, y$; $0, \bar{y}$	—
4	(d)	*m*	$x, 0$; $\bar{x}, 0$	—
4	(c)	2	$\tfrac{1}{4}, \tfrac{1}{4}$; $\tfrac{1}{4}, \tfrac{3}{4}$	As above + $hk: h = 2n, (k = 2n)$
2	(b)	2*mm*	$0, \tfrac{1}{2}$	—
2	(a)	2*mm*	$0, 0$	—

(b) Plane group *p2mg*. See Figure S2.2. If the symmetry elements are arranged with 2 at the intersection of *m* and *g*, they do not form a group. Attempts to draw such an arrangement lead to continued halving of the "repeat" parallel to *g*.

2.7. See Figures S2.3 and S2.4.

4	(e)	1	x, y, z; $\bar{x}, \bar{y}, \bar{z}$; $x, \tfrac{1}{2}-y, \tfrac{1}{2}+z$; $\bar{x}, \tfrac{1}{2}+y, \tfrac{1}{2}-z$	hkl: None $h0l: l = 2n$ $0k0: k = 2n$
2	(d)	$\bar{1}$	$\tfrac{1}{2}, 0, \tfrac{1}{2}$; $\tfrac{1}{2}, \tfrac{1}{2}, 0$	
2	(c)	$\bar{1}$	$0, 0, \tfrac{1}{2}$; $0, \tfrac{1}{2}, 0$	As above + $hkl: k + l = 2n$
2	(b)	$\bar{1}$	$\tfrac{1}{2}, 0, 0$; $\tfrac{1}{2}, \tfrac{1}{2}, \tfrac{1}{2}$	
2	(a)	$\bar{1}$	$0, 0, 0$; $0, \tfrac{1}{2}, \tfrac{1}{2}$	
(100)	*p2gg*	$b' = b, c' = c$		
(010)	*p2*	$a' = a, c' = c/2$		
(001)	*p2gm*	$a' = a, b' = b$		

The two molecules lie with the center of their C(1)—C(1)' bonds on any pair of special positions (a)–(d). The molecule is therefore centrosymmetric and planar. The planarity

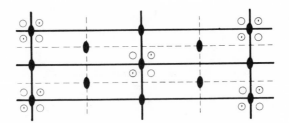

FIGURE S2.1

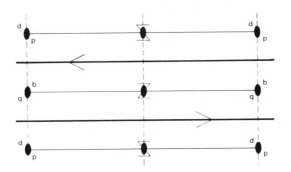

FIGURE S2.2

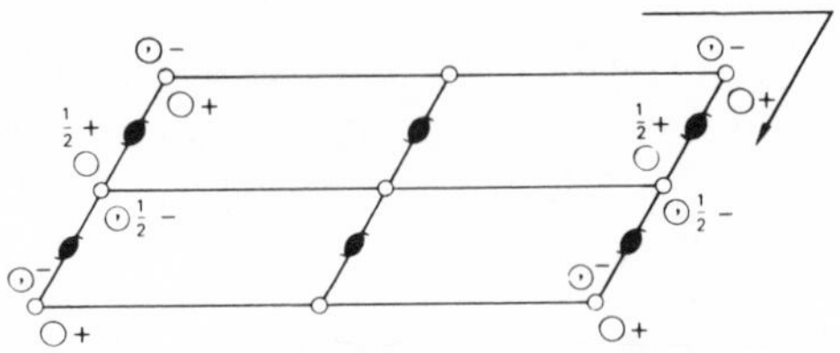

FIGURE S2.3

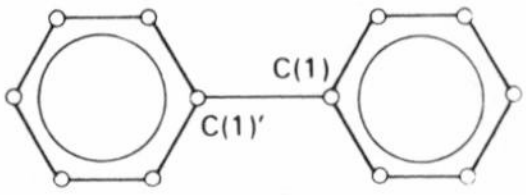

FIGURE S2.4

implies conjugation involving the $C(1)-C(1)'$ bond. (This result is supported by the bond lengths $C(1)-C(1)' \approx 1.49$ Å and $C-C$ (in ring) ≈ 1.40 Å. In the free molecule state, the two rings rotate about $C(1)-C(1)'$ to give an angle of $45°$ between their planes.)

2.8. Each pair of positions forms two vectors, between the origin and the points $\pm\{(x_1-x_2), (y_1-y_2), (z_1-z_2)\}$: one vector at each of the locations

$$2x, 2y, 2z; \quad 2\bar{x}, 2\bar{y}, 2\bar{z}; \quad 2x, 2\bar{y}, 2z; \quad 2\bar{x}, 2y, 2\bar{z}$$

and two vectors at each of the locations

$$2x, \tfrac{1}{2}, \tfrac{1}{2}+2z; \quad 0, \tfrac{1}{2}+2y, \tfrac{1}{2}; \quad 2\bar{x}, \tfrac{1}{2}, \tfrac{1}{2}-2z; \quad 0, \tfrac{1}{2}-2y, \tfrac{1}{2}$$

Note: $-(2x, \tfrac{1}{2}, \tfrac{1}{2}+2z) \equiv 2\bar{x}, \tfrac{1}{2}, \tfrac{1}{2}-2z$.

2.9.

$$x, y, z \xrightarrow{-b} 2p-x, -\tfrac{1}{2}+y, z$$

$$\downarrow{-\bar{1}} \qquad\qquad \downarrow{-a}$$

$$\bar{x}, \bar{y}, \bar{z}$$

$$\left\{ -1 \text{ (or 0)} + 2p-x, 2q-y, 2r-z \xleftarrow{-n} -\tfrac{1}{2}+2p-x, 2q+\tfrac{1}{2}-y, z \right.$$

The points $\bar{x}, \bar{y}, \bar{z}$ and $2p-x, 2q-y, 2r-z$ are one and the same; hence, by comparing coordinates, $p=q=r=0$. Check this result with the half-translation rule.

***2.10.** See Figure S2.5. General equivalent positions:

$$x, y, z; \qquad x, y, \bar{z}; \quad \tfrac{1}{2}-x, \tfrac{1}{2}-y, z; \quad \tfrac{1}{2}-x, \tfrac{1}{2}-y, \bar{z}$$

$$\tfrac{1}{2}+x, \bar{y}, z; \quad \tfrac{1}{2}+x, \bar{y}, \bar{z}; \qquad \bar{x}, \tfrac{1}{2}+y, z; \qquad \bar{x}, \tfrac{1}{2}+y, \bar{z}$$

Centers of symmetry:

$$\tfrac{1}{4}, \tfrac{1}{4}, 0; \quad \tfrac{1}{4}, \tfrac{3}{4}, 0; \quad \tfrac{3}{4}, \tfrac{1}{4}, 0; \quad \tfrac{3}{4}, \tfrac{3}{4}, 0$$

$$\tfrac{1}{4}, \tfrac{1}{4}, \tfrac{1}{2}; \quad \tfrac{1}{4}, \tfrac{3}{4}, \tfrac{1}{2}; \quad \tfrac{3}{4}, \tfrac{1}{4}, \tfrac{1}{2}; \quad \tfrac{3}{4}, \tfrac{3}{4}, \tfrac{1}{2}$$

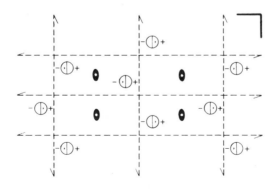

FIGURE S2.5

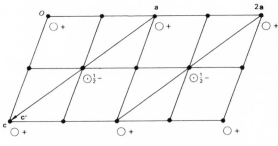

FIGURE S2.6

Change of origin: (i) subtract $\frac{1}{4}, \frac{1}{4}, 0$ from the above set of general equivalent positions, (ii) let $x_0 = x - \frac{1}{4}$, $y_0 = y - \frac{1}{4}$, $z_0 = z$, (iii) continue in this way, and finally drop the subscript:

$$\pm(x, y, z; \quad \bar{x}, \bar{y}, z; \quad \tfrac{1}{2}+x, \tfrac{1}{2}-y, \bar{z}; \quad \tfrac{1}{2}-x, \tfrac{1}{2}+y, \bar{z})$$

This result may be confirmed by redrawing the space-group diagram with the origin on $\bar{1}$.

2.11. Two unit cells of space group Pn are shown on the (010) plane (see Figure S2.6). In the transformation to Pc, only c is changed:

$$\mathbf{c}'(Pc) = -\mathbf{a}(Pn) + \mathbf{c}(Pn)$$

Hence, $Pn \equiv Pc$. By interchanging the labels of the x and z axes (which are not constrained by the twofold symmetry axis), we see that $Pc \equiv Pa$. However, because of the translations of $\frac{1}{2}$ along a and b in Cm, from the centering of the unit cell, $Ca \neq Cc$, although $Cc \equiv Cn$. We have $Ca \equiv Cm$, and the usual symbol for this space group is Cm. If the x and z axes are interchanged in Cc, the equivalent symbol is Aa.

2.12. $P2/c$ (a) $2/m$; monoclinic.
           (b) Primitive unit cell, c-glide plane $\perp b$, twofold axis $\| b$.
           (c) $h0l$: $l = 2n$.

      $Pca2_1$ (a) $mm2$; orthorhombic.
           (b) Primitive unit cell, c-glide plane $\perp a$, a-glide plane $\perp b$, 2_1 axis $\| c$.

      $Cmcm$ (a) mmm; orthorhombic.
           (b) C face-centered unit cell, m plane $\perp a$, c-glide plane $\perp b$, m plane $\perp c$.
           (c) hkl: $h + k = 2n$; $h0l$: $l = 2n$.

      $P\bar{4}2_1c$ (a) $\bar{4}2m$; tetragonal.
           (b) Primitive unit cell, $\bar{4}$ axis $\| c$, 2_1 axes $\| a$ and b, c-glide planes $\perp [110]$ and $[1\bar{1}0]$.
           (c) hhl: $l = 2n$; $h00$: $h = 2n$; $0k0$: $(k = 2n)$.

      $P6_322$ (a) 622; hexagonal.
           (b) Primitive unit cell, 6_3 axis $\| c$, twofold axes $\| a$, b, and u, twofold axes $30°$ to a, b, and u, in the (0001) plane.
           (c) $000l$: $l = 2n$.

      $Pa3$ (a) $m3$; cubic.

(b) Primitive unit cell, a-glide plane $\perp c$, b-glide plane $\perp a$, c-glide plane $\perp b$ (the glide planes are equivalent under the cubic symmetry), threefold axes $\parallel [111], [1\bar{1}1], [\bar{1}11]$, and $[\bar{1}\bar{1}1]$.

(c) $0kl: k = 2n; h0l: (l = 2n); hk0: (h = 2n)$.

2.13. Plane group $p2$; the unit-cell repeat along b is halved, and γ has the particular value of 90°.

Chapter 3

3.1. (a) Tetragonal crystal, Laue group $\frac{4}{m}mm$; optic axis parallel to the needle axis (c) of the crystal.

(b) Section extinguished for any rotation in the ab plane; section normal to c is optically isotropic.

(c) Horizontal m line. Symmetric oscillation photograph with a, b, or $\langle 110 \rangle$ parallel to the beam at the center of the oscillation would have $2mm$ symmetry (m lines horizontal and vertical).

3.2. (a) Orthorhombic.   (b) Axes parallel to the edges of the brick.   (c) Horizontal m line.   (d) $2mm$ (m lines horizontal and vertical).

3.3. (a) Monoclinic, or possibly orthorhombic.

(b) If monoclinic, $y \parallel p$. If orthorhombic, $p \parallel x$, y, or z.

(c) (i) Mount the crystal perpendicular to p, either about q or r, and take a Laue photograph with the X-ray beam parallel to p. If monoclinic, twofold symmetry would be observed. If orthorhombic, $2mm$, but with the m lines in general directions on the film which define the directions of the crystallographic axes normal to p. If the crystal is rotated so that X-rays are perpendicular to p, a vertical m line would appear on the Laue photograph of either a monoclinic or an orthorhombic crystal.

(ii) Use the same crystal mounting as in (i) and take symmetric oscillation photographs with the X-ray beam parallel or perpendicular to p at the center of the oscillation. The rest of the answer is as in (i).

3.4. $a = 9.00$, $b = 6.00$, $c = 5.00$ Å.

$a^* = 0.167$, $b^* = 0.250$, $c^* = 0.300$ RU.

$d(146) = 0.726$ Å; hence $2 \sin \theta(146) > 2.0$. Each photograph would have a horizontal m line, conclusive of orthorhombic symmetry if the crystal is known to be biaxial; otherwise, tests for higher symmetry would have to be carried out.

3.5. (a) $a = 8.64$, $c = 7.51$ Å.

(b) $n_{max} = 3$.

(c) No symmetry in (i). Horizontal m line in (ii).

(d) The photographs would be identical because of the fourfold axis of oscillation.

3.6. Remembering that the β angle is, conventionally, oblique, and that in the monoclinic system $\beta = 180° - \beta^*$, $\beta^* = 85°$ and $\beta = 95°$.

Chapter 4

4.1. The coordinates show that the structure is centrosymmetric. Hence, $A'(hk)$ is given by (4.62) with $l = 0$, $B'(hkl) = 0$, and the structure factors are real $[F(hk) = A'(hk)]$:

$$F(5, 0) = 2(-g_P + g_Q), \qquad F(0, 5) = 2(g_P - g_Q)$$

$$F(5, 5) = 2(-g_P - g_Q), \qquad F(5, 10) = 2(-g_P + g_Q)$$

For $g_P = 2g_Q$, $\phi(0, 5) = 0$ and $\phi(5, 0) = \phi(5, 5) = \phi(5, 10) = \pi$.

4.2. The structure is centrosymmetric. Since $l = 0$ in the data given

$$A(hk0) = 4 \cos 2\pi[ky + (h + k)/4] \cos 2\pi(h + k)/4$$

	$y = 0.10$	$y = 0.15$
$|F_c(020)|$	86.5	86.5
$|F_c(110)|$	258.9	188.1

Hence, 0.10 is the better value for y, as far as one can judge from these two reflections.

4.3. The shortest U—U distance is between $0, y, \frac{1}{4}$ and $0, \bar{y}, \frac{3}{4}$ and has the value 2.76 Å.

4.4. (a) $P2_1$, $P2_1/m$. (b) Pa, $P2/a$. (c) Cc, $C2/c$. (d) $P2$, Pm, $P2/m$.

4.5. (a) $P2_12_12$.
(b) $Pbm2$, $Pbmm$.
(c) $Ibm2$ ($Icm2$); $Ib2m$ ($Ic2m$); $Ibmm$ ($Icmm$)
   hkl: $h + k + l = 2n$
   $0kl$: $k = 2n$, ($l = 2n$), or $l = 2n$, ($k = 2n$)
   $h0l$: ($h + l = 2n$)
   $hk0$: ($h + k = 2n$)
   $h00$: ($h = 2n$)
   $0k0$: ($k = 2n$)
   $00l$: ($l = 2n$).

4.6. (a) (i) $h0l$: $h = 2n$; $0k0$: $k = 2n$. No other independent conditions.
       (ii) $h0l$: $l = 2n$. No other independent conditions.
       (iii) hkl: $h + k = 2n$. No other independent conditions.
       (iv) $h00$: $h = 2n$. No other conditions.
       (v) $0kl$: $l = 2n$; $h0l$: $l = 2n$. No other independent conditions.
       (vi) hkl: $h + k + l = 2n$; $h0l$: $h = 2n$. No other independent conditions.

Space groups with the same conditions: (i) None. (ii) $P2/c$. (iii) Cm, $C2/m$. (iv) None. (v) $Pccm$. (vi) $Ima2$, $I2am$.
(b) hkl: None; $h0l$: $h + l = 2n$; $0k0$: $k = 2n$.
(c) $C2/c$, $C222$.

Chapter 5

5.1. $A(hkl) = 4 \cos 2\pi[0.2h + 0.1l + (k + l)/4] \cos 2\pi(l/4)$. Reflections hkl are systematically absent for l odd. The c dimension appertaining to $P2_1/c$ should be halved; the true cell contains two atoms in space group $P2_1$. This problem illustrates the consequences of siting atoms on glide planes. Although this answer applies to a hypothetical structure containing a single atomic species, in a mixed-atom structure an atom may, by chance, be situated on a translational symmetry element. See Figure S5.1.

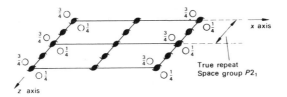

FIGURE S5.1

5.2. There are eight Rh atoms in the unit cell. The separation of atoms related across any m plane is $\frac{1}{2} - 2y$, which is less than $b/2$ and thus, prohibited. The Rh atoms must therefore lie in two sets of special positions, with either $\bar{1}$ or m symmetry. The positions on $\bar{1}$ may be eliminated, again by spatial considerations. Hence, we have (see Figures S5.2 and S5.3).

$$4 \ Rh(1): \pm\{x_1, \tfrac{1}{4}, z_1; \ \tfrac{1}{2}+x_1, \tfrac{1}{4}, \tfrac{1}{2}-z_1\}$$
$$4 \ Rh(2): \pm\{x_2, \tfrac{1}{4}, z_2; \ \tfrac{1}{2}+x_2, \tfrac{1}{4}, \tfrac{1}{2}-z_2\}$$

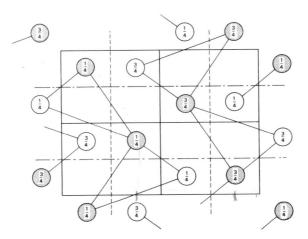

FIGURE S5.2

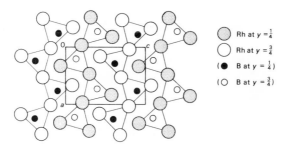

FIGURE S5.3

* R. Mooney and A. J. E. Welch, *Acta Crystallographica* **7**, 49 (1954).

5.3. Space group $P2_1/m$. Molecular symmetry cannot be $\bar{1}$, but it can be m: (a) Cl lie on m planes, (b) N lie on m planes, (c) two C on m planes, and four other C probably in general positions, (d) 16 H in general positions, two H (in NH groups) on m planes, and two H (from the CH_3 that have their C on m planes) on m planes. This arrangement is shown schematically in Figure S5.4a. The groups CH_3, H_1, and H_2 lie above and below the m plane. The alternative space group $P2_1$, was considered, but the structure analysis* confirmed the assumption of $P2_1/m$. The diagram of space group $P2_1/m$ shown in Figure 5.4b is reproduced from the *International Tables for X-Ray Crystallography*, Vol. I, edited by N. F. M. Henry and K. Lonsdale, with the permission of the International Union of Crystallography.

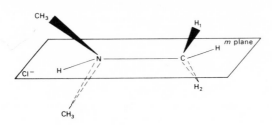

FIGURE S5.4a

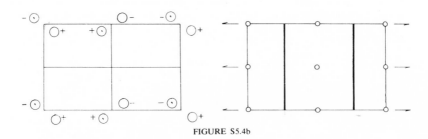

FIGURE S5.4b

Origin at $\bar{1}$; unique axis b

Limiting conditions

4   f 1 x, y, z; $\bar{x}, \bar{y}, \bar{z}$; $\bar{x}, \frac{1}{2}+y, \bar{z}$; $x, \frac{1}{2}-y, z$.

hkl: None
$h0l$: None
$0k0$: $k = 2n$

2   e m $x, \frac{1}{4}, z$; $\bar{x}, \frac{3}{4}, \bar{z}$
2   d $\bar{1}$ $\frac{1}{2}, 0, \frac{1}{2}$; $\frac{1}{2}, \frac{1}{2}, \frac{1}{2}$.
2   c $\bar{1}$ $0, 0, \frac{1}{2}$; $0, \frac{1}{2}, \frac{1}{2}$.
2   b $\bar{1}$ $\frac{1}{2}, 0, 0$; $\frac{1}{2}, \frac{1}{2}, 0$.
2   a $\bar{1}$ $0, 0, 0$; $0, \frac{1}{2}, 0$.

As above +
hkl: $k = 2n$

Symmetry of special projections

(001) pgm; $a' = a, b' = b$ (100) pmg; $b' = b, c' = c$ (010) $p2$; $c' = c, a' = a$

---

* J. Lindgren and I. Olovsson, *Acta Crystallographica* **B24**, 554 (1968).

5.4

hhh	$\|F_o\|$	$X_{Cl}=0.23$		$X_{Cl}=0.24$	
		$\|F_c\|$	$K_1\|F_o\|$	$\|F_c\|$	$K_2\|F_o\|$
111	491	341	315	317	329
222	223	152	143	160	150
333	281	145	180	191	189
		K_1	R_1	K_2	R_2
		0.641	0.11	0.671	0.036

Clearly, $x_{Cl}=0.24$ is the preferred value. Pt—Cl = 2.34 Å. For sketch and point group, see Problem (and Solution) 1.11(a).

5.5. $A_U(hkl) = 4 \cos 2\pi[hx_U - (h+k)/4] \cos 2\pi[ky_U + (h+k+l)/4]$. $x_U \approx \frac{1}{8}$, $y_U \approx 0.20$ (mean of $\frac{1}{4}, \frac{1}{6}, \frac{3}{16}$).

5.6. Since $Z = 2$, the molecules lie either on $\bar{1}$ or m. Chemical knowledge eliminates $\bar{1}$. The m planes are at $\pm(x, \frac{1}{4}, z)$, and the C, N, and B atoms must lie on these planes. Since the shortest distance between m planes is 3.64 Å, F_1, B, N, C, and H_1 (see Figure S5.5a) lie on one m plane. Hence, the remaining F atoms and the four H atoms must be placed symmetrically across the same m plane. The conclusions were borne out by the structure analysis.* Figure S5.5b shows a stereoscopic pair of packing diagrams for $CH_3NH_2 \cdot BF_3$. F_1, B, N, C, and H_1 lie on a mirror plane; the F_2, F_3, H_4, H_5, and H_2, H_3, atom pairs are related across the same m plane.

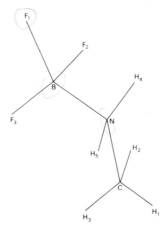

FIGURE S5.5a

* S. Geller and J. L. Hoard, *Acta Crystallographica* **3**, 121 (1950).

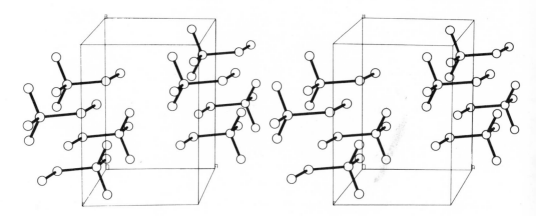

FIGURE S5.5b

Chapter 6

6.1. (a) $|F(hkl)| = |F(\bar{h}k\bar{l})|$
$|F(0kl)| = |F(0k\bar{l})|$
$|F(h0l)| = |F(\bar{h}0\bar{l})|$

(b) $|F(hkl)| = |F(\bar{h}k\bar{l})| = |F(h\bar{k}l)| = |F(\bar{h}k\bar{l})|$
$|F(0kl)| = |F(0\bar{k}\bar{l})| = |F(0\bar{k}l)| = |F(0k\bar{l})|$
$|F(h0l)| = |F(\bar{h}0\bar{l})|$

(c) $|F(hkl)| = |F(\bar{h}k\bar{l})| = |F(\bar{h}kl)| = |F(h\bar{k}l)| = |F(hk\bar{l})| = |F(h\bar{k}\bar{l})| = |F(\bar{h}k\bar{l})| = |F(\bar{h}\bar{k}l)|$
$|F(0kl)| = |F(0\bar{k}\bar{l})| = |F(0\bar{k}l)| = |F(0k\bar{l})|$
$|F(h0l)| = |F(\bar{h}0\bar{l})| = |F(\bar{h}0l)| = |F(h0\bar{l})|$

6.2. (a) Pa; $[\frac{1}{2}, v, 0]$.
$P2/a$: $[\frac{1}{2}, v, 0]$ and $(u, 0, w)$.
$P222_1$; $(0, v, w)$, $(u, 0, w)$, and $(u, v, \frac{1}{2})$.

(b) $(u, 0, w)$ is the Harker section for a structure with a twofold axis along b, whereas $[0, v, 0]$ is the Harker line corresponding to an m plane normal to b. Since the crystal is noncentrosymmetric, the space group is either $P2$ or Pm. If it is $P2$, then there must be chance coincidences between the y coordinates of atoms not related by symmetry. If it is Pm, then the chance coincidences must be between both the x and the z coordinates of atoms not related by symmetry.

6.3. (a) $P2_1/n$ (a nonstandard setting of $P2_1/c$; see Problem 2.11 for a similar relationship between Pc and Pn).

(b) Vectors: 1: $\pm\{\frac{1}{2}, \frac{1}{2} + 2y, \frac{1}{2}\}$ double weight
               2: $\pm\{\frac{1}{2} + 2x, \frac{1}{2}, \frac{1}{2} + 2z\}$ double weight
               3: $\pm\{2x, 2y, 2z\}$ single weight
               4: $\pm\{2x, 2\bar{y}, 2z\}$ single weight

Section $v = \frac{1}{2}$: type 2 vector—$x = 0.182$; $z = 0.235$.
Section $v = 0.092$: type 1 vector—$y = 0.204$.
Section $v = 0.408$: type 3 or 4 vector—$x = 0.183$; $y = 0.204$; $z = 0.234$.

4 S: $\pm\{0.183, 0.204, 0.235\}$ and $\pm\{0.683, 0.296, 0.735\}$

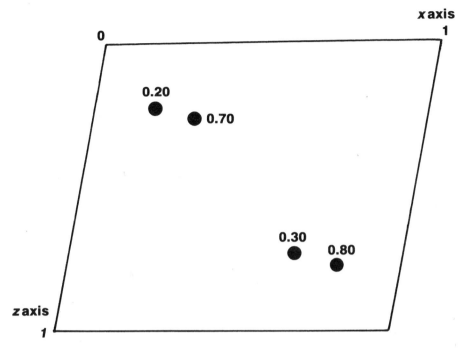

FIGURE S6.1

You may select any one of the other seven centers of symmetry as origin, in which case the coordinates determined will be transformed accordingly. The positions are plotted in Figure S6.1. Differences in the third decimal places of the coordinates determined from the maps in Problems 6.3 and 6.4 are not significant.

6.4. (a) The sulfur atom x and z coordinates are $S(0.266, 0.141)$, $S'(-0.266, -0.141)$.
(b) Plot the position $-S$ on tracing paper and copy the Patterson map (excluding the origin peak) with its origin over $-S$ (Figure S6.2a). On another tracing, carry out the same procedure with respect to $-S'$ (Figure S6.2b). Superimpose the two tracings (Figure S6.2c). Atoms are located where both maps have positive areas.

6.5. (a) $P(v)$ shows three nonorigin peaks. If the highest is assumed to arise from Hf atoms at $\pm\{0, y_{Hf}, \frac{1}{4}\}$, then $y_{Hf} = 0.11$. The other two peaks represent Hf—Si vectors; the difference in their height is due partly to the proximity of the peak of apparent lesser height to the origin peak—an example of poor resolution—and partly due to the particular y value of one Si atom.
(b) The signs are, in order and omitting 012,0 and 016,0, $+ - - + + -$. $\rho(y)$ shows a large peak at 0.107, which is a better value for y_{Hf}, and smaller peaks at 0.05, 0.17, and 0.25. The values 0.05 and 0.25 give vectors for Hf—Si which coincide with peaks on $P(v)$. We conclude that these values are the approximate y coordinates for Si, and that the peak at 0.17 is spurious, arising from both the small·number of data and experimental errors therein.

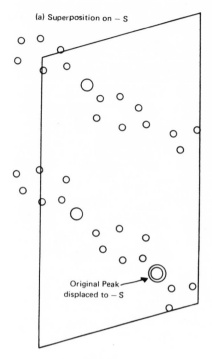

FIGURE S6.2a

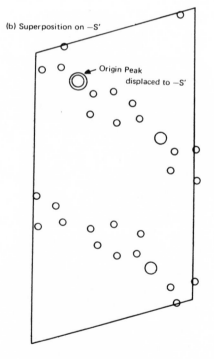

FIGURE S6.2b

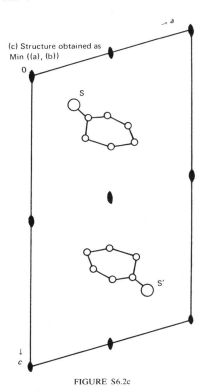

FIGURE S6.2c

*6.6. Since the sites of the replaceable atoms are the same in each derivative, and the space group is centrosymmetric, we may write $F(M_1) = F(M_2) + 4(f_{M_1} - f_{M_2})$

(a)

NH$_4$	K	Rb	Tl	
−	−	+	+	* Indeterminate because $|F|$ is unobserved.
*	+	+	+	
+	+	+	+	
−	†	+	+	† Indeterminate because $|F|$ is small.
+	+	+	+	Omit from the electron density synthesis.
+	+	+	+	
−	−	*	+	
*	+	+	+	

(b) Peaks at 0 and $\frac{1}{2}$ represent K and Al, respectively. The peak at 0.35 is due, presumably, to the S atom.

(c) The effect of the isomorphous replacement of S by Se can be noted first from the increases in $|F(555)|$ and $|F(666)|$ and the decrease in $|F(333)|$. These changes are not in accord with the findings in (b). Comparison of the electron density plots shows that $x_{S/Se}$ must be 0.19. The peak at 0.35 arises, in fact, from a superposition of oxygen atoms in projection, and it is not altered appreciably by the isomorphous replacement. Aluminum, at 0.5, is not represented strongly in these projections.

6.7. $|F(010)| = 149$, $|F(0\bar{1}0) = 145$, $\phi(010) = 55°$, $\phi(0\bar{1}0) = -49°$

6.8. In Figure S6.3, the six intersections 1,1', 2,2', and 3,3' are strongest in the region ∗---∗. The phase angle ϕ_M calculated from (6.95) would lie in this region; the centroid phase angle ϕ_B would be biased slightly toward 1 (see also Figure 6.33).

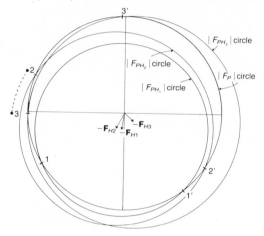

FIGURE S6.3

Chapter 7

7.1. Use 705, 6$\bar{1}$7, 8$\bar{1}$4: 42$\bar{6}$ is a structure invariant, 203 is linearly related to 8$\bar{1}$4 and 6$\bar{1}$7, and 432 has a low $|E|$ value. Alternative sets are 705, 203, 8$\bar{1}$4 and 705, 203, 6$\bar{1}$7. A vector triplet exists between 8$\bar{1}$4, 42$\bar{6}$, and 4$\bar{3}$2.

7.2. $|F(hkl)| = |F(\overline{hkl})| = |F(h\bar{k}l)| = |F(\bar{h}k\bar{l})|$
$k = 2n$: $\phi(hkl) = -\phi(\overline{hkl}) = -\phi(h\bar{k}l) = \phi(\bar{h}k\bar{l})$
$k = 2n + 1$: $\phi(hkl) = -\phi(\overline{hkl}) = \pi - \phi(h\bar{k}l) = \pi + \phi(\bar{h}k\bar{l})$

7.3. Set (b) would be chosen. There is a redundancy in set (a) among 041, $\bar{1}$62, and $\bar{1}$23, because $F(041) = F(04\bar{1})$ in this space group. In space group $C2/c$, $h + k$ must be even, Hence, reflections 012 and 162 would not be found. The origin could be fixed by 223 and 13$\bar{7}$ because there are only four parity groups for a C-centered unit cell.

7.4. From (7.32) and (7.33), $K = 4.0 \pm 0.4$ and $B = 6.6 \pm 0.3$ Å^2. (You were not expected to derive the standard errors in these quantities; they are listed in order to give some idea of the precision of the results obtained by the Wilson plot.) The rms displacement $\overline{(u^2)} = 0.29$ Å.

7.5. The shortest distance is between points like $\frac{1}{4}$, y, z and $\frac{3}{4}$, $\bar{y}$, z. Hence, from (7,41), $d^2(\text{Cl} \cdots \text{Cl}) = a^2/4 + 4y^2 b^2$, or $d(\text{Cl} \cdots \text{Cl}) = 4.64$ Å. Using (7.44), $[2d\sigma(d)]^2 = [2a\sigma(a)/4]^2 + [8y^2 b\sigma(b)]^2 + [8yb^2\sigma(y)]^2$, whence $\sigma(d) = 0.026$ Å.

7.6. By (7.25) $\phi_h = -2.2°$, and by (7.28) $\phi_h = -5.9°$.

Chapter 8

8.1. (a) The I—I vector lies at $2x, \frac{1}{2}, 2z$. Hence, by measurement, $x = 0.422$ and $z = 0.144$, with respect to the origin O.

(b)

hkl	$(\sin\theta)/\lambda$	$2f_\mathrm{I}$	$\cos 2\pi[(0.422h) + (0.144l)]$	F_I	$\lvert F_o \rvert$
001	0.026	105	0.618	65	40
0014	0.364	66	0.995	66	37
300	0.207	82	−0.100	−8	35
106	0.175	86	−0.224	−19	33

The signs of 001, 0014, and 106 are probably +, +, and −, respectively. The magnitude of $F_\mathrm{I}(300)$ is a small fraction of $\lvert F_o(300)\rvert$, and the negative sign is unreliable. Note that small variations in your values for F_I are acceptable; they would probably indicate differences in the graphical interpolation of f_I.

(c) 9.83 Å.

8.2. A simplified Σ_2 listing follows:

h	k	h−k	$\lvert E(\mathbf{h})\rvert\lvert E(\mathbf{k})\rvert\lvert E(\mathbf{h}-\mathbf{k})\rvert$
0018	081	0817	9.5
011	024	035	5.0
	026	035	0.5
021	038	059	0.4
	0310	059	0.4
024	035	059	9.6
038	059	0817	7.2
	081	011,7	6.0
	081	011,9	10.2
0310	059	081	7.9
	081	011,9	9.2

In space group $P2_1/a$, $s(hkl) = s(\bar{h}k\bar{l}) = (-1)^{h+k} s(h\bar{k}l)$, which means that $s(hk\bar{l}) = (-1)^{h+k} s(\bar{h}kl)$. The origin may be specified by $s(081) = +$ and $s(011,9) = +$.

Sign determination

h	k	h−k	Conclusion
011,9	081	038	$s(038) = +$
011,9	081	0310	$s(0310) = +$
038	081	011,7	$s(011,7) = +$
0310	081	059	$s(059) = -$
059	038	0817	$s(0817) = -$
038	059	021	$s(021) = -$
0310	059	021	$s(021) = -$
0817	081	0018	$s(0018) = -$
		Let	$s(035) = A$
059	035	024	$s(024) = -A$
035	024	011	$s(011) = -$
035	011	026	$s(026) = A$

The two indications for $s(021)$ and the single indication for $s(026)$ will have low probabilities and must be regarded as unreliable. Within this limited data set, no conclusion can be reached about the sign of A, and both + and − signs are equally likely. Reflection 0312 does not interact within the data set.

8.3. The space group is $P2_1/c$ (from Table 8.4). Hence, we must recall that $F(hkl) = F(\bar{h}\bar{k}\bar{l}) = (-1)^{k+l}F(h\bar{k}l)$; Fig. S8.1 shows the completed chart. The Σ_2 listing should look something like the following. An * indicates a sign change with respect to the $hk0$ quadrant, and N means that no further relationships were derived by considering the reflexion so marked as **h**.

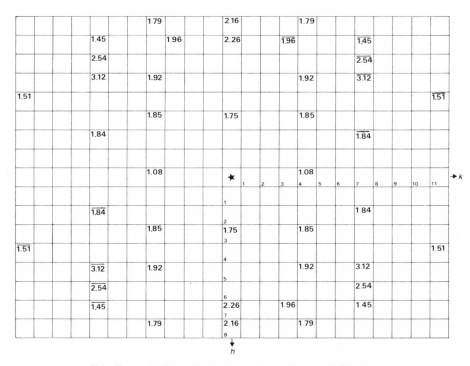

S8.1. The completed chart. Negative signs are shown as bars over the $|E|$ values.

Reference number	$\mathbf{h}$	$|E_\mathbf{h}|$	$\mathbf{k}$	$|E_\mathbf{k}|$	$\mathbf{h-k}$	$|E_{\mathbf{h-k}}|$	$|E_\mathbf{h}||E_\mathbf{k}||E_{\mathbf{h-k}}|$
1	300	1.75	040	1.08	$\bar{3}40$	1.85	3.5
2			840	1.79	$\bar{5}40$	1.92	6.0
3			570	3.12	$\bar{2}70$	1.84	10.1
4	700	2.26	570	3.12	$2\bar{7}0^*$	1.84	13.0
5	800	2.16	670	2.54	$2\bar{7}0^*$	1.84	10.1
6			340	1.85	$5\bar{4}0$	1.92	7.7
7			411,0	1.51	$3\bar{1}\bar{1},0^*$	1.51	4.9
8			040	1.08	$8\bar{4}0$	1.79	4.2
9	730	1.96	$0\bar{4}0$	1.08	770	1.45	3.1
10			$5\bar{4}0$	1.92	270	1.84	6.9
11	040	N					
12	340	1.85	$7\bar{7}0$	1.45	$\bar{4}11,0^*$	1.51	4.1
13	540	N					
14	840	N					
15	270	N					
16	570	N					
17	670	N					
18	770	N					
19	411,0	N					

An origin can be chosen as 0,0 by making, for example, 270 (eoe, and occurring 4 times) and 540 (oee, and occurring 3 times) both +. From the Σ_2 listing we have:

Reference number	Conclusion	Comments
10	$s(730) = +$	
7	$s(800) = -$	$s(411,0) = -s(4\bar{1}\bar{1},0)$
5	$s(670) = +$	
6	$s(340) = -$	Sign propagation has now stopped. Set $s(040)$ equal to symbol A.
	$s(040) = A$	
1	$s(300) = A$	
2	$s(840) = A$	
3	$s(570) = A$	
4	$s(700) = -A$	
8	$s(840) = -A$	
9	$s(770) = A$	
11	$s(411,0) = -A$	

One symbol, A, has been used. In this case, it would be necessary to test, by electron density summation, both possible signs for A. Of course, in a more extended set of $|E|$ values multiple indications could make this test unnecessary. No Σ_2 relationship is noticeably weak, and the above solution could be regarded as acceptable. Alternative results, based on other choices of origin, may be equally correct.

Index